Computer Control of Processes

Computer Control of Processes

M. Chidambaram

Alpha Science International Ltd.
Oxford, U.K.

M. Chidambaram
Department of Chemical Engineering
Indian Institute of Technology, Madras
Chennai, India

First Reprint 2006
Second Reprint 2008
Third Reprint 2008

ALPHA SCIENCE INTERNATIONAL LTD.

7200 The Quorum, Oxford Business Park North
Garsington Road, Oxford OX4 2JZ, U.K.

www.alphasci.com

ISBN 978-1-84265-063-9

Printed in India

Preface

An important objective of the present book is to present details on computers and on the principles in designing digital controllers. This can be used as a textbook for an elective course in 'computer process control' for postgraduate students in chemical, mechanical, instrumentation and control engineering disciplines. Most of the books treat topics either on controller design methods or on details on computer hardwares. This book covers the essentials of both these aspects in a single course. Newer topics like PLCs, smart sensors and fieldbus, identification and design of nonlinear controllers, fuzzy logic controllers are included.

The book is divided into two parts containing seven chapters in each part. Part 1 provide details on computers, interfacing aspects, networking of computers, smart sensors and fieldbuses, and programmable logic controllers. While Part 2 covers identification of linear models, nonlinear models and design of controllers for linear and nonlinear discrete systems.

Beginning with an overall view on personal computers, details on components of a typical mother board, memory, basic input/output interfaces, communications with external devices, serial and parallel communications, plug-in-cards, and important aspects of families of PCs are given.

Important aspects of interfacing of real world (analog and digital) signals and actuators with a PC are discussed in Chapter 2. The signal conditioning of analog input signals is explained. The steps in signal conditioning of digital input signal and digital output signal are described. ADC and DAC are provided in detail.

In Chapter 3, data acquisition and control using a standard add-on-card are explained. The details on industrial signal conditioners are provided. The method for communication using IEEE-GPIB interface card and associated cabling is explained.

Introduction to the evolution of digital computers for process control is given in Chapter 4. Important aspects on supervisory control, direct digital control, distributed digital control are provided. Details on computer networking, topologies, communication protocol, and a brief description of a DCS are provided.

Chapter 5 analyzes six reported case studies on experimental computer control of processes are analyzed. These studies include, computer control of liquid level, a heat exchanger, a batch reactor, a fermenter, plastic injection molding process, on-line optimizing control of a distillation column.

Basic principles in smart sensors such as smart temperature transmitter, differential pressure transmitter, valve positioners are provided in Chapter 6. The HART and field buses are explained. The details on Foundation field bus and PROFIBUS are also given.

Chapter 7 brings out introduction to programmable logic controller for implementing control techniques for sequential processes. The details on ladder logic diagram, boolean algebra, a brief discussion on connecting PLC to a PC and networking of different PLCs are given.

Methods for designing linear feedback controllers for the discrete time system are discussed in Chapter 8 while Chapter 9 gives basic methods in designing feed forward controllers. In Chapter 10, classification of discrete time models and techniques of identifying discrete time parametric model from input and output data set are provided. Chapter 11 provides detail on adaptive control systems that automatically adjust the controller settings in real time to compensate for unanticipated changes in the process conditions. Design methods on model predictive controllers are touched upon.

An introduction to different nonlinear discrete models and identification of some important class of such models with linear in parameters are given in Chapter 12. Essential steps in designing nonlinear controllers based on such nonlinear models are also given. Chapter 13 provides simulation case studies on control of processes and Chapter 14 presents fundamental on fuzzy logic control, design of such controllers, and use of fuzzy logic in improving the performance of conventional PID controllers. Three simulation case studies are also given to illustrate the design methods.

It is a pleasure to acknowledge the contributions of people who helped the author in writing this book. The book owes much to the numerous authors of original research papers and textbooks. The author gratefully acknowledges the contributions made by his students (M.K. Jhunjhunwala, Mrs. S.N. Jyothi, S. Satiskumar, C. Clement Valentine, M.M. Mahesh, N. Bhardwaj, S. Arvind) as part of their project works required for the award of degrees. The author thanks Mr. J. Prakash and other research scholars in the Process Control Laboratory, I.I.T. Madras, Dr. T.K. Radhakrishanan, Assistant Professor, Dept of Chemical Engineering, R.E.C, Trichirappalli, for their comments on the first draft of this book. The author is thankful for the support received from the Department of Chemical Engg, Indian Institute of Technology, Madras and for the support received from the Centre for Continuing Education for providing financial aid for writing this book. The author thanks Mr. A. Pandian for drawing the figures. The author expresses his sincere appreciation to his son Master C. Srihari Prasath for his patience and understanding.

The author dedicates the book in memory of his wife Mrs. C. Sheila.

THE AUTHOR

Contents

Part 1

Computers and Interfacing

1

Introduction to Computers

This chapter gives an overall view on the generic personal computer (PC). The PC consists of: (i) a mother board (also called main logic board) containing central processing unit, external cache, memory, battery for real time control and video memory and other support circuits, (ii) optional expansion card, (iii) mass storage devices and (iv) power supply. Details of these components will be given in this chapter.

1.1 Personal Computer

The personal computer (PC), introduced by IBM, included a feature called open architecture, i.e. computer box contains a microprocessor and other support circuits on a single mother board having several slots. Each slot is a standard connector into a standard bus (bus is a set of conductors controlled by the computer). This architecture is called 'open' because IBM made the information available so that other manufacturers could design circuit boards that could be plugged into the bus (slots). IBM also provided information on the operation of the mother board so that other manufacturers could write IBM PC programs to use the new circuit boards. Boards could be designed and software written to interface the IBM PC to sensors, actuators, or other computers, without IBM having to do the design or programming. The user has to select the appropriate interface boards so as to connect the sensors and actuators to the computers. A computer system mainly consists of four components: an input device, an output device, main memory and a processor (Fig. 1.1). Secondary storage (floppy disc or hard disk) is often used to extend the memory capacity.

1.1.1 Mother Board

The main board in the PC, called the mother board, contains CPU, maths coprocessor, system clock, BIOS ROM (Basic Input/Output System Read Only memory), RAM (Random Access Memory), key board controller and some logic circuit to deal with input/output, interrupts, DMA (Direct memory Access) etc. (Fig. 1.2). The support structure of the printed circuit board (PCB) used in a PC to make the mother board, is made of woven fiberglass material filled and strengthened with green colored epoxy plastic material. On the printed circuit board, the components are soldered and connected with the help of copper tracks. The modern PCB contains many such tracks to make electrical connection between different parts of the system. The board also has the expansion slots as its main component on which expansion cards are connected. We can fit many additional expansion cards like display

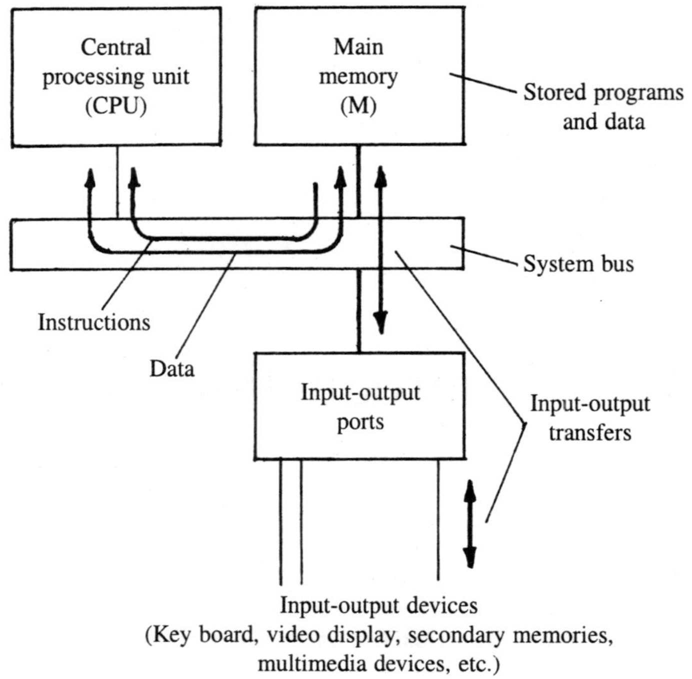

Fig. 1.1 Basic components of a computer

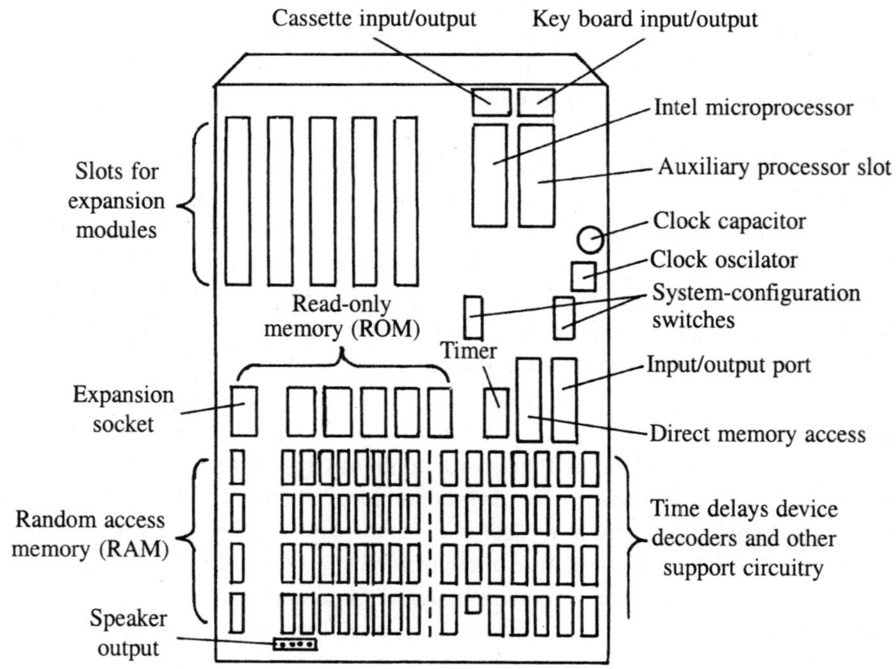

Fig. 1.2 A computer mother board

card, hard disk controller, sound card, network card, data acquisition and control card etc. on these slots (one card for one slot). The mother board is connected to the expansion cards through the data, address and control lines/buses on these slots. A PC may have 4 to 8 such slot expansion which enables consumers to configure computer systems to fit their needs and budgets.

A mother board generally has jumpers or DIP (dual in-line package) switches to set a wide range of options including whether: (i) a coprocessor is fixed, (ii) the amount and type of memory and (iii) color or monochrome video etc. A jumper is a small plastic-encased conductor (shorting plug) that slips over the pins. When settings are often changed, then jumpers cause too much trouble. Banks of DIP switches are generally used in such situations (two types of DIP switches are: slide and push). Some mother board manufacturers provide some common interfaces such as for the display, disk drive, serial/parallel I/O (Input/Output) etc., on the mother board itself. This saves the slot space, but restricts the future upgradability and makes maintenance difficult and costly. Let us now discuss briefly the different components soldered or fixed to the mother board.

1.1.2 Microprocessor

The microprocessor (μP), sometimes referred to as a CPU (Central Processing Unit) or Microprocessing Unit (MPU), controls memory and I/O through series of connections called buses. A bus is a common group of wires that interconnect components in computer systems. The bus may be tracks on a printed circuit board or the wires in a ribbon cable. Components can be soldered or plugged to the bus connectors. Buses *select* an I/O or memory device, *transfer* data between an I/O device or memory and the μP, and *control* the I/O and memory system. The memory and I/O are controlled through instructions that are stored in the memory and executed by the μP. The μP performs three main tasks for the computer system: (i) data transfer between itself and the memory or I/O systems (ii) simple arithmetic and logic operations, and (iii) program flow via simple decisions. The power of the μP is in its ability to execute millions of instructions per second from the software (group of instructions) stored in the memory system.

The buses that interconnect the sections of a computer system *transfer* address, data, and *control* information between the μP and its memory and I/O systems. Three buses that exist for this transfer of information are: address, data and control. Figure 1.3 shows the interconnection of μP, *R/W* memory (RAM), read only memory (ROM), and a few I/O devices.

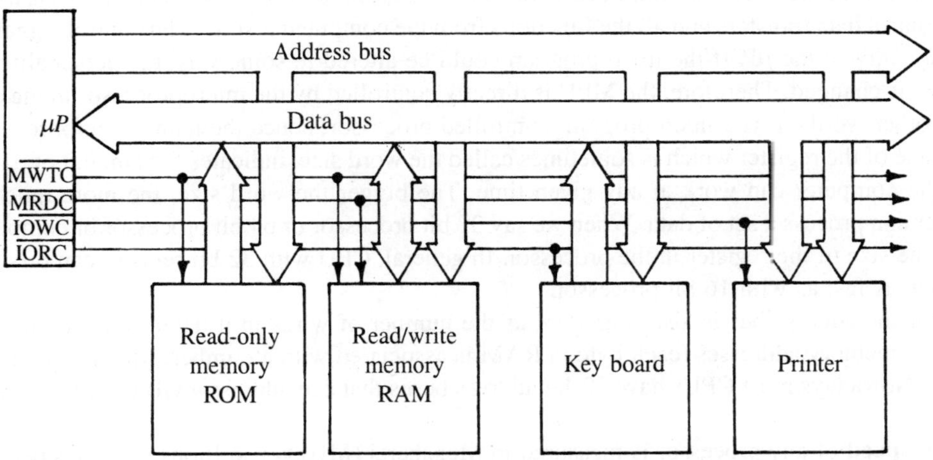

Fig. 1.3 The block diagram of a computer system showing the address, data, and control bus structure

The address bus requests a memory location for the memory or an I/O location from the I/O devices. The data bus transfers information between the μP and its memory and I/O address space. The control bus contains lines that select the memory or I/O and cause them to perform a read or write operation. That is, the address line provides information on where the information is to be sent (or where it is to obtained from); the data lines show what information is; and the control and status lines indicate when it is to be sent. In most computer systems, there are four control bus connections: memory read control, memory write control, I/O read control and I/O write control. The control signal is active low that is, it is active when a logic zero appears on the control line.

The μP is usually packed in a dual-in-line (DIP) plastic or ceramic container. Dual-in-lines means the connections that lead from the integrated circuit inside to the world outside are arranged in two rows on either side of the long axis; they have the same appearance as an IC chip and are numbered in the same way. The CPU consists of a number of basic elements: an arithmetic logic unit (ALU), several accumulators (storage registers for work in progress), general-purpose registers (temporary storage devices), program counter (a register that stores the address of the next instruction). Registers are memory locations inside the CPU. The mathematical operation depends on the register width.

Arithmetic Logic unit (**ALU**) uses logic manipulation, such as 'and, or, exclusive or, and complements', and shifts to perform the arithmetic functions required.

Accumulator is used to pass data into and out of the MPU as well as to hold data for further manipulation.

Program counter (**PC**) keeps track of the location of the next instruction to be executed from program memory. After each fetch from memory, the program counter is automatically incremented by 1.

Instruction register (**IR**) receives each current instruction from memory which holds it until it has been executed; then the next instruction is loaded in, which simulataneously destroys which is already in the register.

Condition codes register (**CC**) contains the flags that are used to indicate the outcome of the previous operation. The contents of this register are used as part of the decision making process of the processor.

Control unit (**CU**) directs the operation of the μP and is actually a computer inside a computer, that is, it has all the functional elements of a classical computer, including read only memory (ROM). The CU executes the user's program instructions by executing a series of its own micro instructions which control data transfers and all the functions from the computed results. The control signals form the personality of the μP. If the microprogram could be altered in some way, the personality of the processor is changed. Therefore, the MPU is directly controlled by the microprocessor in the control unit; in other words, it is a microprogram controlled processor, hence the name 'microprocessor'.

The size of the register which is sometimes called the word size, indicates the amount of data with which the computer can work at any given time. The bigger the word size, the more quicker the computer can process a set of data. When we say 32 bit processor or 64 bit processor or computer we refer to the size of the register in the processor. In general, CPU with 32 bit register can process the data twice as fast as with 16 bit processor.

Size of the address bus is also important as the number of wires in it determines the maximum number of memory addresses (each byte in RAM is associated with a number, which is the memory address). Nowadays most CPUs have 32-bit address buses that can address 4 GB (4 million bytes) of RAM.

Clock speed of any processor is measured in Megaherts (MHz) or millions of cycles per second and it specifies how a processor can perform different internal operations. The CPU is specified in terms of MIPS (millions of instructions per second) or MFLOPS (millions of floating point operations per second).

1.1.3 The Memory

The main memory system (Fig. 1.4) is divided into three parts: (i) TPA (Transient Program Area), (ii) system area and (iii) XMS (Extended Memory System). The TPA (Fig. 1.5) holds the operating system and other programs that control the computer system. The TPA also stores any currently active or inactive application programs. The DOS (Disk operating system) controls the way that the disc memory is organized and controlled as well as the function and control of some of the I/O devices connected to the system. The free TPA area holds application program. The rest includes word possessors, spread sheet programs, CAD program etc. The BIOS (basic input output system) is a collection of programs (stored in RAM) and operates many of the I/O devices connected to the computer. (Kilo bytes: K bytes = 1024 bytes; Mega bytes: M bytes = 1024 × 1024 bytes = 1024 K bytes; Giga bytes: G bytes = 1024 M bytes).

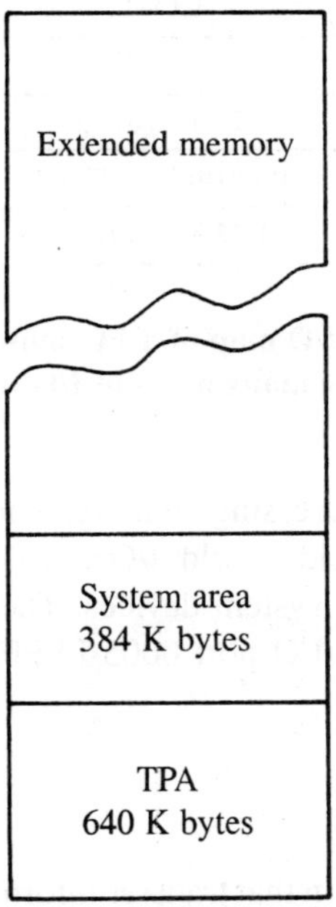

Fig. 1.4 A typical memory map of a personal computer

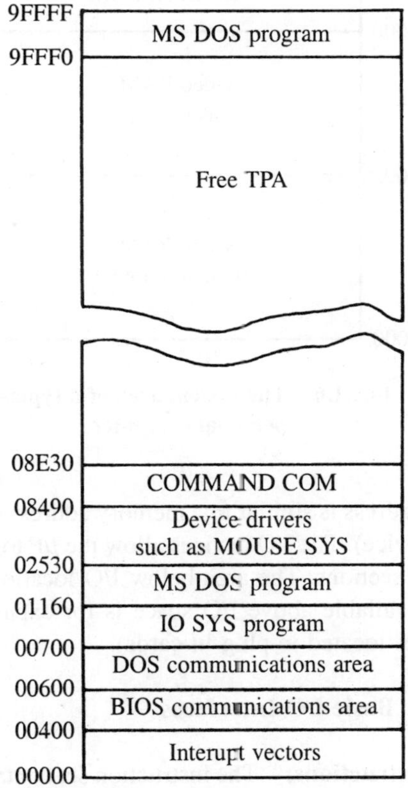

Fig. 1.5 A typical memory map of the TPA in a PC

The system area contains programs on either a ROM or flash memory and also area of read memory for data storage (Fig. 1.6).

I/O space: Input/Output (I/O) space extends from I/O port 000H to port FFFFH (Fig. 1.7). (An I/O

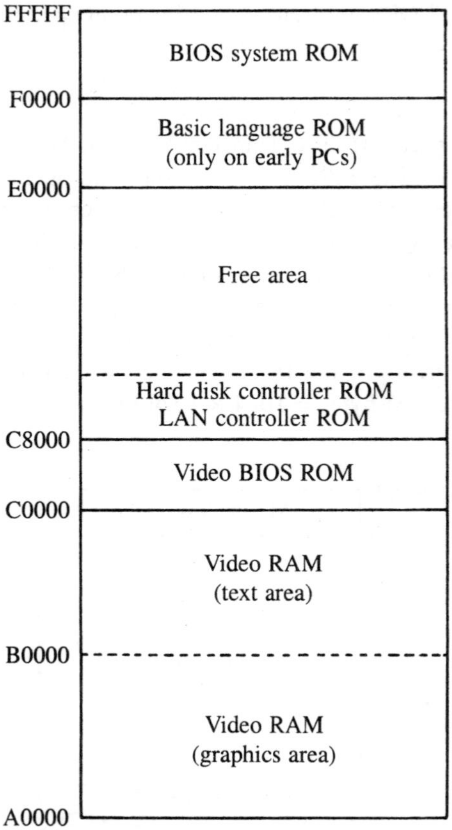

Fig. 1.6 The system area of a typical personal computer

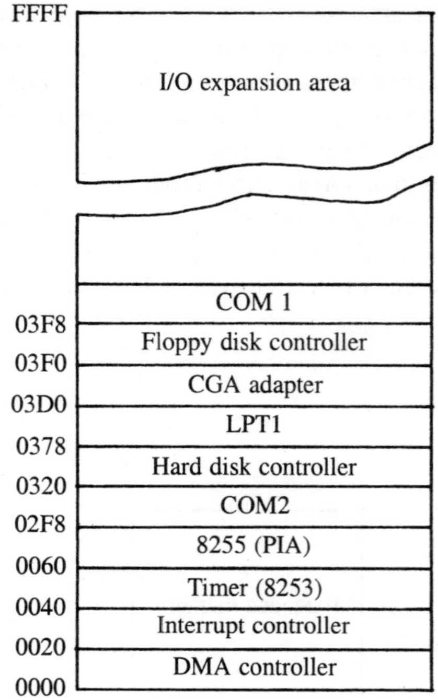

Fig. 1.7 The I/O map of a PC showing some of the many areas of I/O devices

port address is similar to a memory address except that instead of addressing memory, it addresses an I/O device). The I/O devices allow the μP to communicate to the outside world. I/O area contains two major sections. The area below I/O location 0500H is reserved for system devices. The remaining area available above I/O space is for expansion that extends from I/O port 0005-FFFFH (address devices located in plug-in-cards).

1.1.4 Basic I/O Interfaces

I/O instructions: The instruction set contains one type of instruction that transfer information to an I/O device (OUT) and another to read information from I/O device (IN). Both IN and OUT instructions transfer data between an I/O device and the μP's accumulator. Whenever data is transferred using IN or OUT instructions, the I/O address, often called a port number, appears on the address bus.

Isolated and memory mapped I/O: There are two completely different methods of interfacing I/O to the μP: (i) isolated I/O and (ii) memory mapped I/O (Fig. 1.8). In isolated I/O, the IN and OUT instructions transfer data between the μP accumulator or memory and the I/O device. In

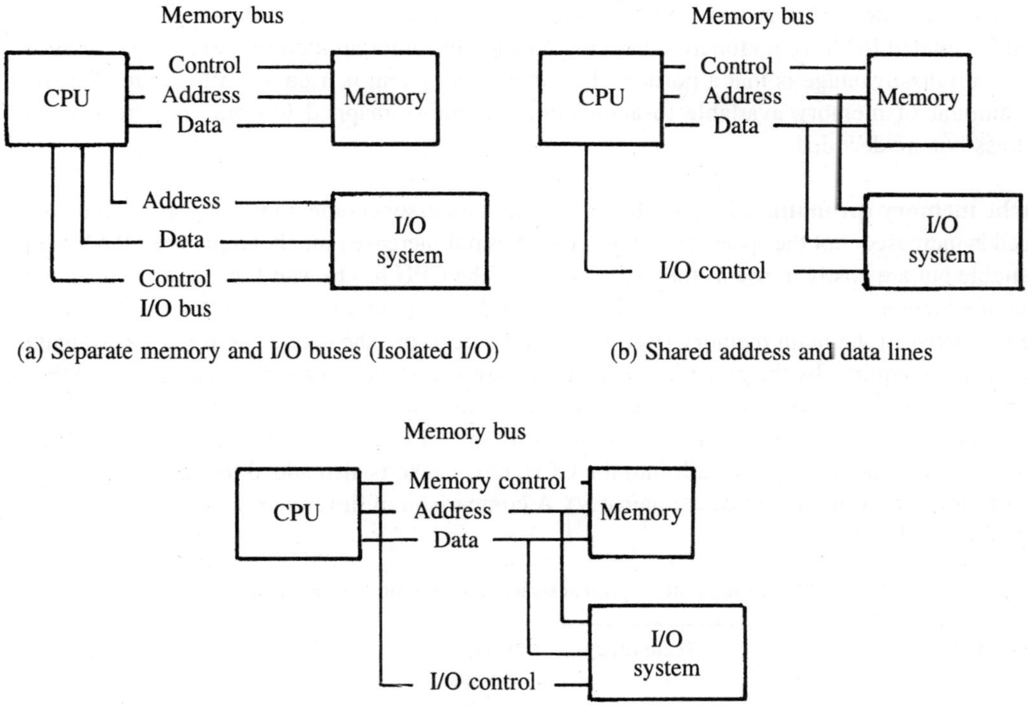

(a) Separate memory and I/O buses (Isolated I/O)

(b) Shared address and data lines

(c) Shared address, data and control lines (memory-mapped I/O)

Fig. 1.8 Independent and shared memory and I/O bus

memory-mapped I/O, any instruction that references memory can accomplish the transfer. Both isolated and memory-mapped I/O are in use. The most common I/O transfer used is isolated I/O. The term 'isolated' describes how the I/O locations are isolated from the memory system in a separate I/O address space. The addresses for isolated I/O devices, called ports, are separated from the memory. As a result, the user can expand the memory to its full size without using any of this space for I/O devices. A disadvantage of isolated I/O is that the data transferred between I/O and μP must be accessed by the IN and OUT instructions. Separate control signals for the I/O space are developed that indicate an I/O read or an I/O write operation. These signals indicate that an I/O port address appears on the address bus that is used to select the I/O device. As a rule, 8-bit port address is used to access devices located on the system board, such as the timer and keyboard interface, whereas a 16-bit port is used for accessing serial and parallel ports as well as video and disk drive systems. The input port is actually a hardware device connected to processor's data bus. When executing the IN instruction, the processor will output the address of the input port on the address bus. The selected input port will then place its data on to the data bus to be read by the processor. Data read from the input port always ends up in the accumulator. The input ports can be used to read data from key boards, A/D convertors, DIP switches, clock chips, UARTs, and other peripherals that may be connected to the CPU.

Memory-mapped I/O: Memory-mapped I/O does not use IN or OUT instructions. Instead, it uses any instruction that transfers data between the μP and memory. A memory-mapped I/O device is treated as a memory location in the memory map. The main advantage is that any memory transfer instruction can be used to access the I/O device. Another advantage is that the separate control signals used in isolated I/O have no function thereby reducing the amount of circuitry required for decoding. The main disadvantage is that a portion of the memory system is used as an I/O map. This reduces the amount of memory available to applications. Memory-mapped I/O may require that the full address bus be decoded.

Cache memory (pronounced as 'cash'): As new processors come up in the market, the processor speed is increased, but the speed of memory chip has not increased much. High speed RAM chips are available but are costly to use as the main memory. The CPU has to wait for the result from memory after any memory read or write command is given. Cache memory is very small but very fast memory used in between the main memory (RAM) and the processor. The cache controller has to make sure that the data required by the processor in the processor's next memory access is available in the cache memory. As the CPU often accesses data in a sequential order, when the memory is read by the CPU, the cache controller reads around the area read by the CPU and stores it in the cache memory. In addition to the cache memory built into the CPU, now cache is also added to the mother board (256 KB or more of mother board cache memory). A comparison of different memory used in computers is given in Table 1.1.

Table 1.1 Comparative characteristics of memories (Rajaraman, 1999)

Memory type	Average capacity in bytes	Technology	Average time to access a byte	Purpose in a computer system	Relative cost per byte in units
Cache	0.5 M	High speed integrated circuits	2 nsec	Instruction and data to be immediately used	10
Main memory	50 M	Integrated circuits	20 nsec	Program and data	1
Disk memory (hard disk)	50 G	Magnetic surfaces on hard disks	10 msec	Large data files and program over flow from main memory	1/100
Disk memory (floppy)	10 M	Magnetic surfaces on thin flexible disks	500 msec	Data entry. As input unit. Data and program files in microcomputers	1/1000
Magnetic Tape	5 G	Long 1/4″ tape wound on cartridge	25 sec	Historical files. Back up for disk. Data and program exchange between installations	1/1000
CDROM	600 M	Laser disk	500 msec	Store large texts pictures and audio. Software distribution	1/10000
DVDROM	8 G	Laser disk	500 msec	Video files	1/100000

One type of removal mass storage devices is the PC card. PC cards also referred to as PCMCIA (Personal Computer Memory Card. International Association) cards are credit card sized expansion devices that are inserted into slots in the PC, particularly Laptop computers. PC card adapters are available for desktop computers. Their storage capacity is 170 MB to 510 MB (equivalent to a hard disk in a PC). Personal hard disks combine high capacity with the convenience of a diskette.

Maths Coprocessor: As stated earlier, a standard mother board contains main processor, maths coprocessor, memory, interrupt controller, timer and keyboard interfacing circuitry and some other support circuits (Fig. 1.2). Maths coprocessors are actually microprocessors themselves. They perform a large variety of mathematical operations, with very high clock speeds with a great degree of precision (80 bit results are common). In addition to the basic add/subtract/multiply/divide operations, coprocessors are capable of finding square root, logarithms, a host of trigonometric functions etc. A math coprocessor is a chip that is separately designed to handle complicated mathematical operations. Newer CPUs have a built-in math coprocessor. The math coprocessor executes arithmetic routines much more quickly than the ALU because it uses floating point arithmetic, a computer technique that translates number into scientific notations. The floating point technique simplifies complex arithmetic because the computer need not store large numbers of decimal places.

Direct memory access (DMA): When a large volume data is to be transmitted between the processor and any external device, routing them through the processor is very slow process. The reading from memory and writing to the external device byte by byte is a very slow process and it slows down fast I/O devices such as disk drive, hard disk, scanner etc. Input and output instructions of the CPU (or the memory load and store instructions, if the I/O device is memory mapped) are used to transfer data to and from the device under program control. The device has to get the CPU's attention, and the CPU cannot be doing anything else while it is transferring data. It is desirable to access the memory directly (Direct memory access). It costs a certain amount of hardware.

1.2 Operating System

An operating system (OS) is a computer program that provides an environment in which other programs can run, allowing them to easily take advantage of the processor and of I/O devices such as disks. The OS shares hardware resources among its users. It divvies up memory and regulate/ access to files and devices. An OS is a master control program that provides an interface for a user to communicate with the computer, manages hardware devices, manages and maintains disk file systems, and supports application programs.

In other words, the operating system provides interfaces for the user, software, hardware, and for file management. The functionality of the operating system can be extended with the addition of utilities. Most modern operating systems employ a graphical user interface (GUI), in which users control the system by pointing and clicking on screen graphics. A GUI is based on the desktop metaphor, icons, sizable windows, menus, and dialog boxes. The applications designed to run under a particular operating system use the same interface elements, so users see a familiar interface or common user access (CUA) no matter what they are doing on the computer. Some older operating systems, such as DOS, use command-line interfaces, which the user controls by typing commands at a prompt. The operating system allows programs to share information. In Windows 3.1, 3.11, NT and 95, sharing is accompalished through the Clipboard and Object Linking and Embedding (OLE). A modern operating system also makes multitasking possible.

An operating system uses interrupt requests to maintain oragnized communication with the CPU and other pieces of hardware. Each of the hardware devices is controlled by another piece of software, called, driver, which allows the operating system to activate and use the device. The operating system also provides the software necessary to link computers to form a network.

Operating systems have gradually included more and more utilities. The most common utilities are for file defragmentation, data compression, backup, data recovery, protection against viruses, and screen saving.

DOS dominated the operating system market during 1980's. Windows 3.0, 3.1, and 3.11 are operating environments that provided GUI for computers running DOS. The strengths of Windows 95 are its simplified interface, 32 bit multitasking, and its ability to run older Windows and DOS programs. Mocrosoft Windows NT offered true 32 bit architecture and excellent networking capabilities. BIOS (basic input output systems) is a set of programs stored inside a programmable read only memory (PROM) and the PROM is put into the mother board. It is used for most basic input output processes, such as reading the keystroke, displaying something on the screen, reading/writing of the serial and parallel port, reading/writing to and from the floppy and hard disk etc. Other than these input/output functions, BIOS contains two more modules. Power on self test (POST) and bootstrap loader. POST is another program stored inside the BIOS RAM to check the processor, memory, support circuits and other devices connected to the computer when the computer is switched on. Once the POST routine is executed successfully, BIOS goes to the next job that is to boot the process. Its job is to load the operating system from floppy disk or the hard disk into the computers main memory (RAM).

1.3 Communication with External Devices

Polling: There are two general methods of programming a microprocessor to send or receive information from external peripheral devices. The first is called polling in which the μP is programmed periodically to check the devices connected to its data bus to see if they are ready to be serviced. Advantages of the polling techniques are that it conceptually simple to program, requires minimal hardware and is synchronous with program execution. The main disadvantage of the polling technique is its inherent inefficiency in that the devices being polled will often not require services. This process causes a loss of control system response capability, which means real-time requirement of the system may not be met.

Interrupt: When a polling technique cannot provide a fast enough response to the system control requirements, then the interrupts method can be used. The peripheral devices will be serviced only when they are to transfer data. When the peripheral devices are ready to transfer data, they send interrupt signals to μP, which will then stop temporarily what it was doing in order to service the peripheral device(s). The major advantage of the interrupt technique of communicating is an increase in control system response time.

Interrupt is one of the most common method for an external device to communicate with the processor. But when more than one device is connected to the processor, the processor must have some facility to find out which of the external device has interrupted. An interrupt controller is used in between the processor and the external device to take care of multiple number of external devices. Interrupt service routines (ISRs) are special programs in the computer memory. Each device attached to the processor will have a specific routine in the computer memory. Whenever the processor is interrupted by any device, the processor stops the execution of the current program and jumps to the interrupt service routine of the interrupted device. The processor saves the counters of all the registers, the status flag, and all the counters before starting to execute the ISR. At the end of the ISR execution, whatever logic was to save the old program's register, counter and status flag values must be executed in reverse order to restore these values.

Each interrupting device connected to the processor is given an interrupt number and each interrupt number is given a fixed priority number. The interrupt priorities determine which device receives the processor's attention when more than one device is simultaneously requiring an interrupt.

1.4 Timing Circuits

Clock generator: The computer needs some type of information to coordinate its various activities. Inside the computer, a clock generator chip, intel 8284A, is used to provide the timing signal to control different components of the system. The clock speed is measured in Hertz (Hz) or cycles per second, which is a unit of frequency. The 8284A chip requires an external timer input to keep its clock signal generation in proper order. A quartz crystal oscillator is commonly used to provide the timing information from the outside to the 8284A clock generator.

Counter-timers: Every microprocessor system, especially the real time one, needs at least one counter or timer. Most counter-timer chips come with several channels, each of which can be configured to count or time external events, generate clock pulses of a given frequency, or provide interrupt at given intervals. Counter timers are most often used as bit-rate generators for serial I/O chips as real time clocks (that is, clocks that support timing and delay functions in real time control systems).

Rate generators: This is the mode most frequently used to repetitively trigger some event. In this mode, a pulse is generated at regular intervals; the interval is determined by the value in the count register. The frequency of the oscillator used as the frequency source is divided by the value in the count register, e.g. if the oscillator is 1 MHz, and the count is 5, then the output pulse generated will be at 0.20 MHz. The output typically changes for only one cycle of the oscillator.

Software-triggered strobe: This mode is similar to the rate generator mode, except that the output is a single-pulse, rather than a repetitive series of pulses. When the count is loaded into the counter, it begins counting down until it reaches zero. At that point, the output is a pulse that lasts one cycle of the oscillator.

Clocks: Many real time systems need to know not only the interval between external events, but the actual date and time of day that the events happen. There are number of clock chips, usually derived from digital watch chips, that are designed to connect to microprocessor systems. They are usually CMOS so that they can run on battery when the system is turned off, and have a binary coded decimal interface in place of the direct digital interface used for watches and clocks.

1.5 Parallel I/O Ports

The vast majority of problems encountered in the use of microprocessors involve connecting the processor to multiple signals. In order to simplify the interfacing task the manufacturer of microprocessors and others have created special parallel I/O devices. Generally, these interface circuits are programmable so that they can be configured to meet a variety of special needs including bidirectional I/O, interrupt functions etc. Figure 1.9 illustrates the general external organization of parallel interface devices. In this figure, the device is connected to the microprocessor on one side. Specific I/O components, such as switches, LEDs etc., are connected to one or more parallel ports on the other side of the device. Each of these parallel ports consists of a number of bidirectional signal lines, typically 8. Usually there are only two or three 8 bit ports per device.

1.6 Serial I/O port

One very common use of computers today is for transmission of data from one computer system to another. In order for this data transfer to occur, the computers must be connected in some fashion.

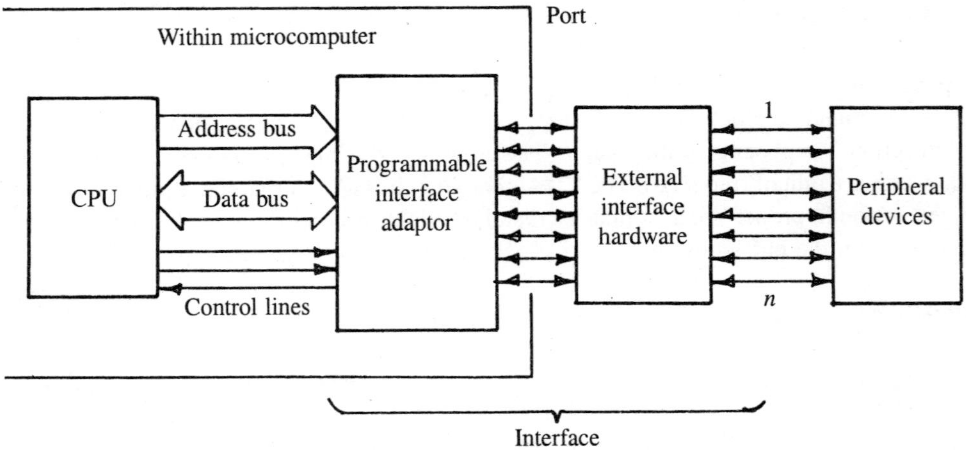

Fig. 1.9 Peripheral interface adapter (PIA)

There are two basic strategies used for this purpose. In the first strategy, computers are connected via parallel ports and can thus transmit and receive data one byte at a time. This strategy, however, requires a large number of wires, 8 for data and several more for controlling the data flow. If the computers are separated by a long distance, then this approach can become quite expensive. The second strategy commonly followed is to use only two wires and transmit the data as a serial bit stream. Obviously the transfer of data will take longer time using this serial approach than with the parallel approach. However, since many fewer wires are needed, this approach will be much cheaper when long distances are involved. The existing telephone network cable can be used as a link between the computers after converting the digital signal into a frequency signal.

The serial data transmission of a byte of data is conceptually simple. The byte is first loaded into a shift register and is then shifted out and transmitted a bit at a time. Reception of this data is basically handled in the reverse order, namely, shift the data into a shift register as each bit is received. There are, however, a number of difficulties in doing this step: determining when the data starts and the location of each bit. Two methods are commonly used to handle these difficulaties: synchronous serial and asynchronous serial communications. Figure 1.10 illustrates the two approaches. In synchronous serial communications, the location of each bit is indicated by a clock signal carried on a third wire (Fig. 1.10a).

Asynchronous serial communications depends on an aprior knowledge of the rate at which the bits arrive. Figure 1.10b shows the operation of this method. The term asynchronous means not clocked, that is, there is no common clock between the transmitting and receiving devices. The total bits including start to stop is called a frame. The start of a data frame is indicated by bringing the signal line low for one bit time, referred to as start bit. The data bits then follow, each being held on the signal line for one bit time. Typically there are either seven or eight data bits, which may be followed by a single parity bit. The parity bit is used to help detect the presence of errors in the transmission. If used, the parity bit may indicate either an even or an odd number of 1's in the transmitted character including the parity bit. Thus, if a single bit error occurs, which is most likely, then the parity of the received character will be different from the parity indicated by the parity bit. Finally, the signal line is brought high again and held for one or more bit times. This last bit is referred to as stop bit. Normally, asynchronous serial communications protocol require one start bit, eight data bits, no parity bit, and one stop bit. Other protocols may be used for specific purposes. Since the bit time, or bit rate, is known, each bit can be identified by simply sampling the signal line at this rate once the start bit is detected. The rate at which serial data is transmitted and received is referred to as the Baud rate, which, in the present case, is simply "bits per second". An I/O support device that can be used for the asynchronous serial communications method just described is generally referred to as a Universal Asynchronous Receiver Transmission (UART).

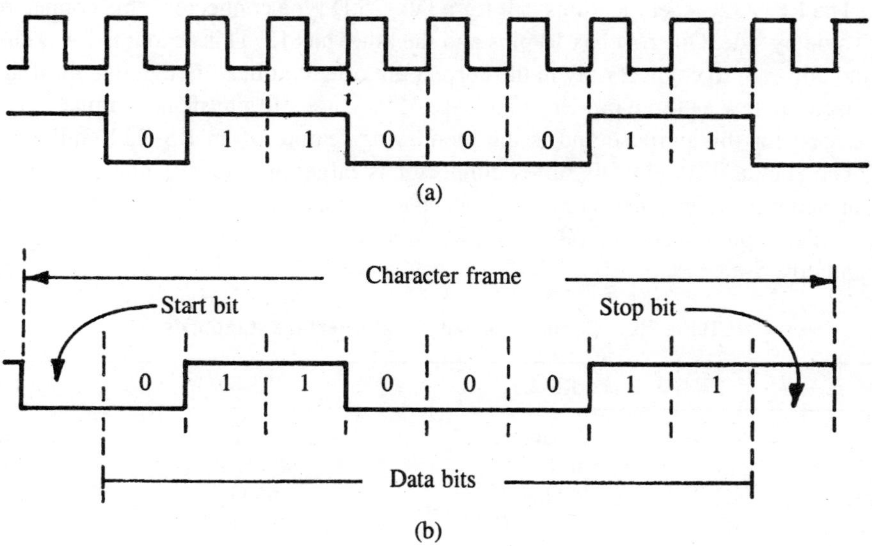

Fig. 1.10 Serial data transmission using (a) clocked synchronous and (b) asynchronous methods

The RS–232C standard: The two major serial data communication standards are the 20–mA current loop and RS-232. The 20–mA current loop uses an electrical current to carry the data; the RS–232 uses voltage levels. The latter is a recommended standard of the Electronic Industries Association (EIA) and is extensively used throughout the computer industry. There are two RS–232 versions, the older RS–232B and the more recent one, RS–232C. The RS–232C standard was issued by the EIA in an attempt to make it possible to interface equipment made by a variety of manufacturers without the need for special engineering for each case. The idea is to use the same electrical connector (i.e., the DB–25 family of D–shell connectors) wired in the same manner all the time, and to use the same voltage levels for the binary digit 1 and 0 all the time. If everyone interprets the standard the same way, it should be possible to connect together any two devices with RS–232 ports without any problem. Modems (modulators/demodulators), CRT video terminals, printers, and other devices all come with RS–232 connectors, atleast as options. The computers provide RS–232 serial output ports.

It is relatively easy to design an RS–232 port, especially if the applications will operate asynchronously at a low data rate such as 110 to 1200 baud. RS–232C receiver and transmitter intergrated circuits are available from various manufacturers (e.g., Motorola). The RS–232 transmitter/buffer chip will accept TTL–level inputs and convert them to the appropriate RS–232C level for transmission. The RS–232 receiver works exactly the opposite: it will accept RS–232 input signals and convert them to equivalent TTL levels. The RS–232C standard assures a data transmission range from 0 to 20 kilobytes per sec with a distance of upto 50 feet. The standard comprises of three different standards: physical, electrical and functional.

The RS–232C standard uses narrower limits between logical 0 and logical 1 in order to make the data transmission speeder. The upper limits for the logical 0 and logical 1 levels are +/–15V, rather than +/–25V as in RS–232B standard. RS–232C standard fixes the load impedance to 3000 to 7000 ohms, and the driver impedance lower than previously. The logical 1 level is –3 to –15V, while logical 0 level is + 3 to +15V.

As stated earlier, the RS–232 standard specifies a standard connector so that all products will be compatible. The RS–232C specifications call for a DB–25 D type connector. This connector has two rows of pins side by side. One row has 13 pins and the other has 12. This accounts for D shape of the connector and allows it to connect only in the correct pin orientation. Although not all of the pins are electrically used (as few as three pins in some cases), they are still physically in place. The D-shell connector is used for this purpose and is the identifying feature of an RS–232 equipped piece of equipment. The standard 25-pin DB series connector is large and clumsy, and it contains a large number of unused pins. After microcomputers became widespred, a new RS–232 connector evolved that uses a smaller 9-pin connector. Table 1.2 gives comparisons of RS–232 with the updated serial interface standards.

Table 1.2 Comparisons of serial interface standards

Parameter	RS–232	RS–423	RS–422
Signal type	Unbalanced	Unbalanced	Balanced
Maximum distance	50 ft	4,000 ft	4,000 ft
Maximum data rate	20 kbps	100 kbps	10 Mbps
Space volatge	+5 to + 15 V	+4 to 6 V	+2 to +6 V
Space threshold	+3 V	+200 mV	+200 mV
Mark voltage	–5 to –15 V	–4 to –6V	–2 to –6V
Mark threshold	–3 V	–200 mV	–200 mV
Source impedence	300 Ohms	50 Ohms	100 Ohms
Load impedence	3 to 7 kOhms	4 kOhms	4 kOhms

1.7 Plug-in-slots

More important to the user is the I/O channel, which is often referred to as the plug-in-slot or just the slot of a computer. It is into these slots that any additional boards are inserted. The slots contain all the necessary electrical connections to the CPU and other parts of the computer. A variety of devices (Figs. 1.11 and 1.12) can be connected to the PC via slots using interfacing cards each for:

(i) diskette drive attachment card
(ii) parallel print port attachment card
(iii) serial port attachment card
(iv) SDLC adapter card (Synchronous data line card), used for connecting terminal and computer together over telephone lines
(v) memory expansion card
(vi) color graphics display adapter card
(vii) data acquisition and control card
(viii) sound card (for multimedia PC)

The maximum number of slots available for these peripherals boards varies with the types of IBM PC (typically from 5 to 8 slots). Additional slots are available if any expansion chassis is purchased. The expansion chassis contain not only additional slots but a separate power supply as well. These add-

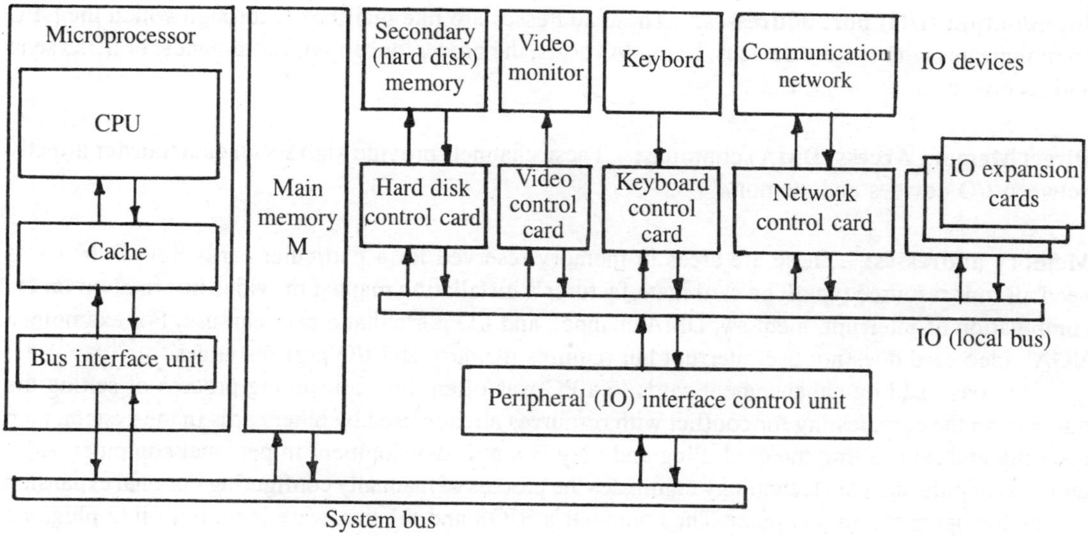

Fig. 1.11 A typical personal computer system showing different adapter cards

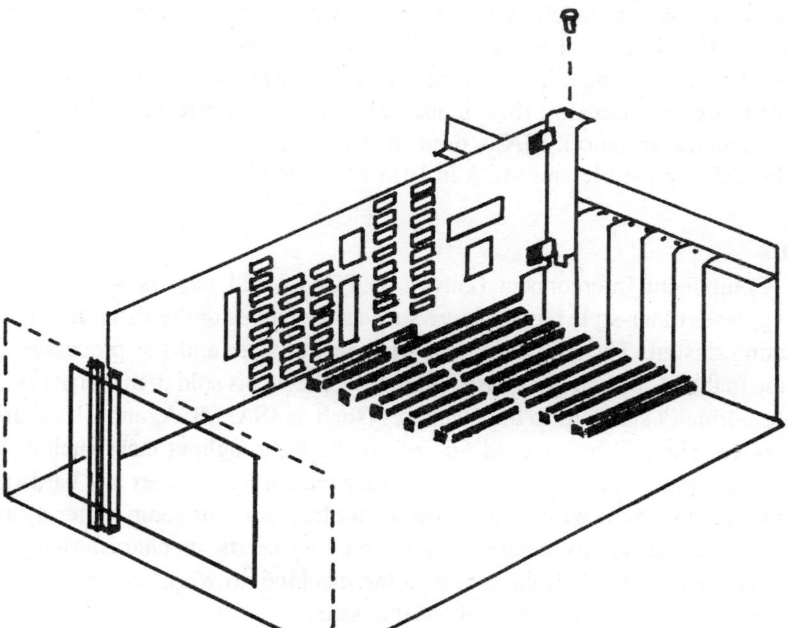

Fig. 1.12 Expansion card fitting

on-card must be configured to indicate the IRQ level, I/O port address, memory address, DMA channel number.

Interrupt request level (IRQs): These are signals that a device uses to communicate with the central processing unit (CPU). Most cards give us more than one choice of interrupt levels.

Input/output (I/O) port addresses: These addresses are like mail boxes through which the CPU communicates with system devices. Using I/O ports, the processor can send commands, or it can send and receive data.

Direct Memory Access (DMA) channels: These channels provide high speed data transfer directly between I/O devices and memory.

Memory addresses: These are areas in memory reserved for a particular card. Not all the cards need all four resource type. The card manufacturer's installation manual provides information on the combination of interrupt, memory, DMA channel, and I/O ports that a card can use. For example, a VGA video card does not use interrupt but requires memory and I/O port resources.

In the past, adding enhancement cards to a PC was often time consuming process of setting the jumpers on the card, testing for conflict with resources already used by other cards in the system, then resetting and reinstalling the card. Plug and play is a new development in personal computer. Add-on-cards of plug and play technology eliminates the process of manually configuring optional expansion boards for operation in computer. The computer's BIOS and the hardware logic built into plug and play add-on-cards work together to automatically assign interrupt reguests (IRQs), I/O and memory addresses, and DMA channels for optimum performance. Plug and play also makes installing non-plug and play expansion boards much easier. However, when installing these cards we must use the ISA configuration utility (ICU) that is supplied with the computer.

The add-on-card is to be compatible with the type of expansion slots in the mother board. There are three types of bus expansion slot: ISA (Industrial Standard Architecture); EISA (extended ISA) and MCA (microchannel structure). ISA is used in IBM PC/AT and most compatible, EISA is for most powerful PC/AT compatible and MCA in IBM PS/2 models.

1.8 PCI Bus

The Peripheral Component Interconnect (known as PCI or PCI bus) is a new type of bus that supplements or replaces older-style bus structure such as the Nubus or the ISA bus with an interactive and interconnecting design. The PCI chip sits between the slots and the processor. To provide a smooth transistion to the PC bus on intel-based platforms, the PCI is sold in machines as a supplement. Bus slots that are normally standard to the machines (such as ISA, MCA, and EISA) are included as the primary interface. The PCI bus is used for specific high throughput tasks such as video or hard disk I/O. The plug and play type card is to give the user the ability to insert any hardware peripheral into a system and begin using it without any special configuration or setup. After a card is plugged in, the PCI asks the card questions, determining its memory needs, its characteristics, and so on, It then configures that card, as well all the others in the machine, to work together so that there is no conflict and no need for the user to worry about the issue.

The PCI requires a set of registers (256 of them) on each card that are used to store configuration information. The content of each register is defined in the PCI specifications to hold such things as the addresses, IQRs, power ratings, and bus mastering priorities for the particular card. These locations essentially replace the need for the configuration of dip switches or jumpers on the card. The PCI specification uses a dedicated line going to each card to read from or write to this area. The PCI bus facilitates configuration of each card through a standard interface. Configuration of the devices present in a PCI system can be done through one of the three means: by the ROM, by the system BIOS through the operating system or by open firmware. Figure 1.13 shows a typical mother board used in a Pentium computer. System controller controls the interface between the processor, memory

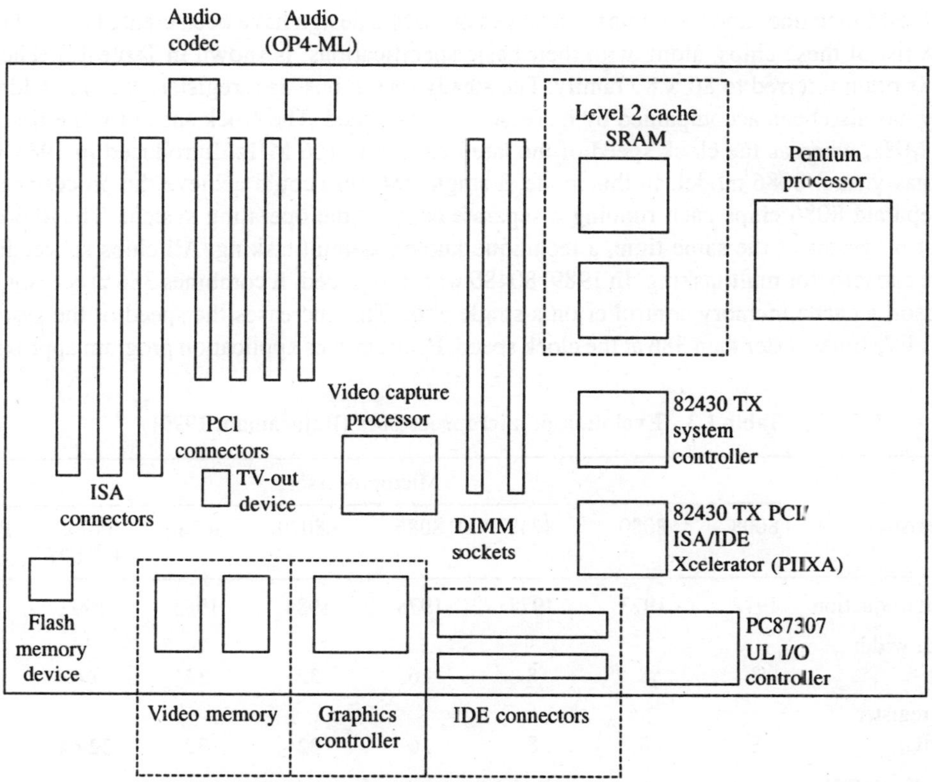

Fig. 1.13 Mother board used in a Pentium computer

and PCI bus. I/O controller controls the interface between the ISA and other buses such a parallel bus, serial bus, floppy disc drives, key board, mouse and infra-red transmission.

PCMCIA (Personal Computer Memory Card International Association) cards are used for storage (mostly volatile) or as modems for portable computers. These cards are new standards for credit-card-size peripherals and are becoming a standard feature on portable and (increasingly) on desktop machines. These cards are sometimes called PC cards (2″ wide and $3^1/_2$″ long and 1/8″ to 1/12″ thickness). With the fast speeds of the PCMCIA storage, data can easily be saved into a PCMCIA card on a portable, and then the card can be taken out and inserted directly into a desk top machine. This technique creates a type of hyper-fast floppy disk for the transportation of data. PCMCIA slots (interface slot that accept the card) are used in desk top machine, portable, digital cameras, and various other devices.

Small computer system interface (SCSI, pronounced as 'Scuzzy'). Instead of plugging the cards into the computers bus via expansion slots, SCSI extends the bus outside the computer by a cable. SCSI is a high speed parallel interface bus. The SCSI can be used to hook up devices such as hard disk, CD–ROM drives, memory storage devices, scanners, and tape drives. We can plug one SCSI device into another to from a daisy chain. To provide personal computer with a SCSI port, SCSI adapter board is inserted into one of the PC's available expansion slots.

1.9 Family of PCs

In 1981 IBM released the first IBM PC, which was based on the microprocessor intel 8088. Since then, all IBM machines and the compatible based on IBM's design have been created around the intel chips. A list of these chips, along with their basic specifications, is shown in Table 1.3. The family of chip is often referred to 80 × 80 family. The steady rise in bus size, register size, and addressable memory has also been accompanied by increase in clock speed. The clock speed for the first PC ran at 4.77 MHz, whereas the clock speed of the late Pentium is 166 MHz. Introduced in 1985 was the 386. It has virtual 8086 model. In this mode, a single 386 chip could achieve the processing power of 16 separate 8086 chips each running a separate copy of the operating system. The 486 can run different programs at the same time, a technique known as multitasking. All chips succeeding 386 have the capacity for multitasking. In 1989, 80486 was introduced. It combines 386 processor, a math coprocesor, a cache memory controller on a single chip. This increases the speed of the system. PC 80386 is five times faster than 386 at the clock speed. Pentium runs application program approximately

Table 1.3 Evolution of microprocessors (Rajaraman, 1999)

	Microprocessor							
Characteristics	8008	8080	8748	8086	68020	80386	Power PC 603	Pentium celeron
Year of introduction	1972	1973	1977	1978	1984	1985	1993	1998
Data path width (in bits)	8	8	8	16	32	32	64	64
Internal register width (bits)	8	8	8	16	32	32	32/64	64
Number of internal registers		8	8	14	16	16	64	32
Directly addressable memory (Bytes)	16K	64K	4K	1M	4G	4G	4G	4G
Estimate of throughput	1	10	20	100	1000	1000	5000	50000
Clock speed MHz	0.75	3	5	5	16	20	80	300
Manufacturer	Intel	Intel	Intel	Intel	Motorola	Intel	IBM/Apple	Intel
No. of transistors in chip (in 1000s)	4	6	20	29	120	200	1600	4000

five times faster than 486 at the same clock speed. Pentium can process more than one instruction in a single clock cycle. In 1995, the Pentium pro was introduced which can process three instructions in a single clock cycle. Similarly Motorola company, another manufacturer of microprocessors for small computer systems, introduced 680 × 0 family of microprocessors (68000, 68010, 68020, . . . , 68060, Power PC MPC 601, MPC 603, MPC 604, MPC 620). In general, computers can be classified as micro, mini and mainframe computers. Table 1.4 gives a comparison of these types of computers.

Multimedia PC: A PC capable of producing high quality text, graphics, animation, video, and sound may include CD-ROM drive, microphone, speakers, a high quality video controller and a sound card.

Table 1.4 Comparisons of micro, mini and mainframe computers

Criterion	Mainframe	Minicomputer	Microcomputer
Volume of data	high	medium	low
Application	universities	colleges	schools
Number of users	hundred	tens	single
Cost (up to)	$ 3.5 million	$ 150,000	$2,000
Mobility	low	average	very high
Mode	batch	both	real time
Main memory	100 Mbytes	50 Mbytes	10 Mbytes
Storgae capacity	80 Gbytes	8 Gbytes	500 Mbytes
Speed (MIPS)	500	100	20
(MFLOPS)	100	75	15
Interfacing	complex	moderate	simple
I/O ports	over 200	15–200	15

MIPS–Million Instruction Per Second
MFLOPS–Million Floating Point Operations Per Second

Work stations: Work stations are also desk top machines. They are more powerful, providing processor speed about 10 times that of PCs, large colour video display (19 inch monitor) and have main memory of 64 MB. Work stations normally use RISC processors and very good graphic facilities. The standard operating system is UNIX. Work stations are used for executing numeric and graphic intensive applications such as arising in CAD and simulations of large scale systems requiring visualization of the results.

Industrial personal computer: PCs are commonly used in the protected environment of offices and laboratories. In an industrial plant the computing power is not the only priority. A greater concern is the reliability of the control system hardware. Manufacturers have introduced rugged versions of popular personal computers. For example, IBM has developed the 5531 and 7531/32, industrial versions of PC/XT and PC/AT, respectively, to withstand the environment of the factory floor. They have the advantage of being able to run any PC-DOS-based software, but are unable to perform real-time control. This problem has been remedied with the advent of the industrial computer disk operating system (IC-DOS), a real time operating system that is compatible with other IBM software. This allows a shop floor computer to provide real-time control while using the software packages previously found only on office models. Although the cost of industrial computer hardware is high in comparison to the desktop machines, the savings in terms of reduced downtime and improved plant performance by virtue of the more advanced control system can result in increased profits for the plant.

Laptop PCs: Laptop PCs (also known as notebook computers) are portable computers weighing around 2 kg. They use a keyboard, flat screen liquid crystal display, and a 80486 or pentium processor. Laptops come with both hard disk (around 200 MB) and floppy disk. They can run on batteries and are designed for energy conservation. Many Laptops can be connected to network. The most common use of Laptop computers is for word processing, and spreadsheet computing while a person is travelling or control engineer would like to test and calibrate smart instruments on plant sites.

1.10 Operators Interface

A mouse is a device that fits in the palm of your hand and glides across a desktop when only a minimal force is applied. It usually has two or three buttons on top that are operated with the fingertips. The computer's operating system and the application program must be written to accept it. When a mouse is used on a system, overall throughput can be greatly increased. The user can physically move the mouse back and forth causing the selection item on the screen to change in accordance with the position of the mouse. Once the item desired is selected, the button on the mouse is pressed, and that function is performed. This method is known as 'point and click'. Another use for the mouse is for drawing. A mouse with a clear plastic guide plate with cross hairs can be placed on a picture surface and trace it. This is how a diagram can be entered into a computer and then modified.

Track balls: A track ball is like an upside-down mouse. It is stationary; the ball (which is much larger then that of a mouse) is moved with your hand. The same type of movement sensors are used as in a ball-type mouse. Track balls are not used as frequently on computer systems as the standard mouse. A track ball is a good alternative to a mouse when one has limited desk space.

Joysticks: Joysticks have rod that sticks up from the main mechanism with a ball attached to the end of this rod. The rod is held in the hand and can be moved forward and backward, and left to right any direction on a two dimensional plane. Some of the uses include drawing, menu selection and game playing. Joysticks are used in arcade type computer games because they allow movement quickly and accurately in any direction.

Touch screens: Touch screens are useful in applications where a keyboard is impracticable, such as in a factory or dirty surroundings. With the touch screens, the user touches the video display surface in response to prompts on the screen. The computer can then read in the responses and act on them. There are two types of touch screens; one that uses wire-mess screen as the sensing mechanism, and one that uses optical sensors.

Light pens: A light pen looks like a writing pen and has a small cable going from it to the video-generating circuitry. Light pens are used in similar circumstances as touch screens, especially where there are just a few responses to every screen prompt. Despite the name, light pens actually do not generate any light rather, they sense the presence of light that is generated by the computer screen.

1.11 Computer Languages

Software is developed using one of the computer languages. Computer languages are either of low level or high level. The main difference between the two is in the extent of programming effort, with high-level languages requiring less effort. In a low-level language, the user programs the computer in 0s and 1s. The low-level language is also called machine language or assembly code, depending on how it is entered. In either case, the program 'talks' directly with the processor, without any intermediate interpreter or compiler. This has both merits and limitations for application programming. The merits are faster response, increased program flexibility, and better control. The limitations are excessive programming time, lack of data structures, required detailed knowledge of the hardware system, lack of programming standards, and no upgrade option. Programming a high-level language, on the other hand, can use English-like commands. The keyword PRINT used in BASIC, a high level language is simple to remember. High-level languages allow powerful operations with fewer programming steps, making them ideal for applications where speed is not important but data processing and other considerations are.

The main differences between a high-level language and machine or assembly language are: (i) High-level languages do not require that users be aware of microprocessor features such as register, I/O channels, internal representation of data, and so forth; (ii) programs written in a high-level language run on any computer provided appropriate compliers or interpreters are available; (iii) High-level language programs are more concise; and (iv) Programs written in high-level language can be designed to use problem-oriented terms. In other words, they may be used to express the problem in familiar words. Computer languages are numerous. The C language is quite popular today. One advantage of C is that it is closer to the machine language, enabling the processor to execute more efficiently. Moreover, programs written in C are easily transportable between computers.

Summary

The components of a PC are mother board containing CPU, memory, expansion slots, and hard disk, monitor etc. In the expansion slots, add-on-cards are added for interfacing with peripheral devices, for example data acquisition control card for interfacing with real world sensors. Computers communicate with the external devices by polling or interrupt.

References

Breeding, K.J., Microprocessor system design fundamentals, Prentice Hall International Edition, New Jersey, 1995.

Brey, B.B., The Intel microprocessors, architecture, programming and interfacing, IV ed. Prentice Hall of India Pvt. Ltd., New Delhi, 1997.

Doyle, C.F., Computer peripherals, Prentice-Hall, Englewod Cliffs, New Jersey, 1986.

Gayakward, R. and C. Sokoloff, Analog and digital control systems, Prentice Hall, Englewood Cliffs, New Jersey, 1988.

Kheir, M., The M68Hcll micro controller: application in control, Instrumentation and communications, Prentice Hall International Edition, New Jersey, 1997.

Lotia, M. and P. Nair, Modern all about mother board, BPB Publications, New Delhi, 1996.

Norton, P., Introduction to computers, Tata-McGraw Hill, New Delhi, 1998.

Rajaraman, R., Fundamentals of computers, 3rd ed. Prentice Hall of India Pvt. Ltd., New Delhi, 1999.

Stechhahn, A.D. and J. Den Otter, Industrial applications for microprocessors, Reston Pub. Co. (Division of Prentice Hall), 1982.

Review Questions

1. Discuss with a diagram the components of a PC mother board.
2. Discuss with a block diagram, the address, data and control bus structures.
3. What are the three parts of memory and discuss them briefly.
4. Discuss on isolated I/O and memory mapped I/O.
5. Give a comparison of different memory used.
6. What is an operating system?
7. What are the steps involved in communication of a PC with external devices?
8. Discuss briefly (i) serial I/O port and (ii) parallel I/O port.
9. Give a comparative study on synchronous and asynchronous communication.
10. Name the different adapter cards used in a PC.
11. Discuss briefly on: (i) PCI bus (ii) PCMCIA card and (iii) SCSI.
12. Give a summary on the evolution of microprocessors?
13. Compare briefly micro, mini and mainframe computers.

2

Computer Interfacing for Data Acquisition and Control

Real-world signals are not compatible with the binary formats used by the microcomputer. Hence, there is a need for signal conversion before and also after processing. In this chapter, the additional circuits required for connecting real world sensors to the computer and for connecting computer to the final control elements are presented.

2.1 Signals

The majority of the signals are analog signals. Analog signals are defined over continuous time interval and assume a continous range of amplitude values. An analog to digital converter (ADC) module changes an analog input signal into a binary output code. For an input analog voltage or current there is a corresponding proportional binary word output. If an 8-bit ADC has a 0 to 2.56 V input signal range, then 0 V input could produce a output word of 00000000, while the +2.56 V level seen at the input would produce an output of 11111111.

As stated earlier, digital signals have only two discrete levels or states. These states are defined as one and zero (1, 0; specifically two levels either *high* or *low*). The control calculations are then carried out by the computer. The signal must be sent to the final control element of the plant. Most of the final control elements we use in the feedback control loop are analog signal based devices, i.e., they require a continuous range of signals such as voltages or current. A digital to analog converter (DAC) produces a current or voltage output that is proportional to some binary word applied to its input. A 8-bit DAC may produce an output signal 0V when the binary word applied to its digital input is 00000000 and say, 2.56 V when the digital input is a word of 11111111. That is, a digital to analog converter receives the binary value (digital signal) from the computer (CPU) and converts into analog voltage (or current) that can be used for actuating external final control element or external devices or for displaying the digital data from the microcomputer to the equivalent analog data. Figure 2.1 shows the block diagram representation of a data acquisition system.

Digital to analog converters and analog to digital converters are available for the IBM PC in a number of forms. The most common form is as one of the components on a general purpose laboratory data acquisition interface card, which plugs into one of the slots of the IBM PC. This card typically contains an analog to digital converter, one or more D/As, and sections for digital input/output and counting/timing.

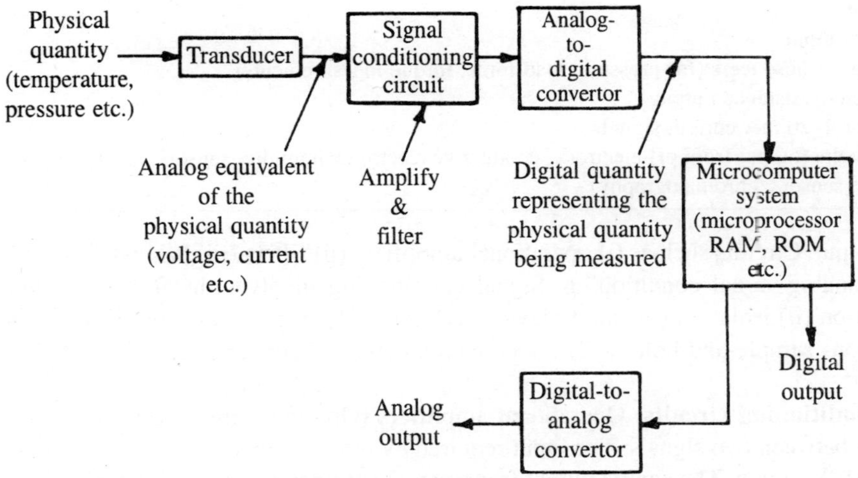

Fig. 2.1 Configuration of a typical data acquisition system

A/D conversion specifications: The A/D specifications are given by conversion time, resolution and accuracy. Conversion time is the time required to complete a conversion of the input signal. It establishes the upper signal frequency limit that can be sampled without aliasing.

$$f_{max} = 1/(2^* \text{ conversion time})$$

Resolution is the number of bits in the converter. The resolution is the smallest analog signal for which the converter will produce a digital code. It may be given in terms of the full scale input signal.

$$\text{Resolution} = (\text{full scale signal})/2^n$$

Resolution relates the smallest signal to the full scale value. Accuracy relates the smallest to the measured signal. Accuracy is given by percentage and describes how close the measurement is to the actual value. The signal is accurate to within

$$(V_{resolution}/V_{signal})\ 100\ \%$$

When converting analog information into digital information, the resolution of the data conversion is important consideration. A/D converters have 8 bit, 10 bit, 12 bit or 16 bit precision. For example, an 8 bit converter with a range of –5V to +5V will slice the total voltage span (10 volts) into 255 steps. Each step or resolution would be equal to 39.2 mV. A 12 bit converter will slice the 10 volt span into 4095 slices. The step size or resolution is 2.4 mV. The higher the resolution, the higher is the cost, however.

2.2 Interfacing Input Signals

2.2.1 Analog Signal Conditioning

Table 2.1 provides a list of transducers that give analog signal as the output. The analog signal from a transducer must be conditioned by a signal conditioning circuit to meet the input requirements of

Table 2.1. Sensors giving output as analog signal form

Thermocouple output
RTD output
Thermistor output
Strain gauges transducers (for pressure, load force, torque measurements)
Potentiometer/resistance output
0–20 mA or 4–20 mA current signals
millivolts/volts (output from pH electrode, conductive electrode, humidity sensors, level sensors, flow sensors, pressure sensors, chromatography)

an ADC input. Circuits such as (i) operational amplifiers (ii) filters (iii) bridges (iv) comparators, are used for analog signal conditioning. Signal conditioning involves the following steps: (i) signal amplification (ii) isolation (iii) multiplexing (iv) noise filtering (v) transducer excitation (vi) use of simultaneous sample-and-hold, (vii) I/V converter (viii) V/I converter and (ix) antialiasing filter.

Signal conditioning circuits: Operational amplifiers (Op-amps) are used to amplify a signal or a difference between two signals. The input required by most of commercially available A/Ds is in the −10V to +10V region. The analog output from many instruments is at low level signals. An amplifier is thus often required. In order to take full advantage of the resolution of the A/D, the amplifier should be designed to provide outputs over the entire range of the A/D. For example, a signal form the instrument that varies over the range of 0 to 10 mV should be amplified by a factor of 1000 to provide 0 V to 10 V output.

There are two basic methods for carrying out the amplification steps: using an external amplifier or buffering an A/D with a built-in-amplifier. Since the external amplifier can be placed inside the instruments or at least very close to it, thereby reducing the effect of noise; they tend to be more expensive than the internal systems, because of added hardware (excluding the power supply unit). They are slower than the internal systems. Even though the data can be collected at a faster rate, transmitter of the digitized information to the host is limited by the transmission speed. The use of amplifier built in an A/D board has three advantages: it is convenient and it avoids possible grounding and impedance matching problem that are encountered with external amplifier. Many of the commercially available A/D boards give the user the option of an on-board amplifier. In some cases, the gain of the amplifier is fixed; in others, the gain can be modified by changing the hardware on the A/D board (i.e., changing the size of the resistor). It is also possible to change the gain in the software between data conversions; this is referred as programmable gain. However, it may slow down data collection because of the need to check the size of the input signal and to reset the gain when appropriate. Its chief advantage is the added flexibility it gives to the user in selecting the gain as the experiment proceeds.

A typical A/D converter for PCs will use a 12 bit of resolution in a fixed analog signal range of −10 V to + 10 V. At best the A/D will be able to distinguish between analog voltage of 5 mV apart (20 V/2^{12}). A signal transducer such as a thermocouple will only produce small voltage; perhaps in the range of 0 to 30 mV. As a result, the 12 bit A/D converter will produce only 6 out of a possible 4096 integers to describe the temperature being recorded by the thermocouple. If the thermocouple range covers 0 to 2000°C, then the A/D can only resolve 333°C (2000/6) differences in temperature. This is not at all acceptable. An external analog amplifier should be used to amplify the thermocouple signal to provide more resolution. An amplifier with a gain of 333 will amplify the thermocouple signal so that the full temperature range (0 to 2000°C) will correspond to 0 to 10 V at the A/D converter input. In this case, with the external amplifier the A/D can resolve 0.98°C (2000/2048) difference in temperature.

2.2.2 Input Signal Buffering and Amplification

Some interfacing applications require impedance matching between the sensor and the input terminal on the DA and C (data acquisition and control) interface board. A unity gain operational amplifier is connected between the sensors and the DA and C board. The high input impedance of the operational amplifier minimizes the loading in the sensor and prevents signal degradation. The low output impedance of the operational amplifier is ideal for load such as motors, recorders or connection to DA and C board.

External amplifiers can also be used as 'buffers' to electrically isolate sensors and transducers from the PC data acquisition systems. This is done so that the noise or signals in the PC A/D converters do not affect the transducer responses. Other sensors that provide large signal such as piezo-electric transducers may damage the PC if they are not buffered or electrically isolated from the PC.

If analog signal produced by transmission are too large for direct conversion to an analog to digital converter, then the signal must be reduced to a lower value. This scale down can be achieved by dropping the required voltage across a suitable resistor.

Offset: An offset occurs with a variety of sensors and transducers. An offset voltage is unwanted voltage that is produced by a sensor though the sensor theoretically should be producing zero volt. Offset occurs due to inaccuracies in manfacturing and calibration of sensors. The offset can be eliminated by using signal conditioning circuit like an operational amplifier connected in a summing application.

Filter: Filter is required in signal conditioning circuits to eliminate unwanted frequencies. There are two types of filters: passive and active. Passive filters are designed using standard discrete component such as resistors, capacitors and inductors. Filters are classified as low pass filters, high pass filters and band pass filters depending on the range of frequencies over which signal is useful.

Bridge circuit: It is used to measure the small resistance changes caused by a resistive transducers (refer to Fig. 2.2). Transducers using wheatstone bridges include strain gauges, thermistors resistance temperature detectors (RTDs) and almost all resistive transducers.

Isolation: In many situations, isolation of the control and controlled signals is desirable or even necessary. For example, the digital output from computer may be used to switch the power to a device on and off. The instrument at the other end may be used to run a furnace or a pump or anything else. While it is quite possible to use the digital output to directly switch thyristor or triac, it is not desirable since a failure of thyristor, may in certain cases lead to feeding the main to the computer with disastrous results. This can be avoided using solid state (optically coupled) relay, or magnetically coupled relay or reed relay circuit. Another situation may be an electrically noisy system where it may become necessary to isolate the controller and the controlled system to avoid feedback of noise into the controllers.

Current loop: If analog sensors are located at a considerable distance from the computer or controller, then the analog signal is transmitted as current rather than voltage signal. The standard

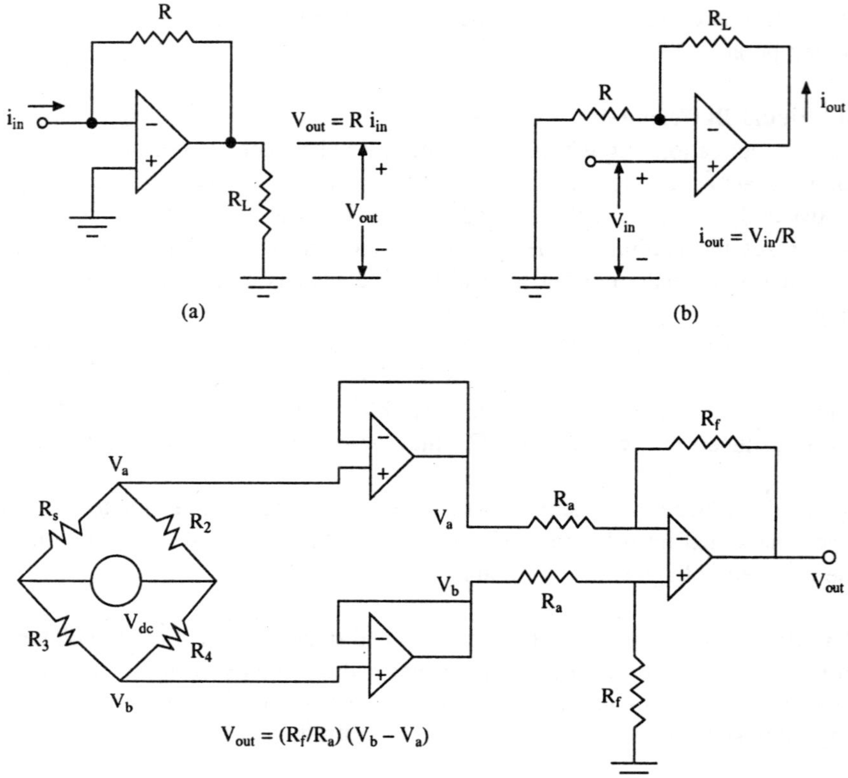

Fig. 2.2 (a) I/V converter (b) V/I converter (c) Unbalanced Wheatstone bridge and instrumentation amplifier

associated with a current loop circuit is 4 to 20 mA circuit. The current ranges from a minimum of 4 mA to a maximum of 20 mA. The minimum current of 4 mA always in the circuit can be used as a check of circuit integrity. A signal above 20 mA indicates the system malfunction. A special module transmitter does the conversion of signal from voltage to 4 to 20 mA current signal (Fig 2.2 and 2.3). The current loop provides a high degree of noise immunity as well as a lack of loading difficulties caused by more than one receiving device in the loop. For example, current loop can be connected to a DA and C board and as well be connected to a strip chart recorder, or other instrumentation devices. Current loop analog transmission is unaffected by the distance between the transmitter and

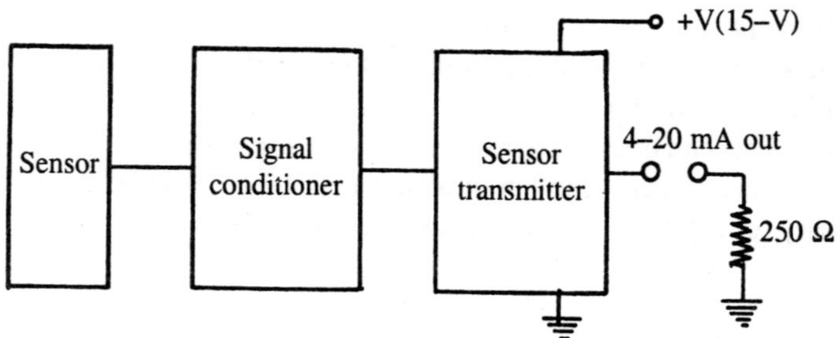

Fig. 2.3 Single conditioning and current loop operation

the receiver. Similarly, most of the actuators (Table 2.2) also work with 4–20 mA control signal. The main advantage of the current loop operation is that the signal is not affected by long cable length. Also, this leads to standardization operation, making it simpler to develop interfaces. Quite often the signal converter that converts the transducer output to a linear 4–20 mA signal induces electrical

Table 2.2 Some typical control system actuations

control action	devices		input
Electric heating	Electric furnace		Voltage or current
Flow adjustment	Pneumatic valve		Air pressure
	Solenoid valve	}	Voltage, current
	Motor-driven valve	}	digital pulses
	Variable speed pump		Voltage or current
	Fluidic control valve		Pressure
Alarms	Lights, bells		Digital pulse
On-off signals	Relays, switches		Digital pulse

isolation of the input from the output. Similarly, many analog elements such as valves are operated by 4–20 mA signals through current to pressure (I/P) converters (Fig. 2.4). For A/D conversion, a short register (normally 250 ohms) is required at the ADC end to convert loop signal in to a voltage. 4–20 mA signal is converted into 0–5 volts, which is easy to handle with single ended inputs.

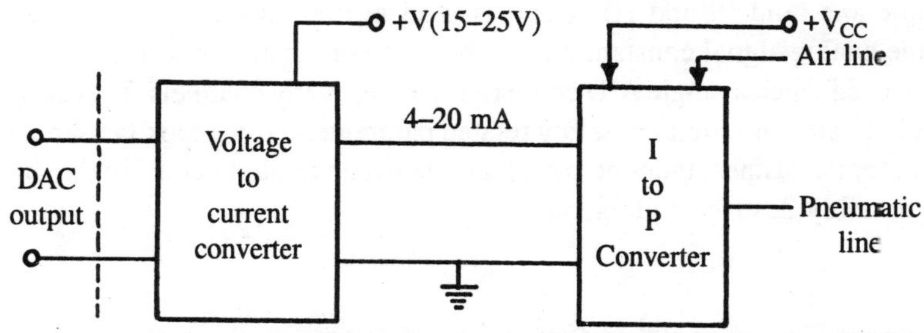

Fig. 2.4 A current loop driving a pneumatic valve through an I to P converter

Analog voltage signals are susceptible to electrical noise and loading of the transmission line between the sensor and monitoring device. If the distance between the sensor and the computer is short then voltage signal can be used. Typical standard analog voltage ranges are 0 to 5 volts, –5 to +5 Volts, –2.5 to +2.5 volts; –50 to +50 mV.

Of more importance is the addition of error introduced by the changes in the analog signal during the ADC conversion time. By definition the digital output of the the ADC should be equivalent to the value of the analog voltage at the instant the start conversion is given. The conversion time of the ADC limits the rate at which the sensor signal can change without producing errors. The solution to the problem is to use either a faster AD (usually expensive) or a sample-and-hold circuit.

Sample-and-Hold circuit

The sample-and-hold (S and H) shown in Fig. 2.5 consists of an electronic switch, a capacitor and a buffer amplifier. The circuit is switched to the sample mode using a control line from the microcomputer

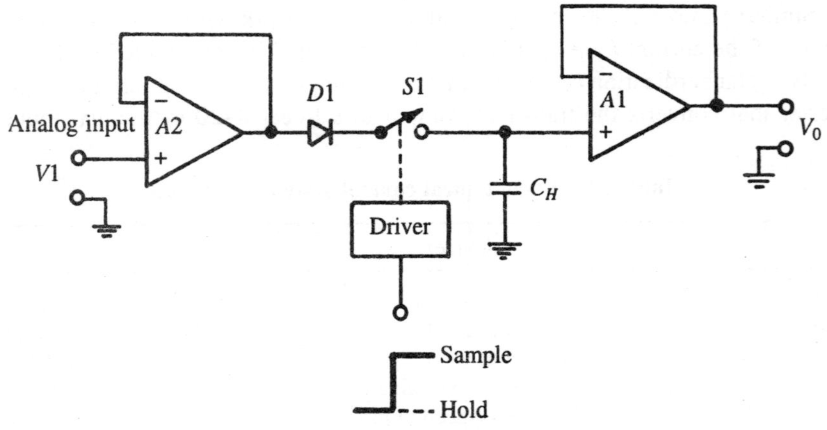

Fig. 2.5 Sample and hold circuit

port and this closes the switch forcing the capacitor to charge to the value of the analog input. After a short interval the circuit is put into the hold mode and the switch is opened. The capacitor then retains the value of the analog signal to which it had been charged. During the hold mode, the ADC performs the conversion. Since a high-quality capacitor and buffer amplifier are used in the sample-and-hold circuit, the voltage signal presented to the ADC will remain constant during the conversion time.

There are two major uses of the S and H circuit in computer data acquisition systems. As stated earlier, the sample-and-hold (S and H) circuit is used at the input of analog-to-digital converters where it keeps the analog signal constant during the data conversion process. Sometimes multiple S and H circuits are used when a single A/D converter handles many channels. This arrangement allows us to take the digital value of several possibly related parameters. If we took the samples at sequential times, instead of identical times, then the correlation between related values in the different channels is reduced, and possibly destroyed altogether.

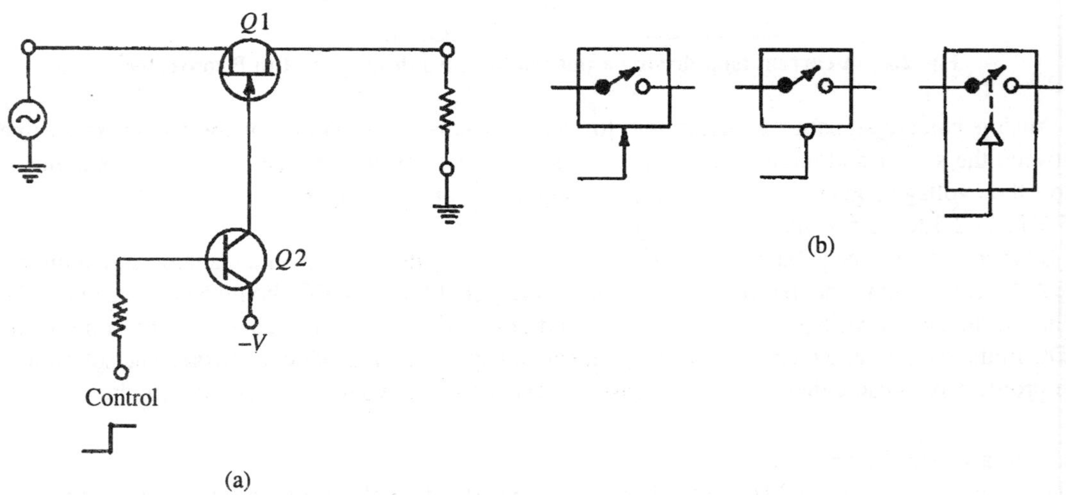

Fig. 2.6 Electronic switch (a) circuit (b) symbols

Analog switches: The heart of the S and H circuit is an analog electronic switch (Fig. 2.6a) while several possible circuit symbols are shown in Fig. 2.6b. In this example, the switching element is a single junction field effect transistor (JFET), while in others (perhaps most) it is either a single metal oxide semiconductor field effect transistor (MOSFET) or a complementary pair of N-channel and P-channel MOSFETs. In any case, the idea is that the transistor is connected in series with the signal path. When the transistor is biased fully off, then no signal passes easily from input to output sides. Typical off/on resistance ratio may be 1000000:1.

The usual field effect transistor analog switch contains a switch driver (such as Q2) that will allow a TTL voltage signal to turn the switch on and off. In the example (Fig. 2.6), the driver transistor is used to control the V-bias to the gate of the JFET. When the base (control input) of Q2 is LOW, then the base is at a lower potential than the emitter, so the transistor will conduct. The gate of the JFET sees a high negative bias that shuts the switch off. Similarly, when the control line is HIGH, then the transistor is reverse-biased and the switch is turned on.

Sampling rate

The next important point is concerned with the sample rate of the system. This is important since between sample, while the microcomputer is carrying out instructions concerned with computing the required output value and outputting this value, the whole system is open loop. The rate at which sampling is carried out will have an important effect on the stability of the system. Too long a period between the samples allows the output to overshoot the desired level since the information from the sensor is gathered too late for the system to take corrective action. If the DAC is an 8 bit type, the manipulated variable can be set to 2^8 (256) different levels which should be more than sufficient to give a smooth control.

The error in reconstructing the signal caused by too low a sampling rate is called aliasing. The effect of aliasing is that it is then possible to construct two signals of different frequencies from the sampled values. The signals from sensors are necessarily complex wave shapes. However, any waveform can be shown, by Fourier analysis, to be made up of a whole series of sine waves of different frequencies and amplitudes. As long as as the sampling rate is at least more than double the frequency of the highest frequency component present in the analog signal, information will not be lost. This follows for Shannon's sampling theorem:

A continuous signal can be represented by, and reconstructed from a set of instantaneous measurements or samples of the signal which are made at equally spaced time. The signal must be sampled at a frequency which is greater than twice the highest frequency component present in the signal.

In practice, the sampling frequency is usually made five to ten times higher than the highest frequency component in the sampling input.

The maximum speed at which the A/D can operate may also be an important consideration for the applications. Experiments can be classified as slow, fast and ultra fast. Slow experiments are those that involve sampling rates less than 1 to 10 Hz, e.g., measurements of weather conditions or plant growth rates and some types of chromatography. Fast experiments are those in the 10 to 100 kHz range, which includes most laboratory experiments. Ultra-fast experiments are those requiring rates above 100 kHz, such as measuring flash photolysis or nuclear chemical reactions.

Multiplexers

A multiplexer is an electronic equivalent of a rotary switch. When in operation the common line is connected to one of the eight channels. The MUX can be disabled, in which case the common is not connected to any of the channels. For a four channel MUX it should only require two binary lines to

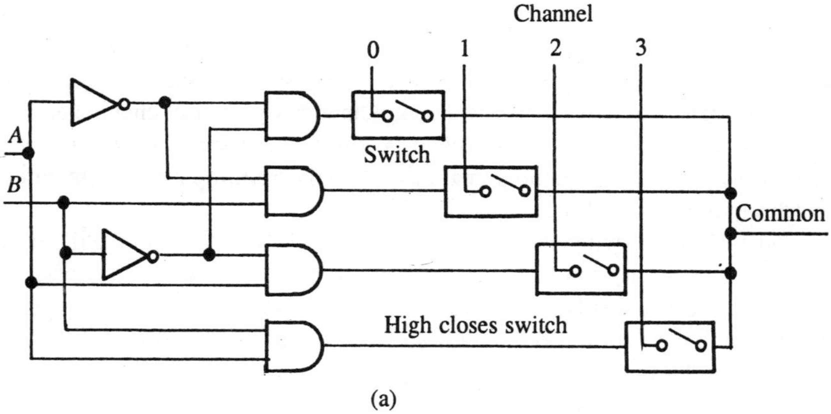

A	B	Channel
0	0	0
0	1	1
1	0	2
1	1	3

(b)

Fig. 2.7 (a) A gate circuit for an elementary four-channel MUX, and (b) channels selected by various combinations of control lines A and B

select any of the four different channels. Figure 2.7a shows a gate circuit that in conjunction with the four switches form a rudimentary MUX. The channels selected by the four combinations of A and B are shown in Fig. 2.7b.

2.3 Digital Signal Conditioning

Table 2.3 provides the list of final control elements that require digital signals and sensors that give output as digital signals. The electrical output voltage and current rating for most DA and C boards is limited to modest electrical ratings. Output ratings are 1.7 mA (sinking) while maintaining ≤ 0.45 V and 200 μA (sourcing) while holding the output voltage ≥ 2.4 volts. These current and voltage

Table 2.3 Digital input/output

TTL or CMOS I/O
Optically isolated I/O
Electromechanical relays
Solid state relays
Frequency input
Proximity sensors
Photo-electrode sensors
Switches/contact . . .
Encoders

values are satisfactory for interfacing to TTL compatible devices such as integrated circuit chips. However, when integrating a DA and C board to real world equipments such as electrical motors connected to pumps, fans, and other rotating devices, electromechanical solenoids, and other heavy duty elctromechanical devices, the TTL compatible outputs simply can not drive these devices.

The output port is said to be sourcing current if it flows out of the port into the load. It is said to be sinking if the current flows through the load and into the output port (Fig. 2.8).

Most actuators require more than 100 mW. Since most types of digital logic circuits source or sink 20 mA (100 mW), the port by themselves are not enough to drive the actuators. The TTL gate will sink 16 mA (i.e., supply a current to ground through a load connected to a +ve supply). Unfortunately it will not source (supply current from a +ve supply through a grounded load) in as much. This means

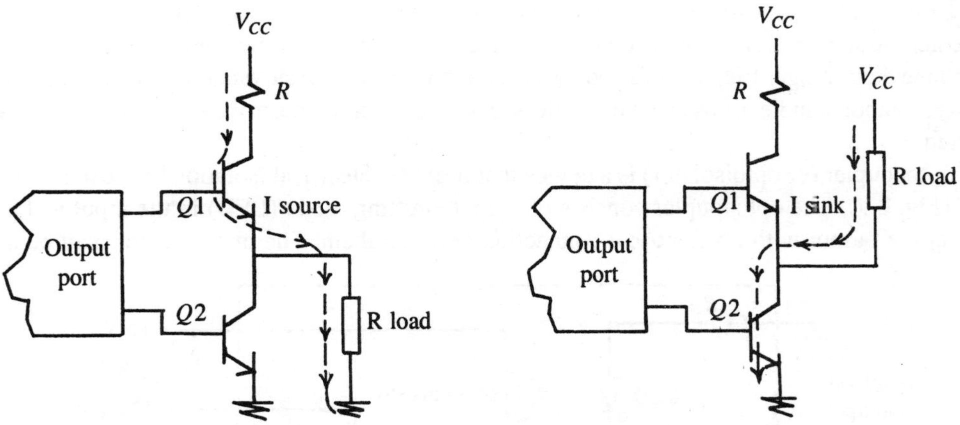

(a) Sourcing current (b) Sinking current

Fig. 2.8 Output with two state actuator

that the device being controlled must be connected between the gate and the +ve supply, and is turned on by a low output, or logic zero.

Driving circuits are: (i) integrated circuits, (ii) discrete solid-state devices, (iii) electro-mechanical relay and (iv) solid state relay.

2.3.1 Electro-mechanical Relays

Electro-mechanical relays have been the primary interface between solid state electric current and heavy duty real world devices such as motors, pumps, and solenoids. If a relay is of high current or voltage rating, then a transmitter must be selected so that the current and voltage ratings are compatible. Proper buffering and current are needed for an interface between TTL level and the real world of electro-mechanical relays.

Electro-mechanical relays have been used to control high DC or AC voltages and currents. Relays allow the sensor to interface with motors, solenoids and other inductive loads. A contact can be either normally open or normally close. The current and voltage requirements of the relays are higher than TTL gate can provide. Transistors buffers can be used to drive relay coil. Contact life depends on the load current and frequency of operation. The relay coil must be shunted by a freewheeling diode to protect it from damage when the driver is turned off.

Advantages of electro-mechanical relays: Switches high currents loads; multiple contacts; switches AC or DC votages; tolerant of inrush current.

Disadvantages: Slow response time (10 to 25 ms); mechanical wear; corrosion of contacts due to arcing; contact bounce; affected by shocks and vibration.

2.3.2 Solid State Relay

The problem of relay can be overcome by replacing relays by solid state relays (SSR) such as triac or silicon controlled rectifier (SCR). They are capable of controlling only AC operated devices such as motors or heaters. The advantages are: fast response time (8 ms) and longer life. The disadvantages are: it can be falsely triggered by large inductive current and can be destroyed by short circuit and it is used only for controlling AC.

Transistor: It is a solid state DC switching device used with low voltage DC powered conductive and capacitive sensors as the output switch. Advantages: instantaneous response; low off state leakage and voltage drop; longer life; not affected by shock and vibration. Disadvantages: low current handling capacity; cannot handle inrush current unless clamped; can be destroyed by short circuit unless protected.

An optocoupler (or optoisolator) is a device that provides electrical isolation between a source and a load (Fig. 2.9). The optocoupler consists of a light emitting diode (LED) in the input and a photo sensor at the output with no electrical connection between them. The photo sensor may be a photo transistor, or a SCR or photo Darlington type or photo triac type. SCRs or triacs are suitable for AC power switching applications (such as motors, solenoids) whereas Darlington output for applications requiring large current gains and photo transmitter output for general purpose applications. Optocouplers allow interfacing systems with different logic levels, different grounds etc. that would otherwise be incompatible with the controller.

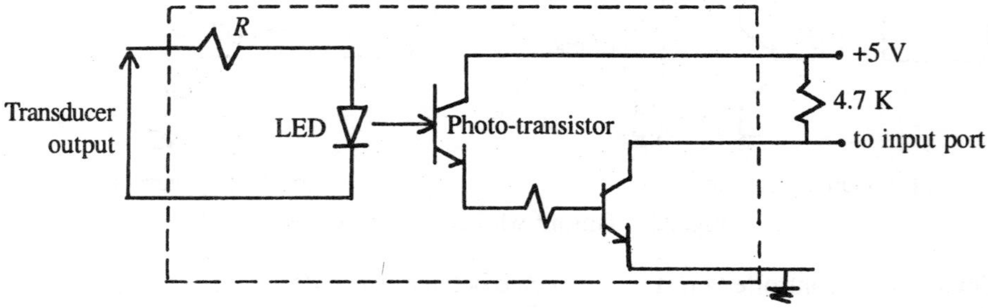

Fig. 2.9 Optocoupler

The current gain of the transmitter switch is improved by using the Darlington connection (Fig. 2.10). This has two (or more) transistors wired so that the base current drive to the output transmitter is supplied from the emitter of the first. This type of connection gives a very high value of current gain (typically 200 to 1000). Darlington can switch a large output current (2A) with a relatively small input current (= 4 mA). In Fig. 2.10 two resisters and a diode are included to reduce leakage current. An output load power of say 300 W can be switched with an input power of about 60 mW. Since collector to emitter current is higher, resulting in to a high heat. Hence, proper air cooling arrangement to be made.

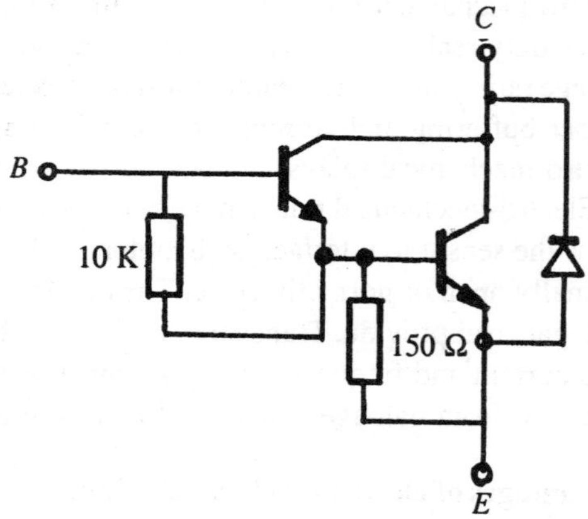

Fig. 2.10 Power Darlington solid state device

Heater control circuit: If the desired temperature is higher than the actual temperature, (output A > B), then the heater is to be on. We also isolate the high voltage AC signal from the low voltage digital signal from the computer. An optically isolated drive MOC3010 satisfies the need (Fig. 2.11). To increase the drive capability of a MOC3010 an additional triac may be connected at the output of the MOC3010 depending on the electrical ratings of the load it drives. The current from the computer

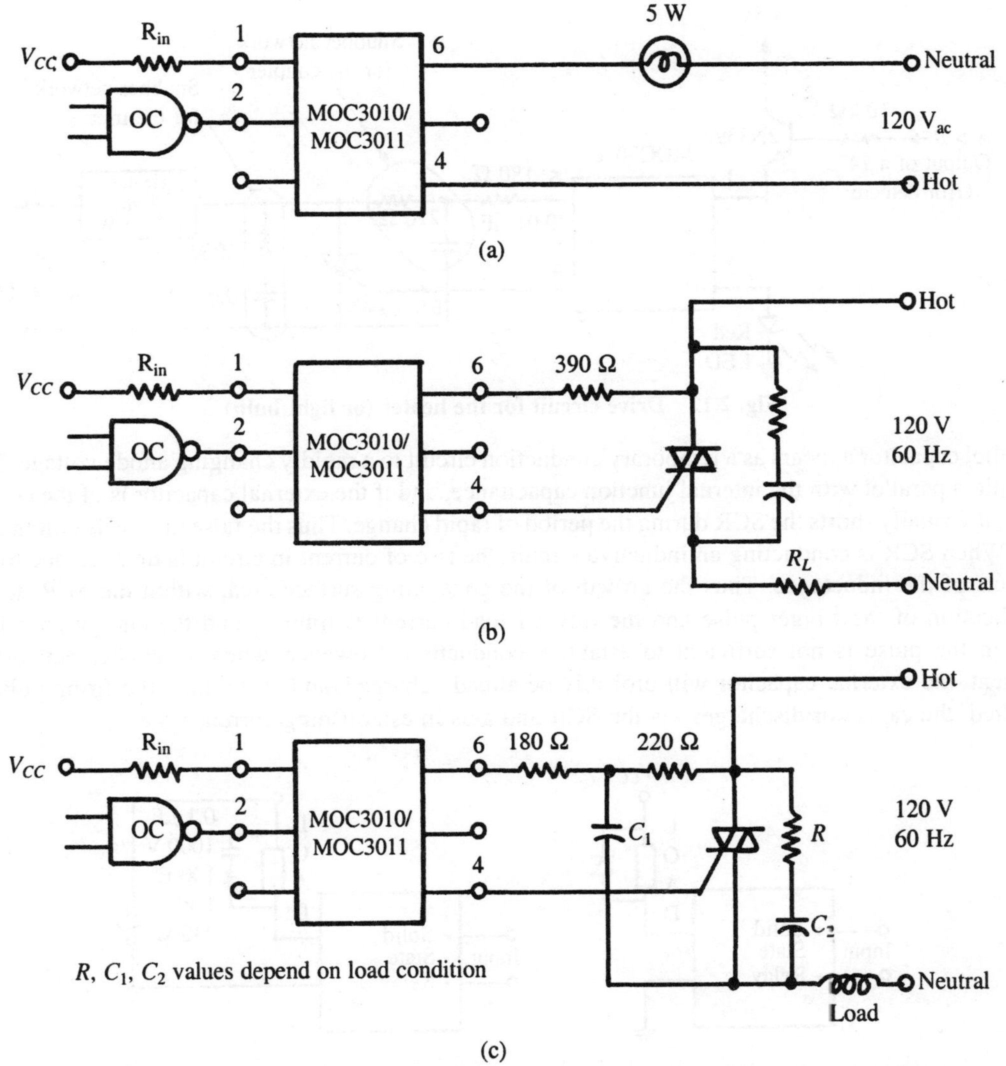

Fig. 2.11 Logic-activated triac driver circuits (a) lamp driver (b) resistive load (c) inductive load

may not be sufficient even to drive the MOC3010. We need current amplification. The schematic diagram of the amplifier, optically isolated triac driver and a triac and a snubber circuit are shown in Fig. 2.12.

Need for snubber circuit: To prevent turn on due to over voltage or line transients, a varistor is connected in parallel with SCR or TRIAC. This also prevents the SCR from being damaged by over voltage during its OFF state. If the voltage applied to the anode changes very abruptly, the SCR can turn on and short conducting (this is due to effect of rapidly changing voltage upon SCR's junction capacitance). To reduce the effect, a combination of series resistor/capacitor network is connected in parallel with the SCR. The parallel branch is called a SNUBBER. The snubber helps because: (i) The

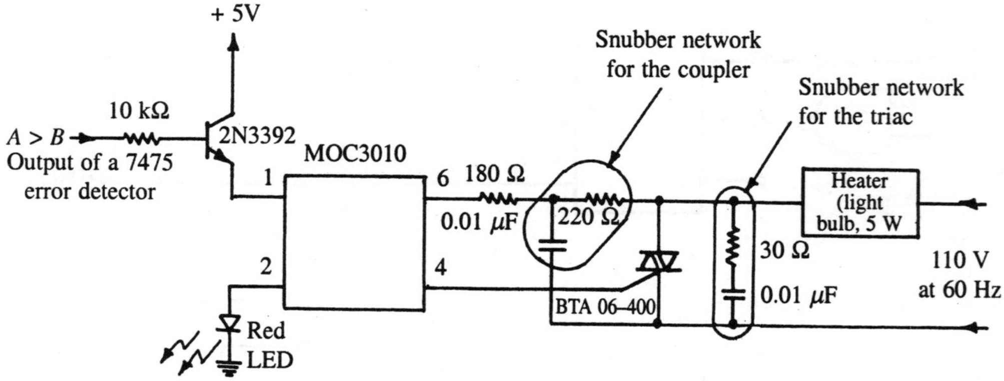

Fig. 2.12 Drive circuit for the heater (or light bulb)

parallel capacitor appears as a temporary conduction circuit to a rapidly changing anode voltage. This circuit is parallel with the internal junction capacitance, and if the external capacitor is of the correct size, it virtually shorts the SCR during the period of rapid change. Thus the false turn on is eliminated. (ii) When SCR is conducting an inductive circuit, the rise of current in circuit is delayed due to the effects of the inductance. Thus the growth of the conducting surface area, within the SCR, to the application of the trigger pulse and the flow of load current is limited, and the energy contained within the pulse is not sufficient to establish conduction. However, when a snubber network is present, the external capacitor will probably be already charged, and at the time the firing pulse is applied, the capacitor discharges via the SCR and aids in establishing current flow.

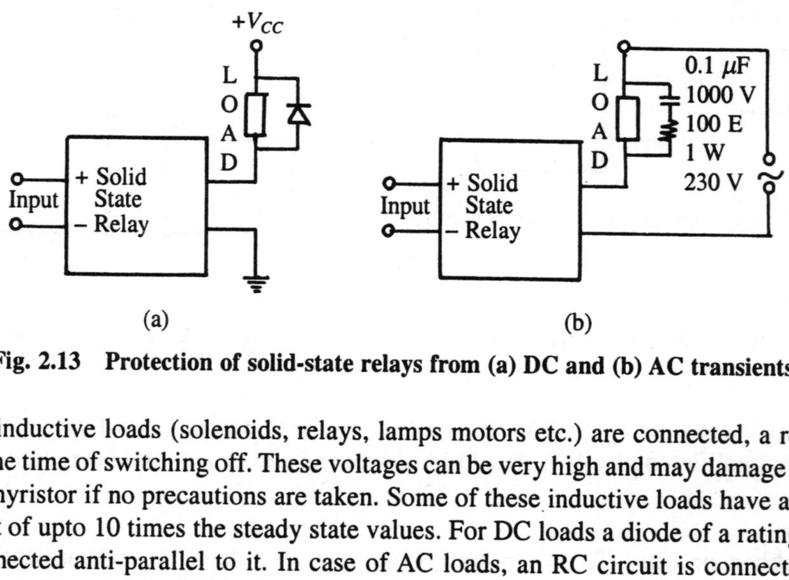

Fig. 2.13 Protection of solid-state relays from (a) DC and (b) AC transients

When the inductive loads (solenoids, relays, lamps motors etc.) are connected, a reverse emf is generated at the time of switching off. These voltages can be very high and may damage the switching transistor or thyristor if no precautions are taken. Some of these inductive loads have a turn-on peak source current of upto 10 times the steady state values. For DC loads a diode of a rating equal to the device is connected anti-parallel to it. In case of AC loads, an RC circuit is connected across the thyristor or triac (Figs. 2.13a and 2.13b).

2.4 Output System with Continuous Actuators

If the DC load driven by the DAC computer combination is somewhat more significant, then a power amplifier is needed to drive the load. Power operational amplifiers with a medium power ratings are used to drive continuous actuators such as DC motors or valves. Other power amplifiers called servo amplifiers have an even higher ratings than that of the power amplifiers. A power amplifier (known as complementary symmetry class B power amplifier) is shown in Fig. 2.14. A complementary pair of power transistor is pair, one NPN and other PNP that are electrically identical except for polarity. When these transistors are connected with their respective bases in parallel and their collector-emitter paths in series, the result is a simple push-pull class B amplifier. When the DAC output voltage V_0 goes to positive, transistor Q1 will tend to turn on, and current flowing through Q1 under the influence of V+ will drive the load also positive. If, on the other hand, the output voltage of the DAC is negative, PNP transistor Q2 will turn on and the load will be driven by current from V– power supply. Since each transistor turns on only one-half of the input signal, the result is full wave power amplification when the two signals are combined in the load.

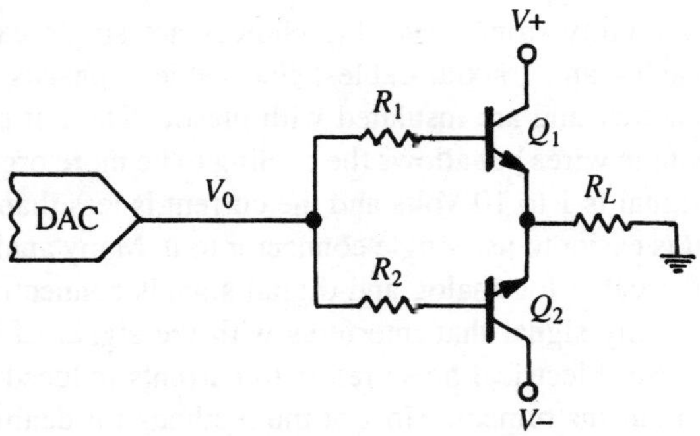

Fig. 2.14 High power DAC output

The load in Fig. 2.14 can be any of several different devices. If it is an electric motor, for example, the DAC output voltage will vary the speed of the DC motor; hence the computer will control the speed because it controls V_0. If we provide some means for measuring the speed of the motor, the

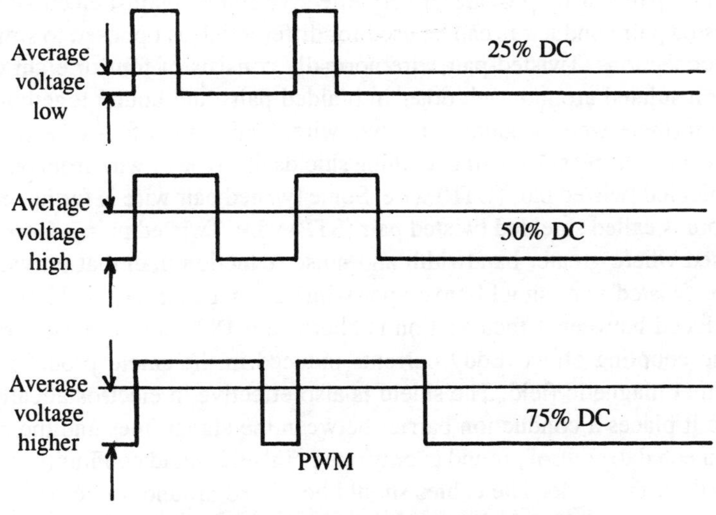

Fig. 2.15 Duty cycle of pulse width modulation

computer can be used in the feedback loop to keep speed constant or change it to follow the set point.

Pulse width modulated amplifiers: Pulse width modulation (PWM) is used to drive motors using PWM amplifiers. These amplifier's power control section consists of solid state switches. These switches are turned on or off at a constant frequency. The duty cycle (refer to Fig. 2.15) of these square pulses can be varied, which will vary the average voltage applied to the load. The output average voltage depends on the pulse duty cycle. The advantages are: low power dissipates, it is easy to implement in a microprocessor.

2.5 Cabling

Several types of cables are commonly used for interfacing tasks. In increasing order of cost and immunity from noise, the choices are single cables, flat cables, twisted pair conductors, coaxial cables and triaxial cables. Flat cables consists of many thin conductors running parallel to one another and are insulated with plastic. The flat cable has almost the same characteristics as that of single wires but allows the cabling to be more organized. Both of these cablings are suitable when the signal is 1 to 10 Volts and the current is less than 100 mA (i.e., low noise situation). For a flat cable it is easier to use single connector to it. Many analog I/O plug-in cards for IBM PC use 25 conductors flat cable for analog and digital signals connections.

Any signal that interferes with the signal of interest is called interference or, more commonly, noise. Electrical noise refers to currents induced by coupling external electric fields and the wiring in our instrument. Most of the methods for dealing with electrical noise are based on the principles of Faraday cup, which is a bing conducting cup (or cage) that surrounds the object of interest and that is connected to ground; objects inside the Faraday cup are effectively isolated from external electrical fields.

Magnetic noise refers to the current in the wiring of our insturments when they are placed near changing external magnetic field. The usual technique with magnetic noise is to isolate physically our instruments from large alternating magnetic fields found in electric motors. For either type noise, the smaller the level of the signal of interest, the more important is the noise prevention.

Twisted, shielding pair wiring provides a very effective shield against electrostatic and magnetic coupling. The twisted pair conductors can be used for differential, as opposed to single ended, analog (A/D, D/A) I/O connections. Twisted-pair wire normally consists of four or eight copper strands of wire, individually insulated around each other in braided pairs and bound togethor in another layer of plastic insulation (there were originally just two wires, rather than four or eight, hence the name twisted-pair). Except for the plastic coating, nothing shields this type of wire from outside interference, so it is called unshielded twisted pair (UTP) wire. Some twisted pair wire is further encased in a metal sheath and therefore is called shielded twisted pair (STP) wire. Twisted pair is more economical than coax. It can be used where greater bandwidth and noise reduction than that provided by flat cables are necessary. The twisted wire should have approximately a 1-inch lay or 12 twists per foot. The magnetic field induced between 1 inch section is alternately 180° out of phase, providing effective cancellation of the coupling effect. Eddy currents induced in the shield produce a magnetic field opposing the original magnetic field. The shield is also effective in electrostatically decoupling the noise source since it places a conduction barrier between the signal lines and the noise source. The shield must be connected to control ground at only one point and shield continuity must be maintained over the entire lenth of the cable. The cables should be routed around, rather than through the high noise area.

Coaxial Cables: A Coaxial cable consists of two conductors: one is a single wire in the center of the cable and the other is a wire mesh shield that surrounds the first with an insulator in between. Current flows in one direction on the inner conductor, and in the opposite direction on the outer conductor. As a result of this bidirectional current flow, the electromagnetic fields generated within the two conductors cancel each other. High frequency analog signals may be transmitted over coaxial cables without causing interference with other laboratory electronics instruments, and with little loses of signal (attenuation). Coax cables are shielded so that laboratory electronic noise does not interfere with coax cable transmissions. BNC connections are usually used with coax cable to provide convenient connections. Impedance matching connectors and terminators are available to improve the signal quality at high frequencies on coax cables. Coaxial cable can carry more data than the older types of twisted pair wiring and are less susceptible to interference from other wiring. They are more expensive and have become less popular as twisted-pair wiring technology has improved.

Terminal stripper connectors: Many commercial PC plug-in-peripherals for analog data acquisition have strips connecting all the signals of interest. Screw mounts on these strips are provided so that bare wires may be interfaced to the A/D or D/A converter. Each screw is isolated with a plastic barrier to prevent short circuiting. The terminal strip connector is mounted on a printed circuit board which usually has a shielded flat cable connecting the PCB with the D-type connector of the PC data acquisition plug-in-card. Terminal strip connectors thus provide convenient connections, but may provide relatively little signal isolation, depending upon the cable attached to the terminal strip.

Digital I/O connections: Digital signals use two voltage levels (or frequencies) that are widely separated from one another to do their signaling. Hence, even rather large amounts of noise result in little degradation in the ability to distinguish between these two signal levels. As a result almost all digital systems use a cheap cable; the most common are flat cable for parallel digital I/O systems and twisted pair for serial digital I/O systems. At very high frequencies the noise may be more of a problem, and shielded wires are required.

Data buffering and Data latching: The term 'buffering' refers to amplifying the signal on the bus. The signal is therefore isolated from the circuit it will drive. If there are many circuits connected directly to the bus, they will draw too much current from the bus. This can load it down to the point that the voltage may not be high enough to be valid or to drive the connected circuitry. Buffering can prevent this from occuring. Likewise, circuitry providing information (signals) to the bus may be loaded to the point that the voltages may be faulty. The circuit is therefore, buffered or amplified. A circuit that performs this buffering in both the directions, to and from the CPU, is known as bidirectional buffer. Buffering is also done to prevent the signals from conflicting with each other. Most circuitry providing information to the bus are dual-state devices. That is, they are either a high or low state. The buffer can act as a kind of a gatekeeper, deciding when the signals should be permitted to or from the bus. When the computer is not addressing the circuitry, the buffer is tri-stated so the circuitry will leave the computer bus alone.

'Latching' is done for the same reasons as buffering and can also be bidirectional. The difference between latching and buffering is that a latch has the ability to hold the information on its outputs for as long as needed. A buffer only keeps the information on its output while the information is on the data bus. Latches are used for output from the bus only, because a latch on the signal going to the bus might 'block' or lock up the data bus. Latches are used to drive many different configuration of circuitry and can even drive devices.

2.6 Digital to Analog Converter

The heart of most digital to analog converters, DACs, is a current summing node where the currents are selected by digital inputs. Consider, for example, the 3 bit D to A converter (Fig. 2.16a). In this circuit, the current I is the sum of the currents in the three resistors on the left and is selected by the switches. This type of resistor network is called binary ladder. Let us discuss how this circuit functions as a DAC. Since the voltage at the summing junction is zero, the current I is given by

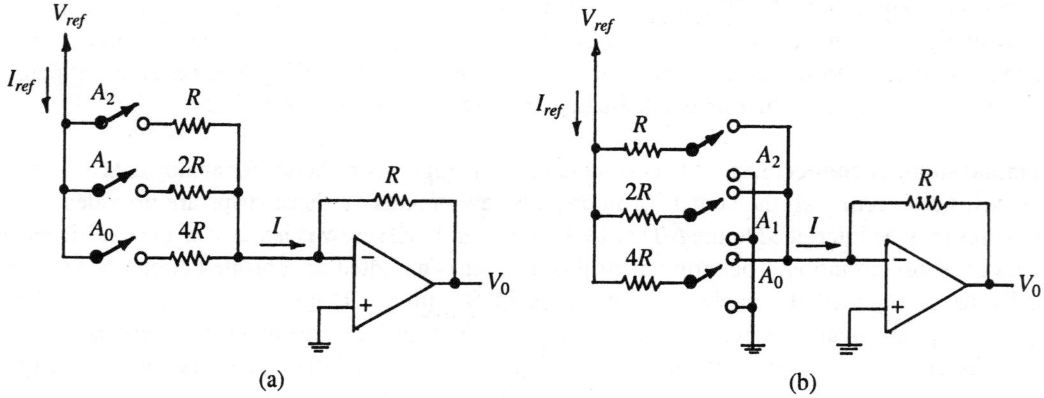

Fig. 2.16 Two versions of the binary ladder diagram DAC

$$I = A_2 (V_{ref}/R) + A_1 (V_{ref}/2R) + A_0 (V_{ref}/4R) \quad (2.1)$$

$$= V_{ref} (4A_2 + 2A_1 + A_0)/(4R) \quad (2.2)$$

where A_1 is 1 if the corresponding switch is closed and 0 otherwise, and V_{ref} is an accurately set reference voltage. In Fig. 2.16, A_0 corresponds to LSB. Since the current must flow through the op-amp feedback resistor, which has been selected to be R, the output voltage, V_0 becomes

$$V_0 = -IR = -V_{ref} (4A_2 + 2A_1 + A_0)/4 \quad (2.3)$$

The magnitude of V_0 ranges from 0 to $-7V_{ref}/4$ volts. This range is not continous but changes in discrete steps of $-V_{ref}/4$ volts.

The particular DAC circuit has two drawbacks.

(i) The current supplied by the reference voltage source changes in steps as the digital inputs change. This can create noise problems in the voltage reference source which may result in poor accuracy at the output. This problem can be avoided by reconfiguring the DAC (Fig. 2.16b). In this case, the reference supply current never changes since no matter which position the switch is in, the current in the associated resistor is the same. Since the switch voltage is zero regardless of switch position, the reference current is given by $I_{ref} = 7\ V_{ref}/(4R)$ amps.

(ii) The second difficulty is that the value of the summing resistors increases dramatically as the number of bits increases. For a 8 bit DAC, the largest value of the resistor is $128R$, where R is the resistance of the smallest resistor. This large variation in the resistor can be avoided by using the circuit in Fig. 2.17. Here, the largest resistor is only twice that of the smallest and hence this network is called an R-$2R$ ladder. The right-hand side of all of the $2R$ resistors is at 0 volts and hence the total resistance of the network is R Ohms. The reference current I_{ref} is V_{ref}/R which is less than the binary ladder (Fig. 2.16b). The current at the summing node I, becomes

$$I = (V_{ref}/2R) (A_2 + 0.5A_1 + 0.25\ A_0) \quad (2.4)$$

$$= (V_{ref}/R) (4A_2 + 2A_1 + A_0)/8 \quad (2.5)$$

Hence, the output voltage V_0 is given by

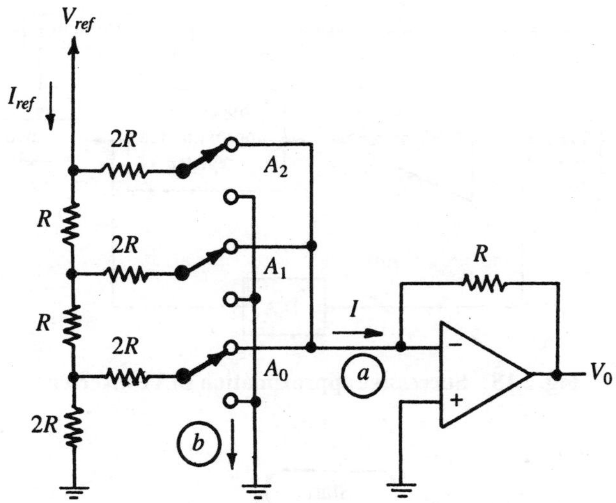

Fig. 2.17 An *R*–2*R* ladder digital-to-analog converter

$$V_0 = -IR = -V_{ref}\,(4A_2 + 2A_1 + A_0)/8 \qquad (2.6)$$

In this case, the output voltage range is from 0 to $-7V_{ref}/8$ volts in steps of $-V_{ref}/8$. This step size corresponds to the least significant bit, LSB.

2.7 Analog to Digital Converter

Counter or Ramp ADC: The counter will start counting from zero and is incremented by a clock pulse. The digital output of the counter is converted back to analog by the digital to analog converter (DAC) and is compared with analog input voltage signal (V_{in}). If V_{in} is higher than the DAC output, the counter keeps incrementing, but if V_0 increases the value of V_{in}, the counter stops immediately and the counter output is the ADC output. The disadvantage of this method is that it is very slow since it depends upon the counter. For example, in 10 bit ADC, the conversion time is about 1024 cycles. The method is also not accurate, especially in converting sudden peaks.

Successive approximation method: This method does not use the counting method but modifies (by trial and error method) the contents of a register (known as successive approximation register, SAR) bit by bit starting with the MSB bit until the register data are the digital equivalent of the analog voltage, i.e. the DAC output matches with the input voltage (Figs. 2.18 and 2.19). The advantages of this ADC are that it is very fast and gives good accurate results. For 8 bit, the conversion takes 8 cycles only. For 10 bit ADC, the conversion takes 10 cycles and so on.

Summary

Computer interfacing for data acquisition and control consists of analog to digital conversion of input analog signals. Prior to ADC, the analog signal has to be conditioned so as to meet the input requirement of an ADC. Signal conditioning consists of amplification (for sensors generating very low level signals), filtering (to limit the amount of noise on the signal), isolation (to protect the sensors from interacting with one another and/or to protect the signal from possible damaging

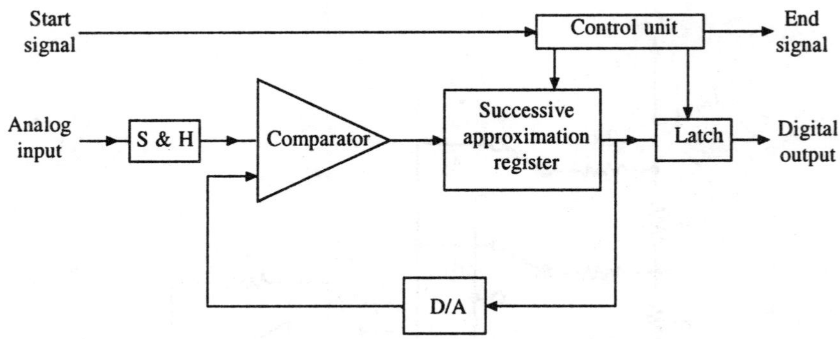

Fig. 2.18 Successive approximation D/A converter

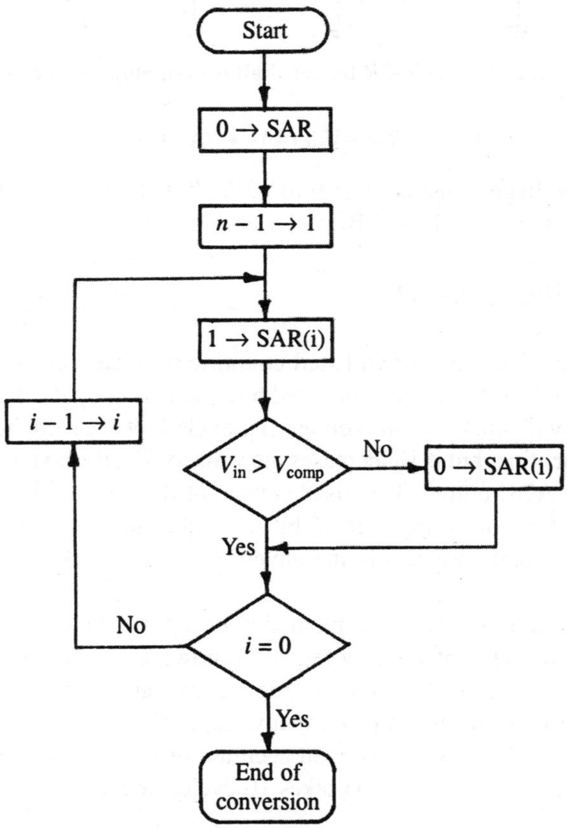

Fig. 2.19 The successive approximation algorithm

inputs). Conversion of digital signal to analog signal at the output is to be carried out if the output signal is sent to a final control element which requires analog signal. The digital signal has to be amplified by a transistor or solid state relay or power amplifier.

References

Breedings, K.J., Microprocessor system design fundamentals, Prentice-Hall International, Englewood Cliffs, New Jersey, 1995.

Brighouse, B. and Loveday, G., Microprocessors in engineering system, Pitman Publishers, 1987.

Carr, J.J., Micocomputer interfacing, Prentice-Hall, Englewood Cliffs, New Jersey, 1991.

Gates, S.C. and Becker, J., Laboratory automation using the IBM PC, Prentice-Hall, Englewood Cliffs, New Jersey, 1989.

Gayakward, R. and Sokoloff, C., Analog and digital control systems, Prentice-Hall, Englewood Cliffs, New Jersey, 1988.

Gupta, S. and Gupta, J.P., PC interfacing for data acquisitions and process control, 2nd ed., ISA Pub., 1994.

Kheir, M., The M68HC11 microcontroller: Application in control, instruments and communication, Prentice-Hall, Englewood Cliffs, New Jersey, 1997.

Rigby, L.H. and Dalby, T., Computer interfacing: A practical approach to data acquisition and control, Prentice-Hall, Englewood Cliffs, New Jersey, 1995.

Tompkins, W.J. and Webster, J.G., Interfacing sensors to IBM PC, Prentice-Hall, Englewood Cliffs, New Jersey, 1988.

Review Questions

1. With a block diagram, explain the concept involved in a data acquisition system.
2. What items must be added to a standard personal computer before it can be used as an industrial controller?
3. How do we interface input analog signals to a PC?
4. Name the sensors giving output as an analog signal.
5. Give examples of digital input/output signals.
6. What are the different control system actuations used in process automation?
7. What is a current loop in a DA and C?
8. What is a sample hold circuit?
9. What are the methods used to carry out digital signal conditioning for DA and C?
10. What is a buffering circuit used in a DA and C?
11. Name the different driving circuit used for interfacing digital output signals.
12. What is an opto-coupler and where is it used in a DA and C?
13. Give circuit diagrams for activating a TRIAC circuit for (i) lamp drive (ii) resistance load and (iii) inductive load.
14. What is a snubber circuit?
15. How do we protect solid-state relays from (i) DC transients and (ii) AC transients.
16. What are the different cables used for interfacing a PC to sensors?
17. What is meant by data buffering and latching?
18. Explain with a suitable circuit how do you implement an ADC.
19. Explain with a suitable circuit how do you implement a DAC.

3

Data Acquisition and Control by Using Standard Add-On Cards

Computer control of processes require a standard personal computer (PC) and additional interface cards and software, We need to know how to select hardware and software, to make the electrical connections and to write the special purpose programs. A PC-based system (Fig. 3.1) can be assembled using purchased interface cards and associated software packages. The PC to be procured can be 'industrially hardened' versions with increased resistance to dirt and vibrations. There is no need to build our own laboratory DA and C board or to code everything in machine or assembly language to save memory space. Computer process control can be as simple as plugging an appropriate data acquisition board into your computer and using the associated software to collect and process the data. In this chapter, the details and availability of standard add-on-cards for DA and C are discussed.

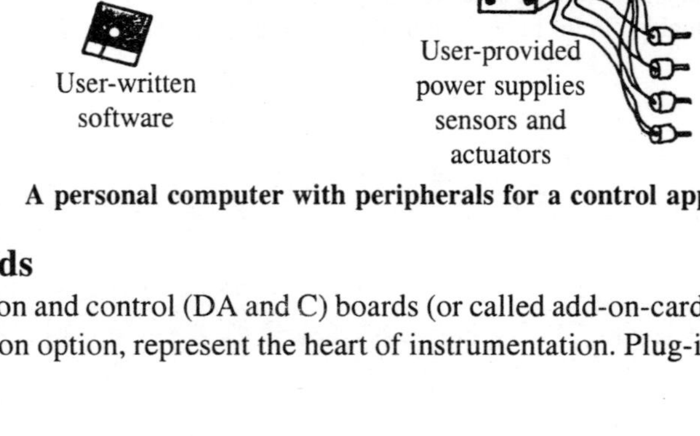

Fig. 3.1 A personal computer with peripherals for a control application

3.1 Plug-in-Cards

Plug-in-Data acquisition and control (DA and C) boards (or called add-on-cards) which are the fastest growing instrumentation option, represent the heart of instrumentation. Plug-in DA and C boards are available for many popular computers. There are also cards for single purpose (i.e., dedicated to a particular function) for example, carrying out only A/D or only D/A conversion. The single purpose card is low cost, simple to use and easy to programme. If we need to use more than one functions, then buying cards for each function will add up to cost and also that many expansion slots may not be available in the PC. Another class of interface boards, known as multi functional cards, have various combinations of analog, digital and pulse inputs and outputs. Many cards have programmable channels sampling and conversion modes; separate gain for each channel. One card often can perform a variety of functions including A/D or, D/A conversion, digital I/O, and counter/timer operations (Fig. 3.2). The multi functional card needs one expansion slot and the cost of the card is cheaper when compared to buying several single function cards. However, the multi functional card requires several consecutive I/O addresses.

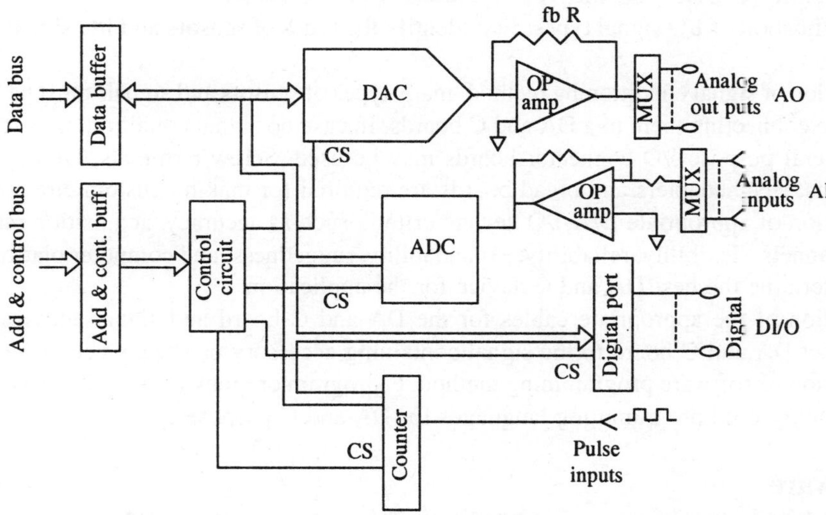

Fig. 3.2 PC add-on card for data acquisition and control purposes

Most of the multi functional laboratory cards are built with 4 to 16 analog input channels; even 4 channel is certainly sufficient. Because the channels are multiplexed (there are 6 channels but one A/D), the sampling rates quoted for A/D is the maximum rate for all the channels combined. If some channels are not used then, the sampling rate is decreased accordingly.

3.2 Input/Output Devices

I/O devices (sometimes called 'port') are the components that the computer uses to exchange data with its peripheral and/or other controller. Functions of an I/O device is to translate data between the form required inside the computer and the form required outside the computer. This functioning is called *interfacing*. Configuring means to setup an I/O device for a special function. Some I/O devices are pre-configured. Pre-configured I/O devices are appropriate for common peripherals such as keyboards, monitors etc. For connecting peripherals such as sensors, actuators etc., it is best to use programmable I/O ports.

Interface cards provide the user the facility to select the exact configuration for the particular card/board. The default configuration is normally so chosen as not to interfere with the hardware present. It is always advisable to procure the software driver with the card. Interface cards with configuration utilities are preferable since some utilities will interrogate the system and then indicate the jumper settings and so on to use. Software configurable cards with auto installation drivers are preferable. The interfacing cards should have the flexibility to allow one to select the I/O port addresses and select or disable DMA channels and/or interrupts. One should carefully read the manual accompanying the cards. If a single interface card is being introduced into the PC slot, configuration set up by the card manufacturer (factory setup) will work. For a subsequent card it may require the disabling of interrupts on one card, DMA on another card.

Configuring of parallel ports consists, partly, of programming the I/O chip to use some lines as inputs and other as outputs. Other configuration options include whether to allow the I/O chip to interrupt the CPU when it receives changed input values.

3.2.1 Guidelines in Selecting the Appropriate DA and C Boards

(i) identification of I/O signal types: first identify the types of sensors and I/O signal types being used.

(ii) selection of signal conditioning method: many types of sensors and signals must be conditioned before connecting them to a DA and C boards. In case no signal conditioning is required then a general purpose I/O connector boards may be used. Screw terminals, DIN rail mounting options, BNC adapters and bread boards are required for making custom circuitry.

(iii) selection of appropriate D/A-I/O device: criteria such as accuracy, acquisition rates, number of channels, flexibility, reliability, expandability, ruggedness, and computer platform are used to determine the best DA and C device for the application.

(iv) selection of the appropriate cables for the DA and C board and signal accessories: cables connect DA and C board to the signal containing accessory or I/O connector block.

(v) selection of software programming method: C program or quick basic (known as QBasic) are commonly used programming languages for DA and C purposes.

3.3 Software

DA and C hardware without software is of little use. The majority of DA and C applications use driver software. Driver software is the layer of software that directly programs the registers of the DA and C hardware, managing its operation and its integration with the computer resources, such as processors interrupt, DMA and memory. Driver software hides the low-level, complicated details of hardware programming while preserving high performance, providing the user with an easy-to-understand interface. Driver functions for controlling DA and C hardware can be grouped into analog I/O, digital I/O and timing I/O. The driver has the following functionality to: (i) aquire data at specified sampling rates; (ii) acquire data in the background while processing in the foreground; (iii) use programmed I/O, interrupts, and DMA to transfer data; (iv) stream data to and from disk; (v) integrate more than one DA and C boards if used; and (vi) integrate seamlessly with signal conditioning units.

Designation of whether an I/O port is configured for input or output is determined by the software. The DA and C board assigns the three I/O ports to the first three addresses beginning with the base addresses. The fourth address is assigned to the control register on the board. The control register retains the binary code that determines the input or output configuration of the three I/O ports. This arrangement, one control register and three I/O ports, accounts for the requirement that four port addresses be assigned to each DA and C board. A multi-port interface chip (known as 8255 programmable peripheral interface chip, PPI) is used in DA and C card. This chip has three operational modes—modes 0, 1 and 2. In mode 0, the 8255 chip provides two 8-bit ports (A and B) and two 4-bit ports (as upper portion of port C, lower portion of port C).

No matter which mode is used and regardless of which configuration of input or output ports is wanted, the control byte (eight bits) is always written to the control register in the PPI chip. The control register is addressed as the base address of the board plus 3. For DA and C board with a base address of 300 Hex, the address of the control register is 303 Hex. Writing a control byte to the 303 Hex address will configure to I/O board.

While the base address of the DA and C board is permanently set via DIP switch, the mode and data direction selection of the I/O ports is not permanent and may be changed by software at any time. Whenever the computer is turned on or the computer is reset, the configuration code must be written to the PPI control register. The control byte should be placed in the beginning of an I/O program. The writing of a byte to the control register should be identified as 'port initialization instruction'.

Let us consider a particular DA and C card CIO-DASS08/Jr-AO multi function interface card supplied by Computer Board Inc. (Massachusets, U.S.A). To receive data from sensors or switches via DA and C boards, the input instruction is written in C language for IBM-PC as

inp (*m*)

where *m* is an integer which represents a number or port address of the input port being selected or accessed through the software. The statement In data = inp (0 × 300) means that a variable name in data is assigned to the data and the data has been input to the computer via the addressed port. The 0X used in this statement show that the port address is provided in Hex format. If the input data is decimal number, then 0X is not included and the following statement is used.

Indata = inp (768)

Executing either of the statements will cause one byte of binary data (8-bit) to be the input into the computer through port 300 Hex (or 768 decimal).

Similarly to output a data from computer though a I/O port is initiated by the out statement: outp (port, byte) where port is the output address in Hex or decimal form and byte is the data in Hex, decimal or binary form that is going to be output through the specified port. Data byte output to a port must be eight bits. For example, in the statement outp (0X301, 0XFF), the output port address on the I/O board is 301 Hex and the digital data 11111111 (eight ones) is delivered to the port. The statement outp (269, 255) means that decimal 255 is written to the port address 269. Each of the above commands tests or controls respectively the operation of a specific register in the card.

I/O port example: Consider an example of a circuit for using an interface card to read DIP switches and to activate LED display. We assign address space from 300H to 303H ports A, B, and C and the control register of the 8255 chip. In this example, port A is connected to an 8-bit DIP switch while port C lower is attached to a 4-bit LED through a driver (74S04). The driver is necessary to provide enough current to the LEDs. A program reads the bit pattern set in the DIP switch to control the LED display. Table 3.1 gives program listing written in C language.

The steps required to make up the analog to digital conversion process are as follows:

1. Choose A/D channel that will be active by writing the channel number to (Base + 2)
2. Write any data value to (Base +1) to start A/D conversion
3. Monitor EOC bit to determine when conversion is finished
4. Read converted data from (Base + 0 to Base +1)

Each of the above steps may require one or more software instructions. The end of conversion (EOC) indicates when the converter is busy calculating the digital representation of the analog voltage. During the A/D conversion process the EOC bit is at a logic 1 level. When the EOC bit falls to zero (0), a logic low level, the conversion is over, and the binary data is ready.

Table 3.1 C program (from Tompkins et al. 1988) for reading the DIP switches and lighting the LEDs

```
#define PORT_A          0 × 300          /*port A on the 8255.*/
#define PORT_C          0 × 302          /*port C on the 8255.*/
#define CTRL            0 × 303          /*8255 control register.*/

main { }
    {
    char input;
    int off_time, on_time = 8

    outp (CTRL, 0×90);                   /*set the 8255 to mode 0 with port*/
                                         /*as an input port and other ports as*/
                                         /*output port*/
    while (! kbhit () )                  /*repeat loop until keyboard is hit.*/
      {
      input = inp (PORT_A);                    /*read from DIP switches*/
      off_time = input and 0×F0;               /*high bits used for off time*/
      off_time = off_time >> 4                 /*shift bits to lower half of*/
                                               /*offtime*/
      outp (PORT_C, input);                    /*Turn LEDs on*/
      delay (on_time);                         /*wait a while*/
      outp (PORT_C, 0×00);                     /*Turn LEDs off*/
      delay (off_time);                        /*wait awhile*/
  }
  }

delay (time)
      int time;
      {
      int i, j;
      for (i = time; i > 0; —i)               /*repeat specified no. of times*/
            for (j = 0; j < 1000; ++ j);
      }
```

The following are the sequence of steps required to output an analog voltage from channel 0:

1. Write the LSB data to (Base +4)
2. Write the MSB data to (Base +5)
3. Perform a dummy read of (Base + 3) to update the D/A conversion

The steps consists of loading the two D/A registers and executing a read of the register at the base address +3. The register at the address Base +3 is the digital input register, however, in this instance it just triggers the D/A output process. The value of the data in the register at the address Base + 3 is not important and is not used in the analog output cycle.

3.4 From Transducer to Control Room

Signal conditioning: As discussed earlier, most active transducers produce an electrical signal that is not usable directly in industrial measurement systems. Moreover, this electrical signal will not be easy to transmit over the several hundredth of meters that might separate the transducer from the control room where it will be processed. The reasons for the transmission problems are many:

(i) very-low-level signal (e.g., thermocouple signal) and (ii) high output impedance (e.g., piezo– or pyroelectric transducers).

As for passive transducers, they have to be powered and their output signal has to be processed before obtaining a useful voltage. Classical examples are the numerous transducer based on the variation of an electrical resistance (strain gauges, potentiometers, resistance temperature detectors, ect.). In most cases, the active resistance is incorporated in a Wheatstone bridge that will be powered by an AC or DC voltage; the unbalance of the bridge gives the measurement. Firstly some transducers have nonlinearities that must be compensated.

3.4.1 Industrial Signal Conditioners

Nowadays most manufacturers of electronic instrumentation devices are producing signal conditioners intended to simplify the use of transducers. These signal conditioners have in-built: (i) power supply of passive transducers (ii) amplifications of the output signal upto the levels of several volts, directly compatible with the industrial analog-to-digital converters; and (iii) filtering either industrial noise or unwanted spectral components in the phenomena measured.

As an example, let us give the family of signal conditioners 5B or 6B from analog devices, including conditioners for strain gauges, thermocouples (with cold-junction compensation) or metallic resistance-temperature detectors (taking into account the nonlinearity of the R-T characteristics). These modules are the rather compact sealed hybrid circuits (size: 57 × 57 × 15 mm); they can be factory adjusted for a lot of different standard measurement ranges and even for customized ranges. They have got a rated precision of 0.05% over the temperature range –25 to –85°C and a galvanic insulation upto 1500 V between input and output. The cold junction compensation of thermocouple is realized by small auxiliary hybrid circuits and the linearization of RTD is implemented in the signal conditioner itself. Several modules can be plugged on a printed circuit back plane and their output is connected in series with an internal analog switch which allows multiplexing the outputs without the need for additional components.

Another example is the hybrid circuit IB32 for the signal conditioning of resistive transducers in a bridge configuration. The circuit features an adjustable excitation voltage for the bridge, an amplifier for the output voltage with a gain ranging from 100 to 5000, a low-pass filter, a common-mode rejection ratio of 140 dB, a linearity of 0.005% and a global precision of 140 bits.

3.4.2 SCXI

Signal Conditioning Extensions for Instrumentation (SCXI) is a high-performance, multi channel signal conditioning and data acquisition system for use with PCI. We can use the SCXI as either a signal conditioning front end with data acquisition (DA and C) boards and modules, or as a complete external system. An SCXI system consists of one or more rugged chassis that can house a variety of signal conditioning modules for most of the I/O needs, such as the following.

Analog inputs: Thermocouples, RTDs and thermistors, strain gauges, voltage sources, 4–20 mA current sources, frequency inputs. Analog outputs: Voltage and current.

Digital I/O: Optically isolated I/O, AC/DC inputs, solid-state relays, electro mechanical relays.

Analog input modules interface the system to a variety of transducers and signals and improve the quality and reliability of the measurements with high performance, low noise signal conditioning such as signal amplification, isolation, multiplexing, filtering, transducer excitation, simultaneous sample-and-hold. For local DA and C systems, we can use the SCXI module to condition I/O signals for plug-in DA and C cards. We can also use the external with suitable DA and C module to digitize the data to the PC via the parallel port or serial port over a distance of 4000 ft. We can use the SCXI as a remote on a RS-485 network upto 1.2 km from the host PC. The expansion system can range from a few channels for desktop computers to a large rack-mount systems with upto 3000 channels.

3.4.3 Remote Input/Output Modules

High quality remote control module products providing cost-effective protection and conditioning for a wide range of valuable industrial control signals and systems are available. Fixed and programmable sensors-to-computer and computer-to-analog output interface modules and associated back planes, accessories and application software are provided (Fig. 3.3). Sensors-to-computer isolated modules provide complete sensor/RS-232C or/RS-485 interfaces with 15-bit measurement resolution. They accept a variety of voltage, current, thermocouple, RTD/thermistors, bridge, timer/frequency, and multi channel digital inputs/outputs. The computer-to-analog output modules are complete, isolated interfaces designed for remote installation and communications with host computers via standard RS-232C and RS-485 serial ports. They offer 12-bit resolutions in a range of analog output voltages.

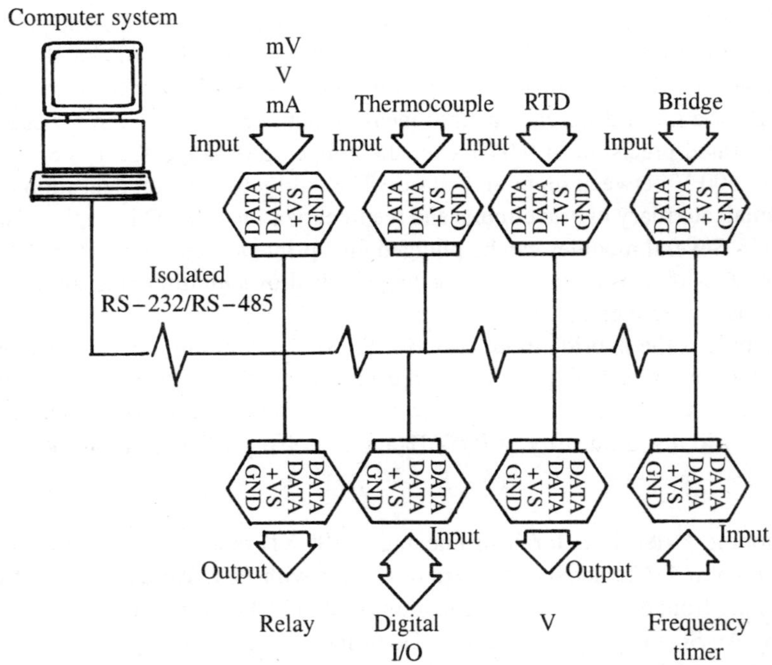

Fig. 3.3 Remote I/O modules (Contec Microelectronics, 1995)

Analog signal conditioning modules: These modules are single-channel analog signal conditioning modules for use with the standard data acquisition interface cards. They provide isolation, noise rejection, and amplification for millivolt source, volt sources, 4–20 mA sources, 0–20 mA sources, RTDs, thermocouples, strain gauges, and frequency inputs. The modules are fully encapsulated and require no field adjustments and can be plugged into a back plane, which is available with 8 or 16 channel capacity. Unconditioned signals connect directly to the screw terminals on the back plane. The modules condition the signals and route the conditioned signals to a 26-pin connector on the back plane. The DA and C card and the back plane can be connected through a suitable cable.

Digital signal conditioning modules: SSR series: The solid state relay (SSR) digital signal conditioning modules are used with DA and C boards to: (i) sense voltages from field devices and convert them to digital signals, or (ii) use TTL digital signals to control field devices and switch signals. SSR modules are high performance, fully encapsulated devices. With the SSR, the PC can interface with field devices with optical isolation. The modules are designed to plug into one of the SSR series back planes, which are available with 32, 24, 16 or 8 module capacity. Each back plane has screw terminals to accommodate the raw signals and LEDs to indicate the on-state of the logic. The conditioned signals are routed to a 50-pin connector on the back plane. The modules can be connected to the DA and C card using a suitable cable.

3.4.4 Configuring Options and Settings

DA and C adapter cards often have configurable options that must be set for the adapter card to function properly. Example include: Interrupt (IRQ), Base I/O port address, and Base memory address. Sometimes it is possible to specify DA and C adapter card settings in software, but these settings commonly must match jumper or dual in-line package (DIP) switch settings configured on the DA and C adapter card. See the DA and C card documentation for DIP switch settings. Many newer DA and C adapter cards use Plug-and-play (PnP) technology, which make manually setting the adapter card option obsolete. The operating system configures hardware device automatically.

Interrupt (IRQ): Interrupt request lines are hardware lines over which devices such as input/output ports, the keyboards, disk drives, DA and C adapter, and computer network adapter cards can send interrupts or requests for services to the computer's microprocessor. Interrupt request lines are built into computer's internal hardware and are assigned different levels of priority so that the microprocessor can determine the relative importance of incoming service requests. Each device in the computer must use a different interrupt request line or interrupt (IRQ). The interrupt line is to be specified when the device is configured.

Base I/O port: The base input/output (I/O) port specifies a channel through which information flows between the computer's hardware (such as DA and C adapter card) and its CPU. The port appears to the CPU as an address. Each hardware device in a system must have a different base I/O port number.

Base memory address: The base memory address identifies a location in a computer's memory (RAM). This location is used by the adapter card as a buffer area to store the incoming or outgoing data. Some adapter cards contain a setting data that will specify the amount of memory to be set aside for storing data. For example, for some cards we can specify either 16K or 32K memory. We have to specify some appropriate part of the main memory as the starting memory or base memory for the card. To set the device or card for this base memory address we have to set appropriate DIP switches in the add-on-card. The base memory address is to be specified in hexadecimal number. The memory required by the add-on-card is given in the manual provided with the card.

A driver is the software utility that enables a computer to work with a particular device. Devices such as mouse devices, disk drives, DA and C adapter cards, network cards, and printers each come with their own driver. The computer's operating system will not recognize a device until its associated driver has been installed, unless the operating system is plug and play compliant. Drivers are included as a disk with DA and C card when it is sold.

3.5 Low Cost Multi-Functional DA and C Card

Let us consider a low cost DA and C card supplied by National Instruments. The Lab-PC-1200 is a low cost, multi-functional play and plug compatible data acquisition (DA and C) I/O boards for ISA computers. The card has 12 bit ADC with eight analog inputs, configurable as eight single ended or 4 differential inputs, 24 lines of TTL-compatible digital input and output, and three 16 bit counter/time channels for timing input and output. The card has two 12-bit DACs with voltage outputs. The board has no DIP switches, jumpers, or potentiometers—the board is configured and calibrated entirely though software. Even the base I/O address is selected by software. The card can be inserted in any ISA computer, replace the computer cover, and completely configure the base I/O, interrupt level, and direct memory access (DMA) channels of the board through software.

A software-programmable gain amplifier has gain selection of 1, 2, 5, 10, 20, 50 or 100. The boards have a 12-bit ADC with analog signal resolution of 2.44 mV at a gain of 1. We can achieve finer resolution down to 24.4 μV by using a higher gain. The 12-bit output of the ADC is automatically sign extended to 16 bits. The board perform both single A/D conversion and multiple A/D conversions of a set of number of samples. A 512-word-deep first-in-first-out (FIFO) memory buffers the data during multiple A/D conversions. Multiple conversion can be carried out either by programmed I/O, Interrupts, or DMA. An on-board counter/timer controls the timing of multiple A/D conversions. The counter/timer generates the sample interval clock with a resolution of 1 μs. Both single A/D conversions and multiple A/D conversions sample sequences are initiated from either software or external timing control signals. Two hardware triggering modes exist: pre-trigger mode and post-trigger mode. By combining the pre-trigger and post-trigger modes, the board can acquire data before and after a trigger condition.

Analog output: The card has two double-buffered 12-bit DACs that are connected to two analog output channels. Each channel can be configured independently for unipolar (0 to 10V) or bipolar (±5V) operations. The resolution of the 12-bit DAC is 2.44 mV in both polarities. The counter/timer generates interrupts and updates signals for the double-buffered DACs.

Digital I/O: The board has 24 digital I/O lines, configurable as three 8-bit ports for input, output, bi-directional, or handshaking modes. Two of the ports can drive Darlington transistors directly for higher-current applications. The digital I/O lines are TTL compatible. The digital output port can sink 1.7 mA on each line.

Counter/timer: The card has two system timing controllers (STCs) for counting and timing. Each STC contains three independent 16 bit counter/timers. One of the STCs, counter A, is dedicated for A/D and D/A timing. The three counters on the other STC, counter B, are available to the user for generating time-related functions such as clock output, pulse output, and event and frequency measurement.

By including a 80C186 processor on a data acquisition board, together with a ROM-resident English-like language such as DAPL (data acquisition processor language) other applications can be addressed. There are more than 100 predefined commands which can be used to allow signals to be analyzed (to reduce noise), linearization for thermocouple, detecting peaks or troughs, set limits or to filter out any of the unwanted frequencies. Faster processing time can be guaranteed because host PC is not involved. Fast Fourier Transforms (FFT), correlation, cross power spectra, amplitude and phase transfer functions can be carried out. On board serial RS 232 communications channel allows the DAP (Data acquisition processor) to run from a terminal so that the board can operate as a stand alone front end system with only a power supply and no host PC. The DAP downloads processed data over the serial link.

3.6 IEEE–488 GPIB

One of the most useful data transmission techniques in the laboratory is a high speed parallel interface (IEEE–488, known by other names such as Hewlett-Packard interface bus (HP-IB), or GPIB-general purpose interface bus). The major advantages of the GPIB standard are that it is very rigidly defined so that various implementations of it are compatible. It allows multiple instruments to be connected via a single adapter in the computer, which lowers the cost. It has a single type of connector, so that there is never any problem with handling the wrong connector. It is a high speed interface, with data transfer rates up to 1 Mbytes/sec. Most common applications of digital to analog converter (D/A) are those in which the D/A is built into the instrument. Many modern instruments use internal D/A's to provide needed voltages. These internal D/As are then often externally controlled using some type of digital technique such as GPIB. A GPIB card installed in the computer that would make it GPIB device and connect the adapter card to the instrument using special GPIB cable.

The major disadvantage of the GPIB standard is that it adds to the cost of an instrument. Its high speed capability is very infrequently used.

The important set of hardware specifications are:

(1) a maximum of 15 devices can be attached to a single network; (2) the total cable length connecting the devices cannot be more than 20 meters without additional buffering and the length of cable to each individual device is also limited to 2 meters; (3) the number of lines in the cable is 24, with the function of each line clearly defined. Eight of the lines are for data transmission, while 8 are for hand shaking and 8 are for grounding and shielding. The 16 data and handshaking lines are collectively the bus of the GPIB; (4) the drives can be connected in almost any convenient arrangements (star, daisy chain, mixed network); and (5) the maximum data rate is 1 Mbyte/sec.

3.6.1 Talkers, Listeners and Controllers

GPIB devices can be talkers, listeners and/or controllers. A talker sends data message to one more listeners, which receives the data. The controller manages the flow of information of the GPIB by sending commands to all devices. The GPIB is an ordinary computer bus, except that a computer has its circuit cards interconnected via a back plane. The GPIB has stand-alone devices interconnected by standard cables. The role of GPIB controllers is comparable by the role of a computer CPU, but a better analogy is to compare the *controller* to the switching station of a city telephone system. The *controller* monitors the communication network (GPIB). When the *controller* notices that a party (device) wants to send a data message, it connects the *talker* to the *listener*.

The *controller* usually addresses (or enables) a *talker* and a *listener* before the talker can send its message to the listener. After the message is transmitted, the controller may address the talker and listener. A controller is necessary when the active or addressed talker or listener must be changed. The controller function is usually handled by a computer.

Each instrument hung on the bus must be capable of serving at least one of the following functions, which can be enabled or disabled at any time.

1. A listener must be capable of being enabled to receive data from other instruments (an example for listener is a printer or digitally programmable power supply).
2. A talker must be capable of being enabled to transmit data to other instruments (an example for a talker is digital voltmeter transmitting its data output).

3. A controller does the enabling and disabling of talkers and listeners. It manages the overall activity on the bus, from initialization at start-up to handling interrupts from instruments. While it often enables itself as a talker or a listener, it can also relinquish both these roles, letting other instruments communicate among themselves.

The interface bus consists of the 16 signal lines (Fig. 3.4). These are:

1. Five control lines used for interface management. Four of these lines are driven by the controller to set up the other instruments on the bus as talkers or listeners. The fifth line SQR, is used by an instrument to interrupt the controller in order to request for use of the bus.

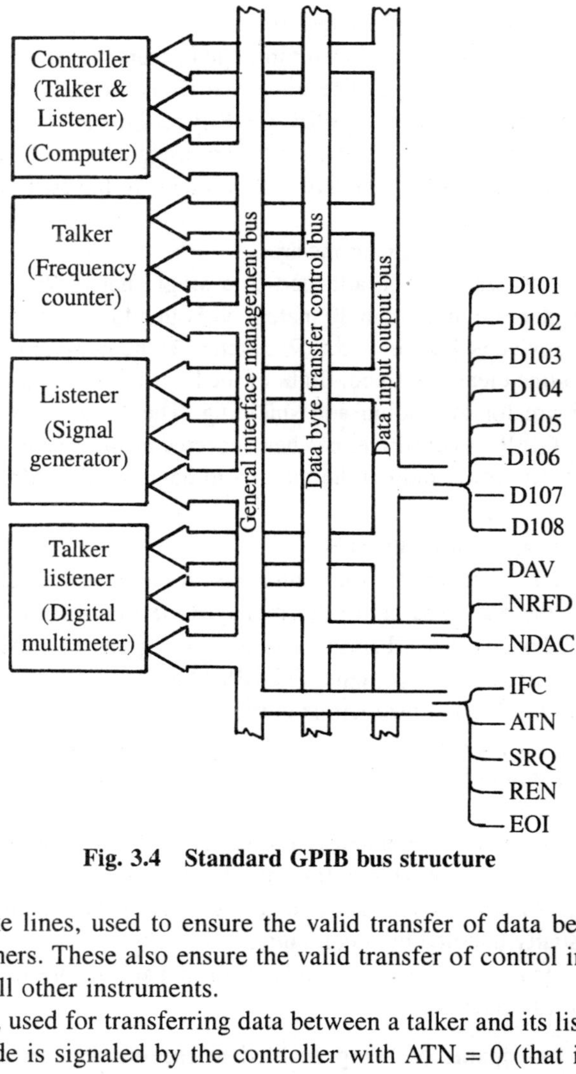

Fig. 3.4 Standard GPIB bus structure

2. Three handshake lines, used to ensure the valid transfer of data between a talker and any number of listeners. These also ensure the valid transfer of control information between the controller and all other instruments.
3. Eight data lines, used for transferring data between a talker and its listeners while in the data mode. This mode is signaled by the controller with ATN = 0 (that is, a high output on the negative-true bus). These data lines are also used for transferring control information from the controller to all the other instruments while in the command mode, signaled by the controller with ATN = 1.

3.7 Standard Add-on-Cards

The range of PC expansion boards currently available from a large number of manufacturers include:

- Analog I/O cards with upto 16 analog inputs (and upto 4 buffered analog outputs)
- Digital I/O cards with direct TTL-compatible inputs and outputs
- Digital I/O cards with opto-isolated inputs and outputs
- Digital I/O cards with buffered I/O lines
- Digital output cards with reed relays
- Digital output cards fitted with relays for AC or DC power control or triacs for AC power control
- Expansion memory card
- IEEE-488/GPIB interface cards
- Network adapter cards
- Modem cards (to permit transfer of data via conventional telephone lines)
- Multi functional I/O cards (offering mixed analog and digital I/O facilities)
- Bus expansion cards (which interface with external card frames or mother boards)
- Thermocouple interface cards
- Solid state RAM/EPROM based disk drives
- Specialized instruments cards (e.g., digital multimeters, counters/digital frequency meters, etc.)
- Stepper motor controllers
- High speed data acquisiton cards
- Serial communication cards with upto 4 serial ports

In addition, the system builder is able to select from a large range of signal conditioning cards which provide the necessary interfacing circuitry for a wide range of popular sensors and output devices. It is easy to construct PC based process control system by selecting 'off-the-self' modules. Only dealing with specialized applications is it necessary to go in for one's own dedicated I/O cards and/ or external signal conditioning boards.

The system designer can select from a range of options depending upon the complexity and individual requirements of a particular application. The following general approaches are available:

- stand-alone PC systems based on internally fitted expansion cards, rack modules, or separately enclosed units
- PC systems based on standard PC expansion cards (and I/O processing cards, where appropriate) fitted into external card frame modules
- Industrial PC system fitted with internal or external expansion cards in a rack or free standing enclosure
- RS-232 based systems with the PC as controller (peripheral hardware connected via an asynchronous serial link)
- IEEE-488 based systems with the PC as controller
- Backplane bus-based (e.g., IEEE-1000) systems with a PC bus master/controller and standard bus I/O cards
- Networked/distributed PC systems (e.g., based on Ethernet or Bitbus) with enclosures and expansion cards to meet local requirements
- Solid-state RAM/EPROM based disk drives
- Specialized instruments cards (e.g. digital multimeters, counters/digital frequency meters, etc.).

Industrial PCs are an obvious choice for use in the harsh environment found in most industrial plants. Typical of such equipment is the popular Hewlett Packard Vectra series of computers. Industrial PCs usually offer the same range of facilities associated with conventional PCs and invariably support the industry standard bus architecture. Industrial PCs may be fitted with bus extenders which are normally based on an external rack assembly. In extremely hard environments, it may become necessary to replace the floppy/hard disk drive system with a 'silicon disk'. Several manufacturers provide RAM/EPROM-based disk emulator cards which are ideal for use in dedicated diskless systems.

Sensors, the devices that are actually used to measure physical phenomena typically require some sort of 'signal conditioning'. A variety of choices are available in packaging of sensors and signal conditioners. These include: (i) data logger (ii) PC-plug-ins (iii) computer backplane PC-plug-ins and (iv) network based systems.

A data logger solution is often a stand-alone instrument but could include a low-cost interface to a personal computer. A PC-based plug-in solution generally uses available low cost slots in the computer to hold measurement modules. A computer back plane solution makes use of a separate main frame holding multiple modules connected to a personal computer with a high speed interface. Their strength comes from being able to accommodate large channel counts; provide an extremely wide range of measurements; and record/analyze the resultant data with high real time speeds. A network-based solution can be physically distributed, inheriting the advantages of LAN technology. Using such standards as Ethernet, small measurement modules can be connected to PCs, sometimes at great distances. This new architecture provides an information bandwidth greater than that typically available from a standard data logger, but less than that available from computer parallel-back plane architecture, such as VXI. The network based system encourages to digitize the measurement data as close to the point of measurement possible, then brings the data back to the controller digitally. It clearly reduces the cabling. This lead to the development of 'smart sensors', those that can be plugged directly into such a 'measurement network'.

3.8 Backplane Bus

The term backplane bus refers to the board (Fig. 3.5) on which connectors are mounted and into which printed circuit boards containing a particular function, e.g., memory, can be plugged. The backplane provides the data, address and control bus signals to each board, thereby enabling the

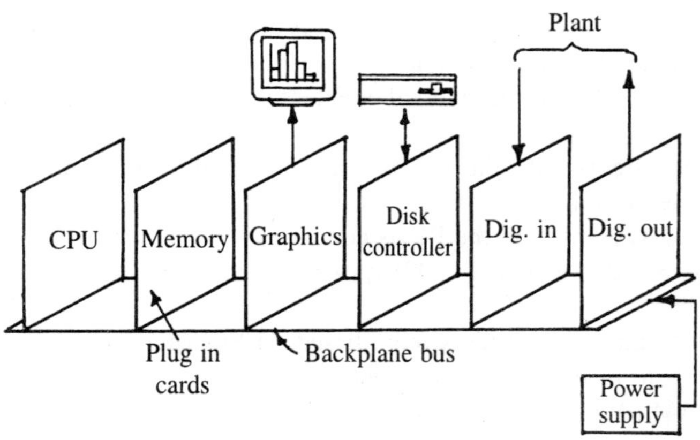

Fig. 3.5 Backplane bus

system to be easily expanded by the use off-the-shelf boards. It is into such computer buses that the data acquisition boards and boards used for interfacing instruments and other peripheralsl have to interface. Data acquisition and instrument boards are usually available in various configurations, depending on the computers with which they are to be used. For example VME bus is the bus designed by Motorola for use with its 32 bit 68000-microprocessor-based system. Such a bus is now, however, widely used with other computer systems as the bus for use with instrumentation systems.

A backplane provides the data, address and control bus signals, and various cards can be plugged onto the bus to configure the system as required. The advantages of a bus system are, the standardization, off-the-shelf cards, ease of expansion and a DIY (do-it-yourself) approach. Each and every microprocessor manufacturer devised their own standards, with different sized data words (8 bit, 16 bit or 32 bit) different address ranges and different edge connectors and pin layouts. Fortunately some common standards do seem to be emerging, notably the VME and STE Bus standards.

With a high speed clock (24 MHz data transfer rate), 32 bit data bus and an enourmous address range, the VME bus is arguably the most powerful industrial control system available. It has a high speed of performance. It is however, sophisticated and expensive.

The STE bus is an 8-bit data bus system with origins in the earlier STD bus. It has 20 address lines (allowing over 1 Megabtyte of memory space) and 4 kilobytes of addressable I/O. It has been formalized under the IEEE-1000 standard. This ensures compatibility between different manufacturers, and has led to it becoming a general purpose acceptable standard where the higher performance of VME bus is not required. STE bus has many attractive features. The cards are based on the compact Eurocard size (100 × 160 mm) and connections between the cards and the backplane are made by a robust two-part connector (DIN 41612) which is resistant to vibration and shock loads. The STE bus has been designed around an interface between a master processor and a wide range of I/O boards. The microprocessor type is not specified, and any microprocessor can be used provided the interface between CPU board and the backplane meets defined standards. The bus can support up to three master CPU boards on the bus, although, of course, only one of these can be active at any time. A well defined procedures is laid down for the selection of control of the bus when connection occurs between the different masters.

The VL bus (or VERSA bus) established by the Video Electronic Standards Association as a means of improving the graphics performance of modern PCs, is useful in any graphic intensive environment (e.g., Windows) as it not only provides an extension to the standard 16-bit ISA bus so that 32 bus data transfers can take place, but is also capable of operating at speeds upto 33 MHz.

3.8.1 VME Bus

The VME bus is a bus structure designed originally to allow the interconnections of various printed-circuit board modules that made up the Motorola's lines of VERSA modules. The VERSA modules are products designed to take full advantage of the capabilities of the 16- and 32-bit MC68000 family. A more compact version of the boards called VERSA module Euro cards was developed in Motorola's Munich factory and became a standard in Europe. In 1981, motorola announced the VME (VERSA Module Europe or Eurocard) bus as a non-proprietary standard in the public domain. The VME bus system is designed to do the following:

1. Allow communication between devices without distrurbing the internal activities of the other devices interfaced to the bus
2. Specify protocols that define the interconnection between devices

3. Specify the electrical and mechanical system characteristics required to design devices that will reliably communicate with other devices interfaced to the VME bus
4. Allow a broad range of design latitude so that the designer can optimize cost and performance without affecting system compatibility.

Mechanical structure: The mechanical structure consists of subracks, backplanes, and a plug-in board. The subrack (usually 19″ wide) includes I/O modules, the VME bus module, the MVME820 mass storage module, chassis, backplane and power supply. Backplanes may have upto 21 slots. Each backplane includes several 96-pin connectors and signal paths that bus their pins. All VME bus system have a J1 backplane that provides all the signal paths needed for 24-bit addressing and 16-bit data. In system that utilizes a 32-bit address and data path, a J1/J2 backplane is provided. The VME bus boards could be either single or double height. A single height board is 100 mm (high) × 160 mm (deep). A double-height board is 235.35 mm (high) × 160 mm (deep).

3.8.2 VXI Bus

As discussed earlier, the VME bus is the bus designed by Motorola for use with its 32-bit 68000-microprocessor-based system. The VXI bus (VME extension for Instrumentation) is an extension of the VME bus which has been designed for instrumentation applications such as automatic test equipment where higher speed communications are required than those possible with the GPIB bus. It also provides better synchronization and triggering and has been developed by a consortium of instrument manufacturers so that inter-operability is possible between the products of different companies. The system invloves VXI boards plugging into a mainframe. Figure 3.6 shows a number of possible system configurations that can be used. In Fig. 3.6(a), a VXI mainframe is linked to an external controller, a computer, via a GPIB link. The controller talks across this link using a GPIB protocol to an interface board in the chassis which translates the GPIB protocol into the VXI protocol. This makes VXI instruments appear to the controller to be GPIB instruments and enables them to be programmed using GPIB methods. Fig. 3.6(b) shows the complete computer embedded in the VXI chassis. This option offers the smallest possible physical size for the system and allows the computer to directly use the VXI backplane bus. Fig. 3.6c uses a special high-speed system-bus-on-cable, the MXI bus, to link a computer and the VXI chassis, the MXI being 20 times faster than the GPIB.

3.9 Micro Controllers

Perhaps more than any other factor, the development of microprocessors has been responsible for the explosive growth of the computer industry. As stated in Chapter 1, microprocessors are ‘computer on a chip’. While early microprocessors required many additional external components in order to perform any useful work, the increasing use of large scale integration (LSI) or VLSI semiconductor fabrication techniques has led to the production of microcomputers where all of the required circuitry is embedded on one or atmost a small number of integrated circuits. A further extension of this integration is the single chip micro controller which adds analog and binary I/O, timers, and counters so as to be able to carry out real-time control functions with almost no additional hardware (Fig. 3.7). Examples of such micro controllers are Intel 8051, 8096 and Motorola MCH68HC11. These chips, developed largely in response to the automotive industry’s desire for computer-controlled ignition, emission control, and antiskid systems, are now widely used in process industries. Table 3.2 gives a comparison of the micro controllers.

The major types of micro controllers are: (i) embedded 8-bit micro controllers, (ii) 16–32 bit micro controllers and (iii) digital signal processors.

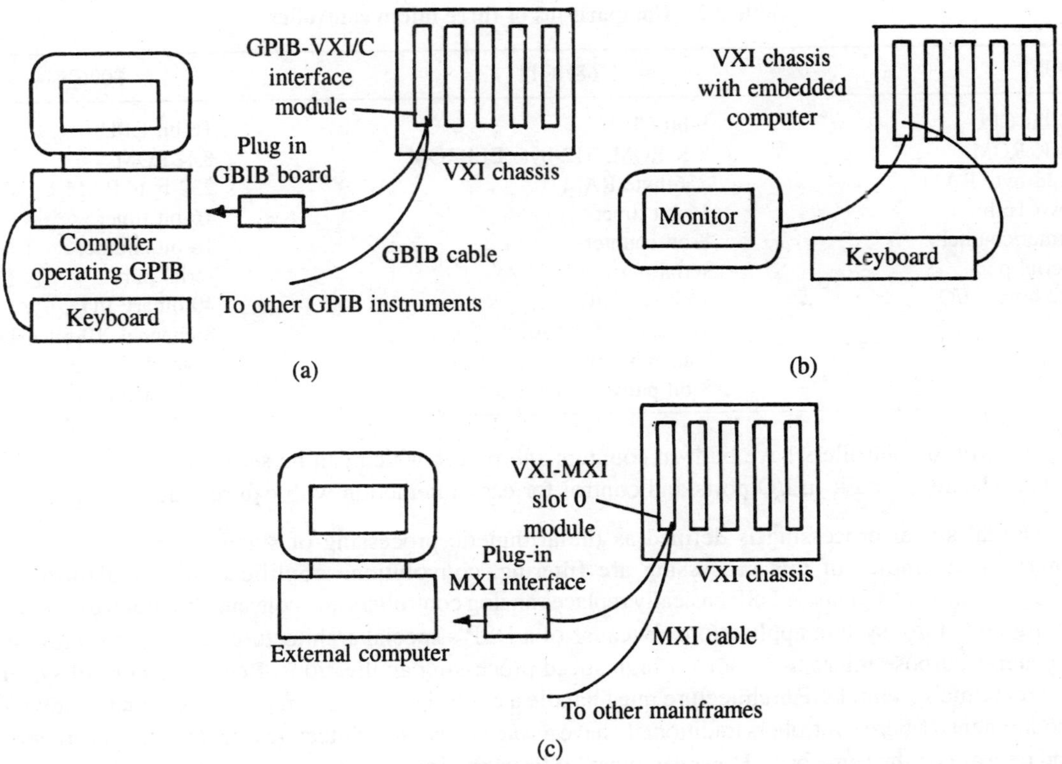

Fig. 3.6 VXI options

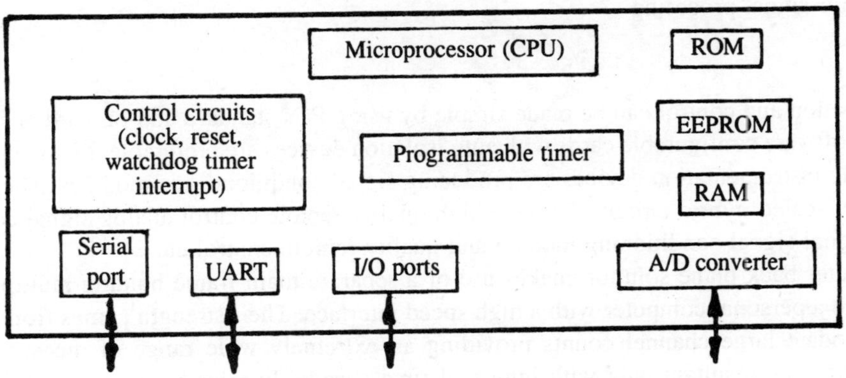

Fig. 3.7 Simplified block diagram of a microcontroller

The basic characteristics of a micro controller are:

(i) It has a built-in ROM (4 Kbytes) within the chip to store the control program

(ii) It has a small built-in RAM (128 bytes) for temporary data storage

(iii) The CPU uses single bit instructions so that the limited program memory (ROM) is effectively used

(iv) Many micro controllers have a Boolean co-processor along with the CPU to simplify implementing Boolean expression occurring often in control applications

Table 3.2 Comparisons of three micro controllers

8051	68HC11	8096
8-bit CPU	8-bit CPU	16-bit CPU
4 K ROM	8 K ROM, 512-byte EEPROM	8 K ROM
128-byte RAM	256-byte RAM	232-byte RAM
two 16-bit	16-bit timer	16-bit timer
timer/counters	8-bit counter	16-bit counter
serial port	serial port	serial port
32 binary I/O	16 binary I/O	40 binary I/O
	8-channel, 8-bit A/D	8-channel, 10-bit A/D
	watchdog timer	watchdog timer
	8-bit pulse accumulator	pulse width output

(v) Micro controllers have bulit-in counters and timers which can be set by users
(vi) There are built-in I/O ports and control for easy interaction with external devices

Digital signal processing is defined as the arithmetic processing of signals sampled at regular intervals. Examples of this processing are filtering, convolution, amplification, modulation, and transformation of signals. A DSP basically replaces analog controllers and conventional microprocessors in digital control system applications. Because of a DSP's special architecture, it is more useful than a general-purpose microprocessor for high speed processing applications. Because a control system is a real time system, DSP architecture must handle a control system's numerical tasks and bandwidth requirements. Micro controllers traditionally have a *von Neumann* architecture (meaning that instructions and data are on the same bus). However, most DSPs use a *Harvard* architecture (meaning that the data is separated from the instruction bus to increase speed) or a modified Harvard architecture that is optimized for signal procesing.

Summary

Data acquisition and control can be made simple by using PCs standard add-on-card and associated software. Software configurable cards with auto isolation devices are preferable. Most manufacturers of electronic instrumentation devices are producing signal conditioners as modules. These modules are compact sealed hybrid circuit. We have high quality remote control analog signal conditioning module, digital signal conditioning module and quality remote control etc.

A computer back plane solution makes use of a separate main frame holding multiple modules connected to a personal computer with a high speed interface. Their strength comes from being able to accommodate large channel counts providing an extremely wide range of measurements and record/analyze the resultant data with high real time speeds. In microcomputers, all the required circuitry is embedded on one or at most a small number of integrated circuits. A further extension of this integraion is the single chip micro controller which adds analog and binary I/O, timers, and counters so as to be able to carry out real-time control functions with almost no additional hardware.

References

Bolton, W., Mechatronics, 2nd Ed., Addision Wesley Longman Ltd., Singapore, 1999.
Coombs, C.F., Electronic Instrument Handbook, III Ed., McGraw Hill, New York, 2000.

Gates S.C. and J. Becker, Laboratory automation using the IBM PC, Prentice-Hall, Englewood Cliffs, New Jersey, 1989.

Gupta, S. and J.P. Gupta, PC interfacing for data acquisition and process control, 2nd (Ed.), ISA Pub. 1994.

Instrumentation Reference and Catalogue, 1997, National Instruments corporation, Austin, 1997.

Kheir, M., The M68HC11 micro controller, application in control, instrumentaton and communication, Prentice Hall, 1998.

Morris, S.B. Automated manufacturing systems, McGraw-Hill Int. Ed., Singapore, 1995.

Product Guide, Contec microelctronics U.S.A. Inc., Vol 3, p22–24, 1995.

Rigby, W.H. and T. Dalby, Computer Interfacing: A practical approach to data acquisition and control, Prentice Hall, Englewood Cliffs, New Jersey, 1995.

Tooley, M., PC-based instrumentation and control, 2nd Ed., Newnes, Oxford, 1995.

Tompkins, W.J. and I.G. Webster, Interfacing sensors to IBM PC, Prentice Hall, Englewood Cliffs, New Jersey, 1988.

Review Questions

1. What are the guidelines to be followed in selecting an add-on-card?
2. What are industrial signal conditioners?
3. What is a SCXI?
4. What are remote I/O modules used in DA and C?
5. What is meant by configuring an add-on-card?
6. What does an RS 232 interface do?
7. What is IEEE-GPIB? When will you use RS 232 and when will you use GPIB?
8. Discuss briefly bus Talkers, Listeners and Controllers in GPIB?
9. What is a VMS?
10. What is a VXI bus?
11. What is a micro controller?

4

Distributed Digital Control Systems

In this chapter, the evolution of digital computers applications in process control is discussed. The development of computer networking and communication among computers in the network are also discussed.

4.1 Evolution of Digital Controllers

The evolution of digital computers, having higher speed, higher reliability, smaller size and reduced cost, has brought about increased use of digital computers in process control. One of the first computer applications was data logging (Fig. 4.1), where the computer was used both to store a vast amount of measurement data produced in a complex process and to display the data for review (by process engineer) to determine the condition of the process. Gradually, the computer performed certain kinds of reduction of this data using control equations and even indicated the type of action, if any, which should be taken to tune a process for maximum operating efficiency. All of the loops in the process were still analog and for the most part, independent except for manual adjustment of set point under guidance of a process engineer.

A natural extension of this concept led to the development of a technique where the computer itself performs adjustments of the loop set points and provides a record of process parameters (Fig. 4.2). The loops are still analog, but the set points that determine the overall process performance are set by a computer on the basis of equations solved by the computer, using measured values of process parameters as input. Such a system is called supervisory computer control.

The ultimate result of computer control in process control has been to use the computer to perform controller functions. In such a system (Figs. 4.3 and 4.4), called direct digital control (DDC), the only analog elements left in the process control loop are the measurement function and the final control element.

Digital controllers can replace several analog controllers. They usually require an analog-to-digital converter at the input because of the wide use of analog sensors, transducers and signal transmission, and a digital-to-analog converter at the output to drive actuators designed for analog techniques. It is possible in the near future (refer to Chapter 6), digitalization will extend to sensors and actuators. This would not only save A-D and D-A converters but would also circumvent certain noise problems, permit the use of sensors with digital output and reprocessing of signals in digital measuring transducers (for example, choice of measurement range, correction for nonlinear characteristics, automatic failure detection, etc.). An example of actuators which accept digital signals are stepping motors. Digital controllers can replace not only just one or several analog controllers, but can perform additional functions previously exercised by other devices or even perform new functions. These additional functions are such as program sequence control of set points, automatic switching to various controlled and manipulated variables, feed forward adjusted controller parameters as function of the operating points, additional monitoring of limit values, etc. Examples of new functions are: communications with other digital controllers, mutual redundancy, automatic failure detection and failure diagnosis, the possibility of choice between different control algorithms and, in particular, adaptive control algorithms. Entire control systems such as cascade control systems, multivariable control systems with coupling controllers, control systems with feed forward control which can be easily changed via software at commissioning time or later, can be effected with a digital controller.

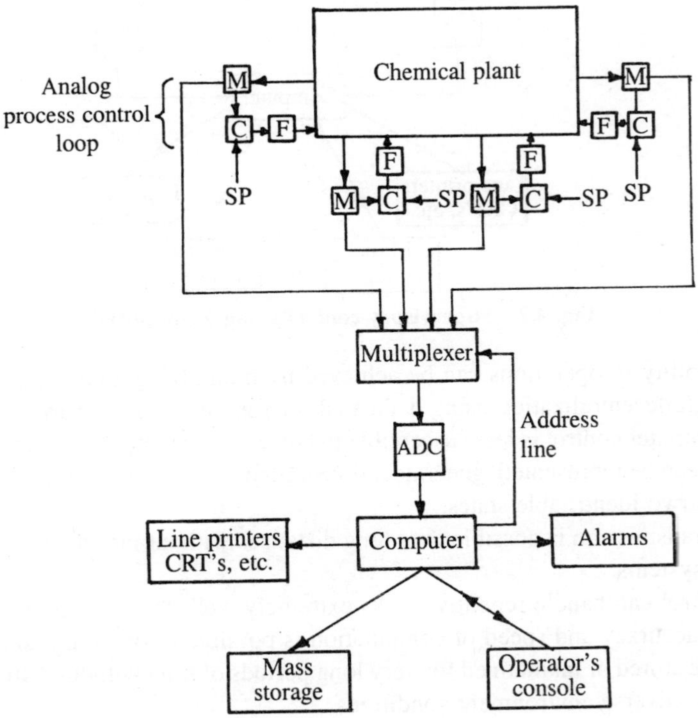

Fig. 4.1 Data logging using a computer

Distributed control systems are those systems that have intelligence in various nodes, are connected by some media and networked together, perform peer to peer communications, and perform a portion of the overall set of tasks at each node.

4.2 Advantages of Digital Computer Control

The important advantages of computer control of processes are:

1. Complex control laws and signal conditioning methods, that might be impracticable to implement using analog devices, can be programmed.

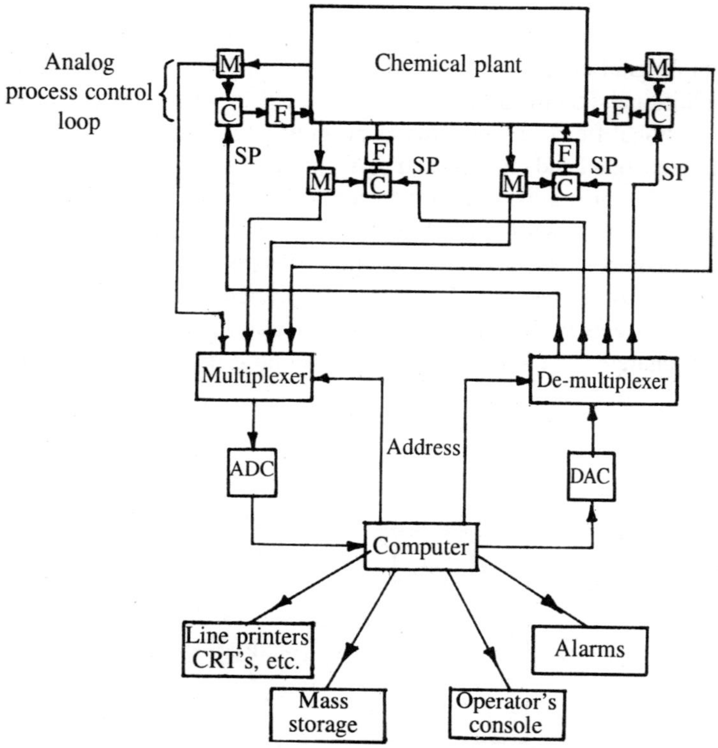

Fig. 4.2 Supervisory control using a computer

2. High reliability in operations can be achieved by minimizing analog hardware components and through decentralization using dedicated microprocessors for various control tasks.
3. Digital computer control is less susceptible to noise or parameter variation, in instrumentation since data can be represented, generated, transmitted and processed as binary words, with bits possessing two identifiable states.
4. Fast data transmission is possible over long distances without introducing dynamic delays, as in analog systems.
5. Digial control can handle repetitive tasks extremely well, through programming.
6. Very high accuracy and speed of computation is possible through digital signal processing.
7. Data can be stored or maintained for very long periods of time without drift and without being affected by adverse environment conditions.
8. Large amount of data can be stored using compact, high density data storage methods.
9. Digital control has easy and fast data retrieval capabilities.
10. Digital processing uses low operational voltages (0 to 12 V DC).
11. Low overall cost.

Sampling is necessary in real time digital processing (i) to allow for the processing time in ADC and in control law calćulations (ii) expensive computer or equipments can be shared among many signals.

4.3 Process Control Requirements of Computers

Industrial process control exists in any of the following forms:

(i) Plant automation/computer integrated control (CIM): Total plant automation is already widely used in the automotive industry, but not yet in many small-size production facilities. The automation of a whole plant which includes integration of several production control units, allows for optimization of production and improved production scheduling.

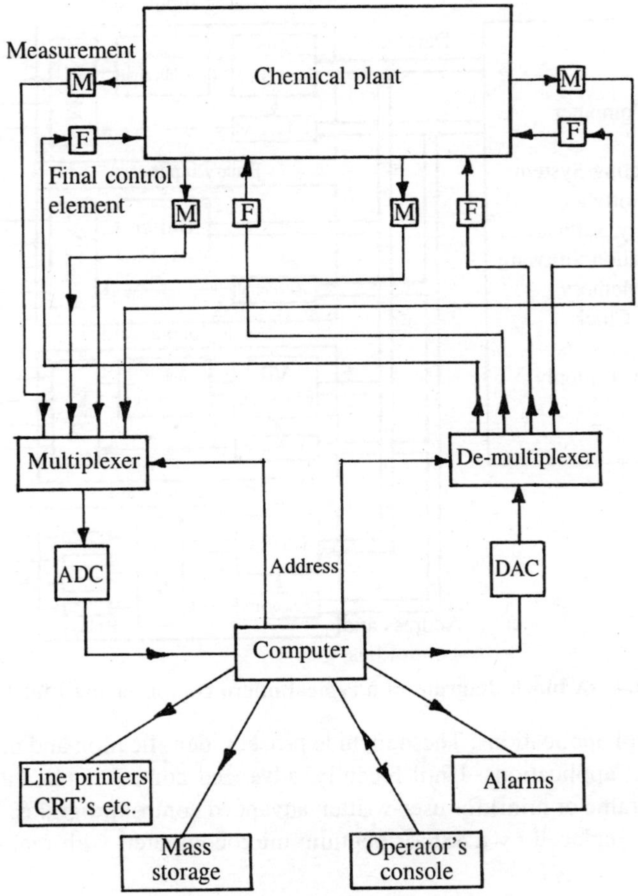

Fig. 4.3 Direct digital control (DDC)

(ii)Programmable logic controllers (PLCs) integration: PLCs are microprocessors with other interfacing circuits used to automate sequencing jobs. A plant that has several PLCs without good operator interface could benefit from a PC-based process control system that displays data from many PLCs, with the objective of upgrading the production facilities. Data from different PLCs can be used to set and monitor production rates. Plant data collected from different PLCs can be combined into one production report.

(iii)Regulatory process control: Distributed control systems (DCS) is the state-of-the-art regulatory control system, which can be configured to the requirements of nearly every application. The DCS controls many concomitant loops and has advanced display capabilities. However, DCS is usually only cost-effective in large applications. Small- and medium-sized plants that are equipped with several single-loop PID controllers can now enjoy most of the benefits of a system by using a PC-based process control system that functions like a DCS system.

(iv)Batch applications: Batch applications are widely used in the food and chemical industries, where mixing or reacting of components to a final product is frequent. Batch applications, especially when many different recipes are used or when recipes are developed, can ideally be done by a flexible PC-based process control system.

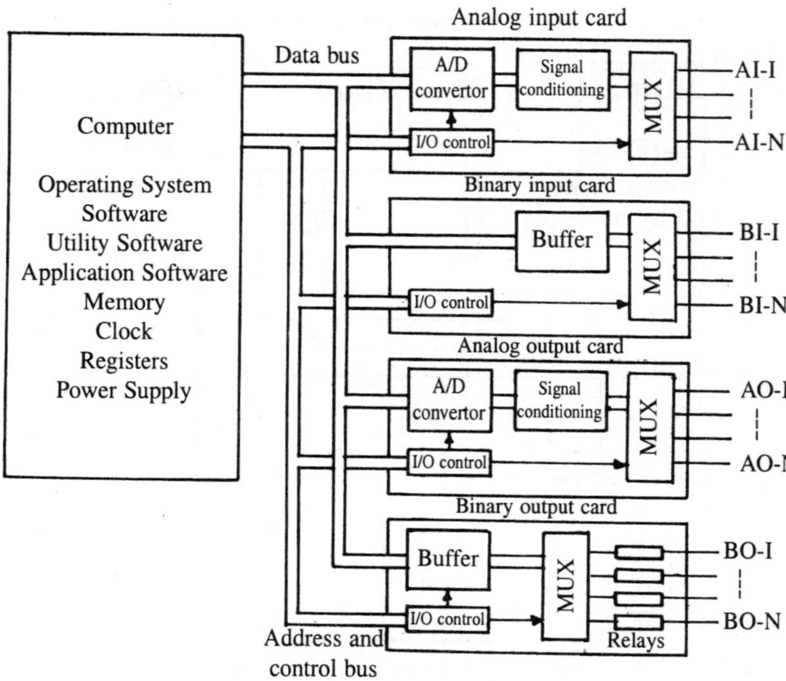

Fig. 4.4 A block diagram of a typical micro computer used for DDC

(v)Advanced control applications: These include process identification and model development in more complex control applications. Until recently, advanced control applications required a large host computer (mainframe or mini) for user-written advanced control programs. In certain cases, this computer can now be replaced by a 486 or pentium microcomputer, with real savings in hardware costs.

4.4 Computer Network

Perhaps the most successful deployment of computers in process control involves a network (known as distributed control systems) consisting of all three computer sizes namely micro, mini, large scale computers (Fig. 4.5). In a network, the micro computers is used to control one to perhaps ten loops in a process. Several such units would provide control of all loops requiring computer control in the process. Each of the micro computers is supervised by a mini computer which also provides the required mass storage and evaluation of over all performance. In a large plant with a number of major processes, the network would include a large-scale computer in communication with the mini computer supervisors.

The advantages of distributed computer system are:

- increased performance through resources sharing
- increase reliability
- the system can be expanded in modular fashion
- reduced cost of wiring.

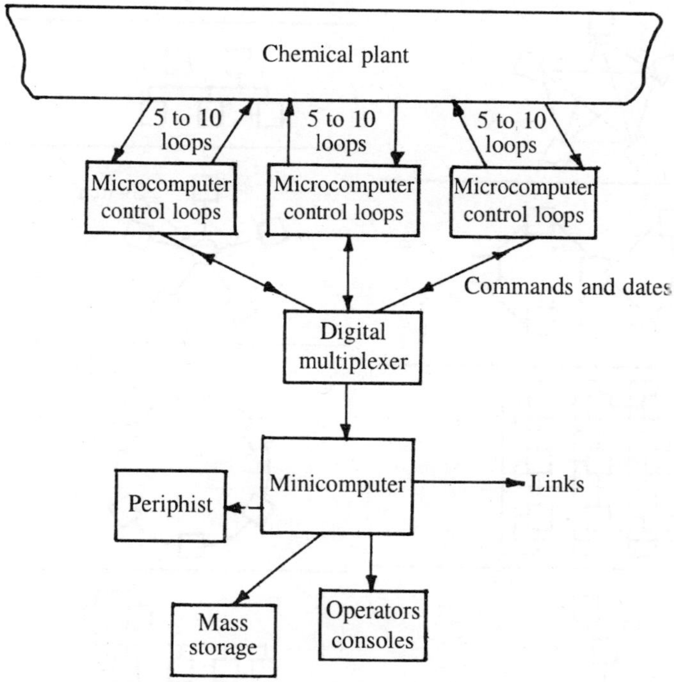

Fig. 4.5 Distributed computer control

The disadvantages are:

- increased software complexity
- difficulty in system test and failure diagnosis
- unique expertise needed during design and development phases.

Communication among the plant computers would not only provide a total picture of corporate production but would also facilitate transferring raw materials and finished product, smoothing loads and demands, and compensating for unexpected bottleneck and shutdown.

The impact of distributed digital control system on control aspects of process is straight forward. Centralized controllers racks drawer full of screw drivers for control loop tuning, and miles of twisted pair cabling and associated shielding gone, replaced by remote controller-nests and digital coaxial cabling. The control room itself can be scaled down significantly with shared display. The vast arrays of indicating/controller panels, strip chart recorders, gauges, dials, process flow diagram, alarms, and so forth can be condensed into an integrated console.

4.4.1 Multi-Mini Computer Architecture

Variations in multi-mini computer architecture (Fig. 4.6) are generally determined by several characteristics such as message rate and size, data paths between mini computers, and switching element routing and controlling messages flowing from source (transmitting) mini computer to destination (receiving) mini computer. There are six basic multi mini computer architectures available.

To a significant degree, distinctions between these configurations can be related to the level of geographical dispersion. A shared memory-based multi computer system is usually connected in one or two adjacent racks, whereas a bus or loop-based system may be distributed between several plants

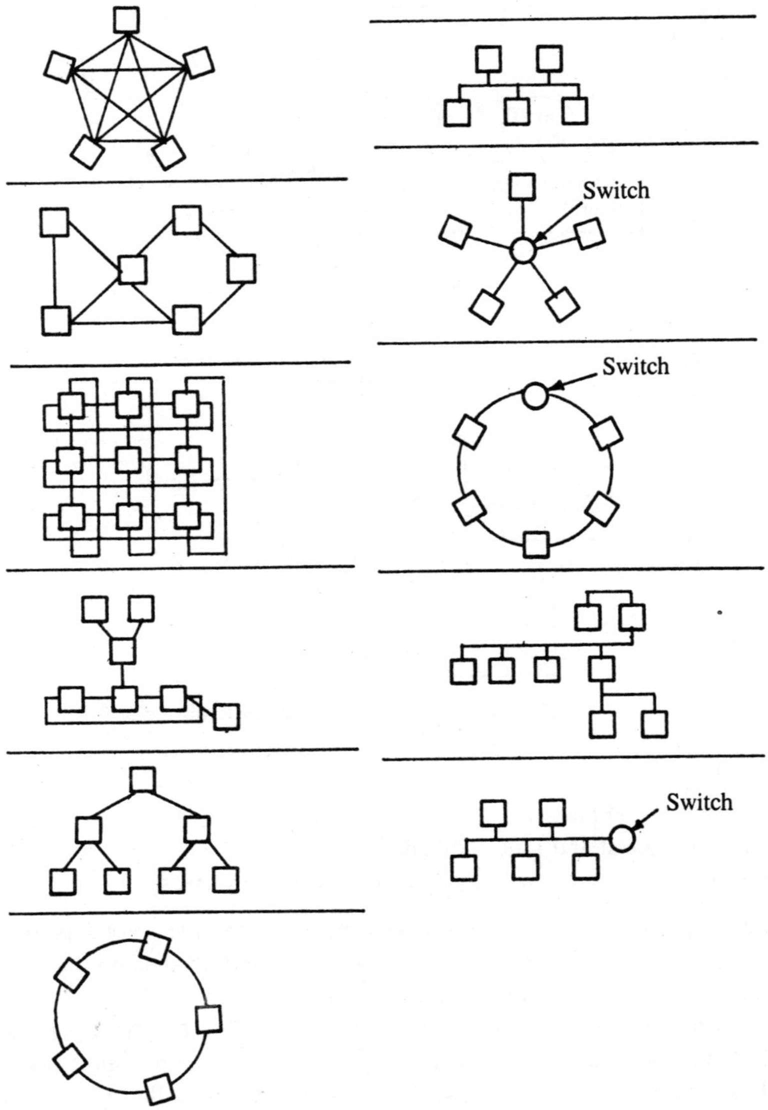

Fig. 4.6 Computer network topology

compltely interconnected
packet switch network
regular network
irregular network
hierarchy
loop or ring

global bus
star
loop with switch
bus window
bus with switch

or buildings, while a point-to-point partially or fully inter-connected system may have one or more mini computers located in a longer distance between them.

The topology of a network describes the cable layout. A network's 'topology' can be considered in two ways: physical topology and logical topology. Physical topology describes what a human sees.

Logical topology describes what the network software thinks is happening. The two can be quite different.

Each topology has certain attributes which are related to cost, reliability, responsiveness, speed, throughput capability, ease of development, modularity, reconfigurability, logical complexity and physical dispersability. The total development cost of a multi mini computer system is divided among the cost of mini computers, mini computer interfaces, communication links including switches and buffers (when needed) and the communications software.

The multi computer system reliability can be measured in terms of mean-time-between failures, fault tolerance, graceful degradation, mean-time-to-repair, and failure reconfiguration time. Mean time between failure is typically measured in hours and can be related to component reliability and level of redundancy. Mean-time-to-repair can be minimized with built in redundancey, built in real time self-check and diagnosis, and the use of on site highly trained maintenance people, coupled with an extensive spare parts inventory.

Failure reconfiguration generally requires redundant paths, and/or mini computers that can be activated, as soon as a failure occurs and is detected. The speed can be improved by proper balance in processing and communication, type of main memory and peripherals, selection or design of operating systems and communications protocols. The modularity can be assessed in terms of the incremental cost of adding a mini computer to the system and the system ability to make incremental changes in system capability by reconfiguring the system. Reconfiguration of a star based may be fairly simple compared to reconfiguring a shared memory system.

4.4.2 Peer-to-Peer and Server-Based Network

A network can be divided into two broad categories: (i) peer-to-peer and (ii) server-based networks. In peer-to-peer networks each computer can act as both a client and server. Servers are computers that provide shared resources to a network user. Clients are computers that access shared network resources provided by the server. The media is the way the computers are connected.

Peer-to-peer networks make sharing data and peripherals easy for small groups of people. However, a consistent security is difficult in a peer-to-peer environment because administration is not centralized. If there are 10 users or fewer and the users are also located in the same place, then the security is not an issue. Server-based networks are best for networks sharing many resources and data. An administrator overseas operation on the network and ensures that security is maintained. The performance of the network is also improved. This type of network can have one or more servers depending on the volume of traffic, number of peripherals and so on. For example, we may to have a print server, a communication server, and a database server all on the same network. The server runs completely on a different operating system and rarely accommodate a direct user at its keyboard.

Server-based networks appear to mimic the old central mainframe system. The difference is that the relatively 'dumb' terminals connected to a mainframe have been replaced by very sophisticated personal computers or work stations. These PCs are just as capable of running software as in the server. A work station can operate without a floppy (diskless work stations) if an extra memory chip is added to the network adapter card. This would contain just enough code to initiate the transfer of the DOS files and any other required files from the server. Diskless work stations are not cheaper but are often used for security reasons so that the user cannot copy data into floppies for personal or illegal use or introduce any unwanted and possibly damaging software viruses into the network. A network design needs two, somewhat separate, purchasing decisions—what server software to use or which adapter card to use.

4.4.3 Network Topology

Bus topology: A bus network is a single conduit to which all the network nodes and peripheral devices are attached to the cable at tap points. Nodes on a bus network transmit data at any time, regardless of any data being sent by the computers. Keeping the data transmission from colliding requires extra circuitry and software. Computer on a bus topology network communicate by addressing data to a particular computer and putting that data on the cable in the form of electronic signals. The information is accepted only by the computer whose address is encoded in the original signal, only one computer at a time can send messages. Because of this the network performance is affected by the number of computers attached to the bus. The bus is a passive topology. Computers on a bus only listen for data being sent on the network. They are not responsible for moving data from one computer to the next. If one computer fails, it does not affect the rest of the network. In an active topology, computers regenerate signals and move data along its network. Because the data, or electronic signal, is sent to the entire network, it will travel from one end of the cable to the other. If a signal were allowed to continue uninterrupted, it would keep bouncing back and forth along a cable and prevent other computers from sending signals. Therefore, the signals must be stopped after it has a chance to reach the proper destination address. To stop the signal from bouncing, a component called a terminator is placed at the end of the cable to absorb free signals.

Star topology: Groups of data are routed through a central hub to their destinations. A hub is the central computer in a star toplogy. It monitors the traffic and prevents data collisions. A broken conection does not affect the rest of network. If the hub fails the entire network fails. More expensive hardware is required because of hub. Hooking of an additional computer is easy because only one line is to be made to the node from the hub. Most hubs are active in that they regenerate and retransmits the signals. Many active hubs contain diagnostic capabilities to indicate wheter or not a connection is working. Active hubs require electrical power to run. Some type of hubs are passive and act as connection points and do not amplify the signal, the signal passes through the hub. They do not require electrical power to run.

The size of the hub can range from an expansion card in a PC, through a small box, to a large cabinet in a large company. Intellignet hubs can provide diagnostics, which can often be investigated from a work station with suitable software. At the very least, this will be in the form of LEDs on the front panel which tell you when data is flowing.

Ring topology: The ring topology connects the nodes of the network in a circular chain in which each node is connected to the next. The final node in the chain is connected to the first node completing the ring. Each node examines the data that is sent through the ring. If the data is not addressed to the node examining it, that node passes it along to the next node in the ring. There is no danger of data collisions because data always flow in one direction. Each station acts like an amplifier. If a connection is broken in the ring then the entire network goes down. A great amount of traffic exists through each node, thus it slows down the network performance. Since each station is needed to complete the ring, either each station will have to be always turned on or the two ends of the ring will have to be 'jumpered' when a station is turned off or removed.

Chain or wave network: A variation on the bus is the 'daisy chain'. Each adapter card has two identical connectors that are directly connected within the card. They are arbitrarily labelled IN and OUT. The next machine in a line will have a cable connected from its IN connector to the OUT of the previous machine. The end machine usually has a terminating resister connected to its OUT connector. Interconnected machines can be turned off without upsetting the operation since the internal connection is direct. AT&T's star Lan and the Macintosh's localtalk connectors use this arrangement. Data packets are passed down the daisy chain from computer to computer until they reach the end of the network. Some computers like Macintosh have built-in-capability to have this type of networking. This network is cheap and easy to set up. A great amount of traffic prevails. Termination on the beginning and end of the chain is citical for it to be able to function.

Broadband bus topology: A broadband bus topology is similar to the bus topology. The bus allows only one communication to occur at a time and is known as 'baseband' operation because the entire bandwidth of cable is used for each message sent. This is accompained by breaking up the frequency spectrum of the cable into several divisions. Each of these divisions can then be used to carry separate messages. This process is known as frequency-division multiplexing (FDM). Because of the high speeds and multiple messages being sent, the transmission medium is critical. An open, short, or potentially, any damage to the cable can result in degradation of these data channels.

Advantages of broadband bus topology: (i) multiple messages can be sent at the same time and (ii) the throughput of this network is the greatest of all topology types.

Disadvantages: (i) smart and more costly controllers are needed and (ii) the transmission medium is more critical than for the other topology types.

4.4.4 Network Adapter Card

Role of network adapter card: A network adapter card acts as the physical interface or connection between the computer and the network cable. The cards are installed in an expansion slot in each computer and server on the network. After the card has been installed, the network cable is connected to the card's port to make the actual physical connection between the computer and the rest of the network.

The card contains hardware and firmware (software stored in read-only-memory) programming that implements the logical link control and media access control functions

The function of the network adapter card is to prepare, send and control data on the network. The card uses transceiver to reform data from parallel to serial transmission. Each card has its unique address which allows it to be distinguished from all other cards on the network. The network adapter card also receives incoming data from the cable and translates into parallel form that the computer CPU can understand. The component responsible for this is called transceiver (transmitter and receiver). Some cards come with an external and onboard transceiver. The appropriate one should be selected by jumper settings or through the software. Each card is built to accept particular type of cable such as coaxial, twisted pair or fiber optics. Some network cards have more than one interface connector.

The network adapter card has configurable options that must be set. These options include the interrupt request level (IRQ), the base I/O port address, and base memory address and transceiver. There are several ways of enhancing the network performance. Some adapter cards have enhancement designed into them. These include direct memory access (DMA), shared memory access, shared system memory and bus mastering. Performance can also be enhanced through RAM buffering and with the use of an on-board microprocessor. There are also network adapter cards for specialized environment such as wireless and fiber optic networks, and work stations without disk drives in a high security environment.

4.4.5 Software

A simple processor is controlled by an operating system that consists of two types of software: (i) control programs and (ii) processing programs. The control programs typically manage the use of system resources, provide easier access to and more efficient use of physical resources (the users interfaces) and perform data management. The processing programs include application and support software where the latter contain text editors, language translators (assemblers and compilers), file management utilities, diagnostic software etc. The group of programs that manage the resources on the network is often called network operating systems or NOS. The most popular NOS is *netware* (from Novell) which can be used to run networks with different protocols, including Ethernet, token ring and Arcnet.

The multi mini computer system software differs from that of single processor in both the control and processing softwares. Often control programs must not only manage resources in the local mini computer but also in the inter-connected computers. The degree or level of management 'responsibility' depends to a large extent on the total system architecture.

The distributed mini computer software must be able to provide features such as diagnosis and recovery procedures beyond those of a single processor-based system and such software often becomes a sizable part of the total system software. In case one of the mini computer in a system either stops functioning or starts generating faulty messages, other processors must detect this and take appropriate action, such as working around the failed elements and reporting the failure to a system operator thereby allowing him/her to reconfigure the system.

4.5 Selection of a Suitable DICS

Some of the important distributed control systems (DICS) available in the market are:

S. No.	DICS	Manufacturer	Loops/controller
1	TDC 2000	Honeywell	8
2	Spectrum	Foxboro	40
3	DCI-4000	Fisher and Porter	16
4	Q-Line	Westinghouse	25
5	Encon-D	EMC control	32
6	ODAC	Reliance Electric	16
7	MV 8000	Beckmen Instruments	16

The differences in the above said DICS are mainly due to the type of processors used, the local station parameters like number of analog inputs, binary outputs, number and types of control algorithms and control schemes, and dataway and data traffic parameters i.e., of the architecture, protocol used and data rate used etc. The selection of a suitable DICS is determined by reliability, operator interface, maintenance, software functional capacity, configuration, vendor capability, delivery, price etc.

4.6 Interconnection of Networks

Let us discuss how different networks can be interconnected (Fig. 4.7). The diagram contains a backbone cable, two ring networks, two star networks, FDDI (Fiber Distributed Data Interface) networks, some hubs, some bridges, a gateway, some concentrators, and a few servers. The thick line to the left of the diagram labeled backbone provides a common link for an organization, building, or geographical area. The backbone is a high speed transmission line (usually fiber or thick cable) that

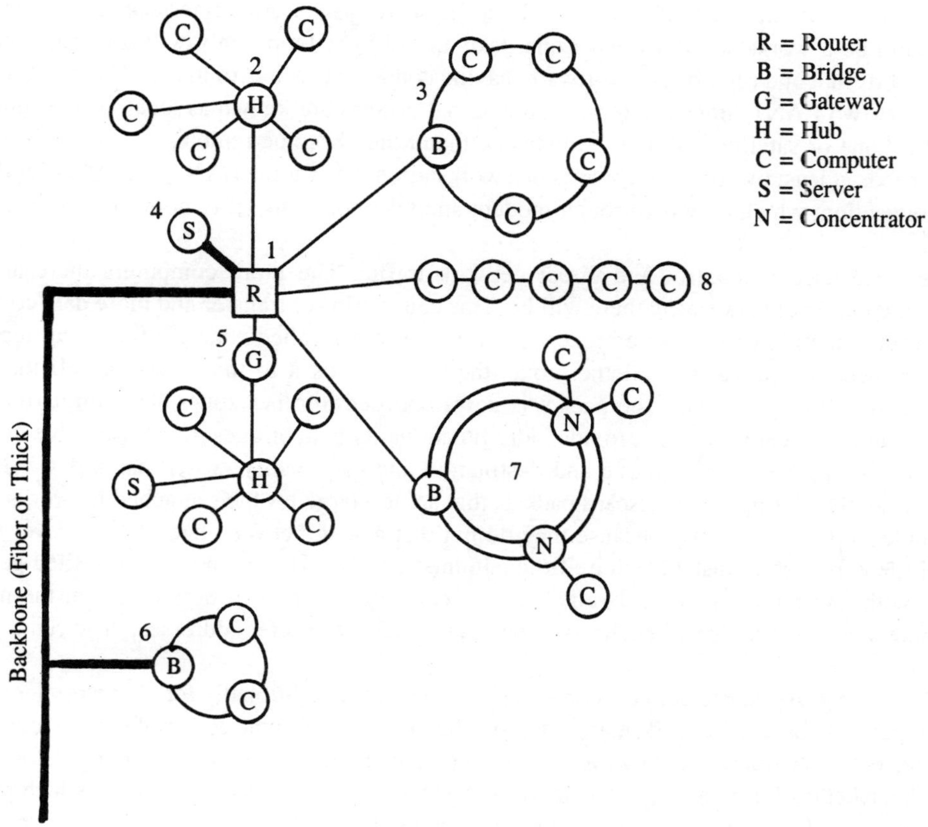

Fig. 4.7 LAN network

allows different networks to be hooked together effectively. A building, for example, might run a backbone upto several floors so that the networks on each floor can communicate with the rest of the networks in the building. On a university campus, a fiber backbone is often used to connect the entire campus together. Rarely will any one network have all of these features and devices. A network, however, will commonly have a number of them.

Hardwares in network: Let us see some of the hardwares used in the network. In the network diagram (Fig. 4.7) there are two junctions off the backbone. This is known as 'tapping' the backbone. One of the taps goes directly into a bridge to a ring topology network. The second tap interfaces with a router. The router distributes packets to the various network hooked to it. As the diagram shows, a router can interface with all of the various topologies.

The router interfaces with a star topology network. It also connects to a ring topology, an FDDI network, a chain or wave network, and a gateway to another star network. A server may also be directly connected to the router with a high speed line for quick access.

The bridge enables connection of networks that have the same network operating system, but different hardware. Gateways are similar to bridges except they can also do protocol translation. Gateways can connect networks that use different NOS (network operating software) and that may not have similar hardware. A gateway may be a special purpose computer, a work station with associated software (e.g., a PS2 with gateway software), or a software module which runs as a task in a mainframe. An internal bridge lives inside a file server as a second interface card. An external bridge could be a workstation. To make one, take an old PC (floopy-only will do), put 2 network interface cards in it and run bridging software, having connected each card in the PC to a LAN. When they connect two LANs directly, internal and external bridges are known as local. When something like a telephone or satellite link is used between them, they become remote. As with a bridge, the gateway needs at least two interface boards in a workstation; one for the LAN and another for the host systen connection, which may be through modem and telephone link (if remote) or a coaxial cable, if direct.

Routers are used to control and direct network traffic. The more computers there are on a network, the more network traffic there will be. This causes slower transfer and more data collisions. Using a router, the network may be broken into little networks called zones. If the router receives a packet that needs to proceed to another zone, the router sends it to the other zone. If the packet destination is in the same zone, the packet is never exposed to other zones. This limits the traffic across the entire network. Routers provide more functionality than bridges: routers are able to use the best available path between a source and destination and can readily switch paths if a better one becomes available. Bridges will discard packets that are too large for their attached networks as they are not able to prioritise traffic because they do not distinguish between packets. However bridges require little expertise to install; require little administrative overhead, and once installed generally function with minimum attention. Bridges are generally transparent to other hosts in the network while routers usually require others hosts to be configured with router addresses, or to run a passive routing protocol.

A hub is a central place to connect computers to a network, traditionally for a star topology. There are two types of hubs: passive and smart. A passive hub merely sits at the center of the star, connecting the various nodes. A smart hub, however, can be remotely controlled by the system operator and can send back packet traffic information for analysis. A server can be a simple file server, which provides a central location for all of the nodes to store files, mail, data and so on. A server may also monitor and govern network traffic, exchange data formats, print serving, do data queries, and so forth.

4.7 Communication in Network

For two computer systems to communicate, they must share a common set of rules for generating and interpreting the messages they send or receive. This can be done by establishing modularized set of system-interconnection rules that define a hierarchical series of layers of functions. The same set of layered functions must exist on each system in order for the two systems to be able to communicate. The communications between peer layers on two systems is achieved via protocols. A protocol is a set of rules or conventions that govern the format and control of information that is transmitted through a network or that is stored in a database.

Let us discuss an example of functioning layer of OSI (Open System Interconnections). It is a seven layer architecture (Fig. 4.8). The bottom three layers of this model (the physical, data-link, and network layers) apply specially to the architecture of the individual local networks. Their concern is the transmission, framing, and routing of messages between adjacent machines. Protocols at this level consists of different transmission technologies (such as coaxial cables and optical fibers), a variety of network topologies (such as star, ring, and bus configurations) and different accessing methods (both deterministic and random access schemes). Layer 4, the transport layer, serves as the boundary of the data communication functions of the upper three layers. Its functions include error detection and recovery, message sequencing, end-to-end addressing, and multiplexing. The lower three levels are technology-independent. The session layer (No. 5) initiates, ends, synchronizes, and

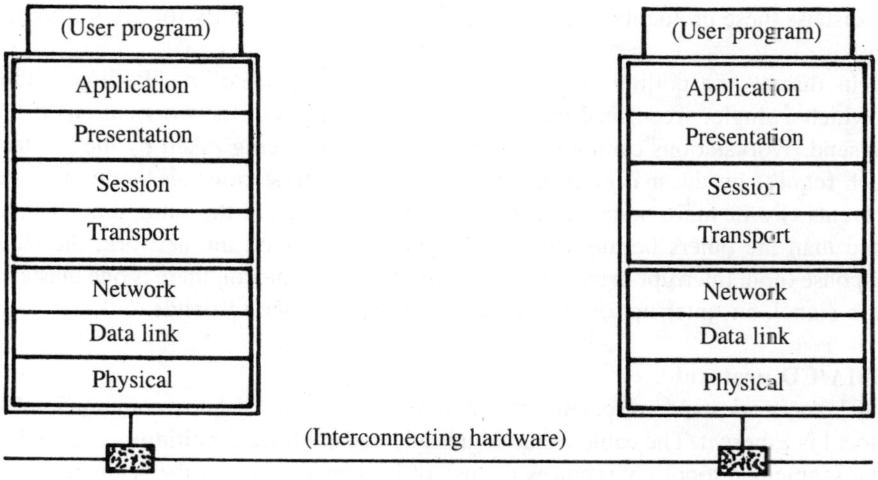

Fig. 4.8 OSI seven layer model

coordinates communications and end-user tasks. At the presentation layer (No. 6) the syntaxes of the different machines on a heterogeneous system are converted into a common syntax for information being exchanged among communication machines. This eliminates the need for each machine to know the formats of all others. The top layer (No. 7) is the application layer which interprets the meaning of information transferred between application processes.

A *repeater* operates at the physical link layer
A *bridge* interconnects *LAN*s at the data link layer
A *router* functions at the network layer
Gateways handle higher-level internet protocols.

To transmit a message over a network, the message is represented as a string of binary symbols (0s and 1s) known as bits. When transmitting these messages bits across a network additional bits must be added to it to ensure reliable communication, correct routing of the message, and prevention of network congestion. Long messages are usually broken down into shorter bit strings called packets. Packets, rather than one complete unit are transmitted to reduce message delay, better buffer management and to reduce congestion control. Packets are transmitted through the network as individual units and are reassembled into complete messages at the destination station.

4.8 Communications in Distributed Control Systems

As stated earlier a computer network is a system of computers–also called network nodes–interconnected by data transmission links, which allow for information exchange between nodes. To provide error-free and maximal convenient information transfers, the network operation is regulated by a set of rules and conventions called network protocol. The protocol defines connectors, cables, signals, data formats, and error-checking techniques as well as algorithms for network interfaces and nodes. The requirements for communications in distributed control systems include reliable transmission of data, deterministic delivery of messages under all communications media load conditions, end to end message confirmation, and message and delivery integrity. Apart from the two simple protocols like polling and interrupt driven there are three important communications network protocol available: (i) carrier sense multiple access with collision detection (CSMA/CD) (ii) token ring and (iii) token bus. Let us discuss these protocols with special reference to its suitability for use for process control applications.

First let us discuss the polling and interrupt driven protocols. Polling: Polling is the protocol method in which a single, predefined master computer asks each of the workstations if they have a message to send. Workstations cannot send a message without being asked by the master to do so. Usually each request by the master comes back from the workstations as 'nothing to send'. This method prevents two or more nodes from trying to communicate at the same time. This method is more limited than the others because the more there are nodes on the network, the slower is the network response time. Interrupt-driven: In the interrupt driven protocol, there is one master computer. It waits for a signal, an 'interruption' from a workstation and then responds to it.

4.8.1 CSMA/CD protocol

The IEEE 802.3 network standard specifies the CSMA/CD protocol. The most common implementation of this protocol is Ethernet. The cable technologies allowed are linear, multidrop, tree and segmented layouts. The segmented topology requires the use of bridges to connect the segments.

The CSMA/CD technique is a contentious scheme, in which network nodes compete for the right to use the medium. A winner can transmit one packet and then has to release the medium for other nodes. If a node is already using the medium, all other nodes must defer their transmission until the medium becomes idle. The activity on the network is detected (carrier sense), and a lack of activity means that the medium is idle and available. Several nodes may then start transmitting to the medium. If only one node starts transmitting, then the medium becomes busy, and all other nodes that were to start will have to wait until the medium becomes idle again. If several nodes start transmission more or less simultaneously, a collision occurs. While each station is transmitting it must also listen to detect a collision of messages. On detecting a collision, all transmission have to abort their transmissions and back off for some time before trying again. The back off time is random to avoid repetitive collisions. The 'carrier sense' means that every node must listen to see that the cable line is quiet before sending message. 'Multiple access' means that each user has a random chance of using the channel at any given time. 'Collision detection' means that any user immediately stops if another signal is detected—there is no sense in continuing because the data will be already be garbled.

Advantages: The advantages of the CSMA/CD protocol are: a simple algorithm for operation of the network and almost no delays in stations access time to network load. This is due to low media access overload. No communication bandwidth is used to gain access to the network; if a station is ready to transmit and the network is silent, the station can send its message.

Disadvantages: The protocol is limited to a cable length (2.5 km with repeaters). Priority levels of messages are not supported. The media access time is non deterministic; this can cause a station that needs to transmit to wait an undetermined amount of time before it is able to send a message. Since data for alarms, interlocks, and control valves can not be guaranteed to arrive at the proper destination in a timely fashion, the network is inappropriate for real time work. The CSMA/CD protocol does not offer a low-level acknowledgement of transmitted messages. Each station must be able to detect a signal from the weakest station, even while transmitting a message. This forces the collision detection circuitry to be analog technology, which is more complex than all digital implementation. Network efficiency drops as the network speed increases, and the network is not well suited to fiber optics due to difficulty of installing taps on the network cable. At high network load, message collisions are a major problem because they affect data throughput greatly. Even under short term burst conditions, a station's message can run into a series of collisions, and the station can discard the message after a number of attempts. That is, the network can loose a message without ever transmitting it.

4.8.2 Token Ring

The token ring is built on the physical ring topology (Figs. 4.9 and 4.10). Signals transmitted by a node of a network based on physical bus topology propagates across entire medium (broadcast). In the physical ring topology, signals propagate over unidirectional point-to-point paths between network nodes. The nodes and unidirectional links are connected in series to form a physical ring. The nodes play an active role in signal propagation. In order to reach a destination node, signals originating at the transmitter port of a source node have to be re-transmitted by all nodes located along the ring between the source and the destination nodes. Each node transfer network signals internally from the receiver to the transmitter port. Hence, the signals are reinforced during re-transmission and therefore, the total length of the physical ring is not limited by signal attenuation in the medium.

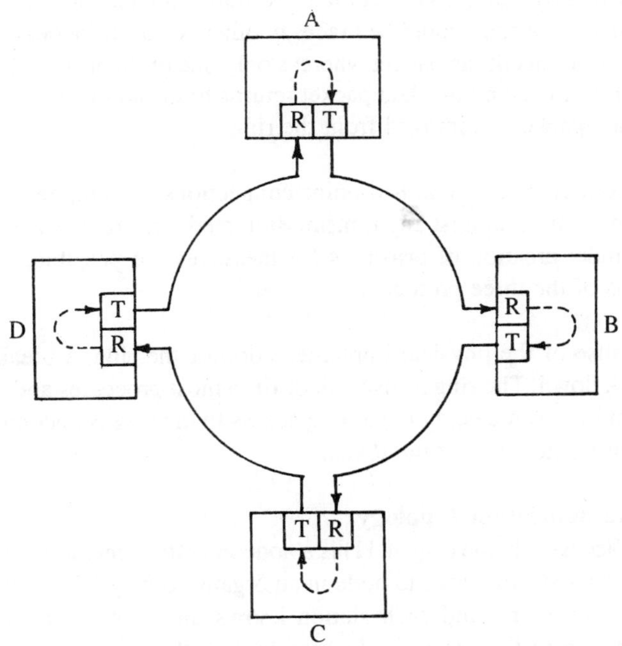

Fig. 4.9 Physical ring topology

Special, short packets called "tokens" are constantly passed around a logical ring of work stations to invite each to transmit in turn. The tokens in the token ring scheme are not addressed. Instead a token can have a free or busy forms. As long as none of the nodes on the ring has a data packet to send, a free token message circulates along the ring. Note that at a given instant only one token is present in the entire ring. A node that has a data packet to send has to wait until it receives a free token. At this moment, the node changes the token to busy, transmits it further along the ring, appends the data packet to the busy token. The busy token, followed by the data packet, is then re-transmitted around the whole ring. Only the node that changed the token to busy can change it back to free. The data packet contains a destination address field in its header. The destination copies the packet as it passes through the node. For example, if the data packet originated at node A and was sent to node

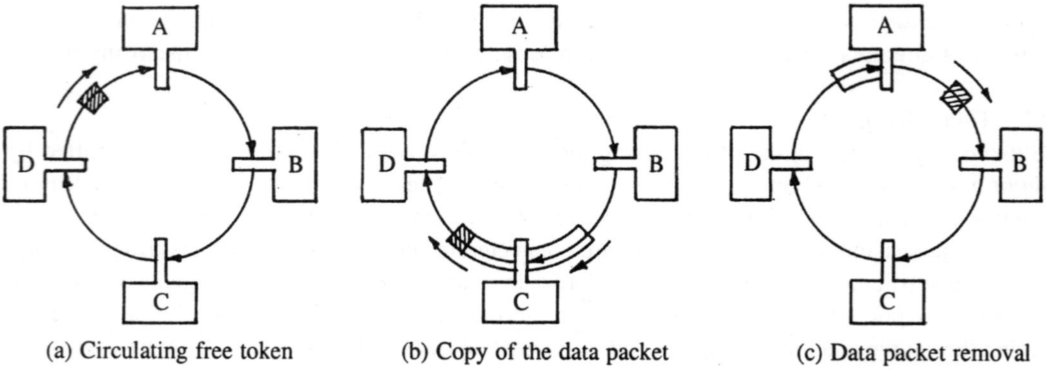

Fig. 4.10 Token and packet transfer in a token-ring network (Reiss, 1987)

C, the busy token (followed by data packet) would be re-transmitted by nodes B, C, and D. However, at node C, a copy of the data packet would be made. In other words all nodes on the ring, except for the packet source node, re-transmit the packet whereas only one of them-the destination node accepts it. When the busy token followed by the data packet returns to the source node, the token is changed back to free and the data packet is removed from the ring.

Advantages: The token ring uses point-to-point connections, so engineering is easy and fully digital. A ring can be made from almost any transmission medium, from twisted pair to fiber optics. Other advantages include: creation of priorities for messages are possible, token ring allows the longest network lengths of the three protocols.

Disadvantages: Because of the physical implementation of the ring, a break in the cable would bring the whole network down. The ring is also a poor fit to most processes and assembly lines found in manufacturing operations. Some delay in gaining access to the ring is encountered at low network load because the station has to wait for the token.

4.8.3 Token Bus Communication Topology

The token bus medium access scheme (Fig. 4.11) is a noncontentious method in which the right to use the network medium is passed from node to node in an organized way. The stations on the token bus ,are arranged logically into a ring and each station knows the address of its predecessor and its successor. During operation of the network, the station with the token transmits data frames until either it runs out of data frames to transmit, or the time it has held the token exceeds the token holding

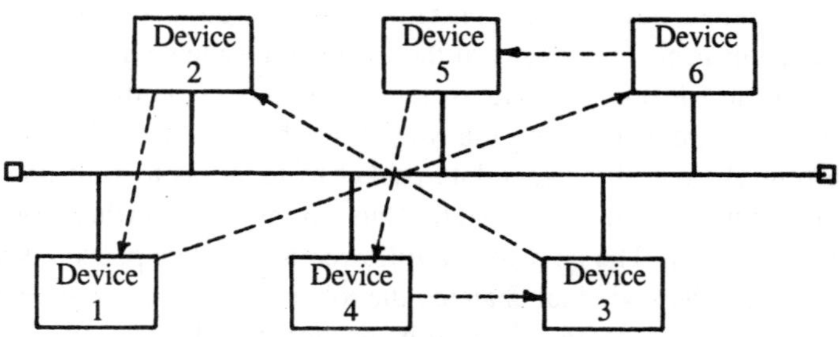

Fig. 4.11 Token bus (Wheelis, 1993)

time. Then the station regenerates the three-byte token and transmits it to its logical successor on the network. The physical location of the successor is not important; the token is sent to the logical neighbour. In this manner the token propagates around the logical ring. Collisions of data frames do not occur, as only one station at a time can transmit. Priority levels for messages are also allowed. The protocol defines four priority classes (0, 2, 4, and 6) for the messages, with a 0 denoting the lowest priority and 6 denoting the highest priority.

Advantages: The token bus protocol can support message priorities, which ensures that a station's high-priority messages, such as alarms, interlock data, and critical process variables, can be given access to the network first. The protocol has excellent throughput and efficiency at high loads, can handle short data frames, and can dynamically add devices to the network. The protocol allows the receiving station to immediately acknowledge receipt of a Send Data with acknowledgement message, providing true end-to-end confirmation of the delivery of a data frame.

Disadvantages: The token bus is a complex protocol, which can raise the cost of the communications equipment. Similar to CSMA/CD protocol, the token bus uses a substantial amount of analog circuitry. It also is not as well suited to use fiber optics as the token ring.

Comparison for process control application under low to medium network loading, CSMA/CD has a distinct advantage, because it uses no media overhead to determine which device has access to the network. DCS vendors cannot guarantee that the system communications loading will remain under 40% under plant upset conditions, and since an overloaded CSMA/CD network can collapse totally, it is not appropriate for process control applications.

In comparing the two token passing schemes, the token ring wins marks for its capability to more easily use fiber optics to extend network distances and its ability to use many network media in its implementations. The token bus, on the other hand, gains ground because its message priority scheme is more fair than the token ring's scheme, devices can be dynamically added to the bus and end-to end confirmation of messages is available.

4.9 Logical Topology

What the network software thinks is happening can be quite different from what the physical layout of the wires might suggest. Logical topology does not always match the physical toplogy. For example the more common form of ARCnet is a physical star. However, when the 'token' is passed, it follows a very definite order from one machine to the next. All stations actually receive the electrical signal, but only the one with the matching address reacts to it. It is, therefore acting as a logical ring. The order of the token passing is determined by the network addressed and not by physical proximity.

Token ring could look, at first glance, like a physical star. However, when the individual wires are examined, a physical ring will be seen to exist. When tokens and messages are passed, they follow this ring, and so the network is also a logical ring. The passing of the token is controlled by physical order.

4.10 Ethernet Card

Let us discuss one of the network in detail. Ethernet is a CSMA/CD network with 10 Megabits/sec transmission rate. Each station 'contends' for the use of the ETHER (the connecting cable) using CSMA/CD. The original, and still common cable system is a linear bus that uses a fairly large diameter coaxial cable. This is often referred to as Thick Cable Ethernet. Although offering a very high performance, the original design with external transceivers is relatively expensive for personal computer use. Cabling modifications (Thin Cable Ethernet, twisted pair Star Lan and twisted pair Ethernet) have been created to reduce the cost. Except UTP Ethernet, all other versions of Ethernet use a common, shielded cable to both transmit and receive. The UTP Ethernet in contrast uses only a pair of wires to transmit and a separate pair to receive. The following codes are used for the Ethernet:

10 base 5	Thick Cable Ethernet	500 meter bus
10 base 2	Thin Cable Ethernet	185 meter bus
1 base 5	Twisted pair Star LAN	500 meter diam. Star
10 base T	Twisted pair Ethernet	100 meter diam. Star

Twisted pair Ethernet uses an unshielded twisted pair (UTP) in a star network. The length of any one arm of the star is 10 meters. The ethernet adds a central concentrator with a dedicated cable running to each work station. The dedicated station cable contains two pairs of unshielded, twisted wires, a transmitter pair and a receiver pair. The two wires of each pair are twisted independent of each other to reduce cross talk and maintain a balanced state. Should a connector fail it can be disconnected via software without disturbing other users. The adapter cards usually have four visible LEDs on the mounting bracket to assist with trouble shooting. The polarity and link integrity lights should be on constantly. The other two will flash as data is transmitted or received.

Each Ethernet reciver includes a "squelch" circuit that keeps small amplitude noise out. This circuit monitors the average dc voltage on the cable to see when a station is transmitting. As indicated, a cable voltage of about –1.05 indicated that a single transmission is in progress. As the average voltage moves down (negative voltages) towards this value, the receiver is turned on. All stations monitor all packets at least to the end of the destination address. At that point, a station will have determined if the packet is of interest or not. If the packet is to be received, it is moved into a RAM buffer on the circuit board. Then, if CRC (Cyclic Redundancy Check) and few other checks are valid, the host computer is notified that a packet has been received, usually via an interrupt.

Many Ethernet card manufacturers use a set of three integrated circuits produced by National Semi-conductor (Fig. 4.12). These chips look after the entire protocol implementation, encoding and

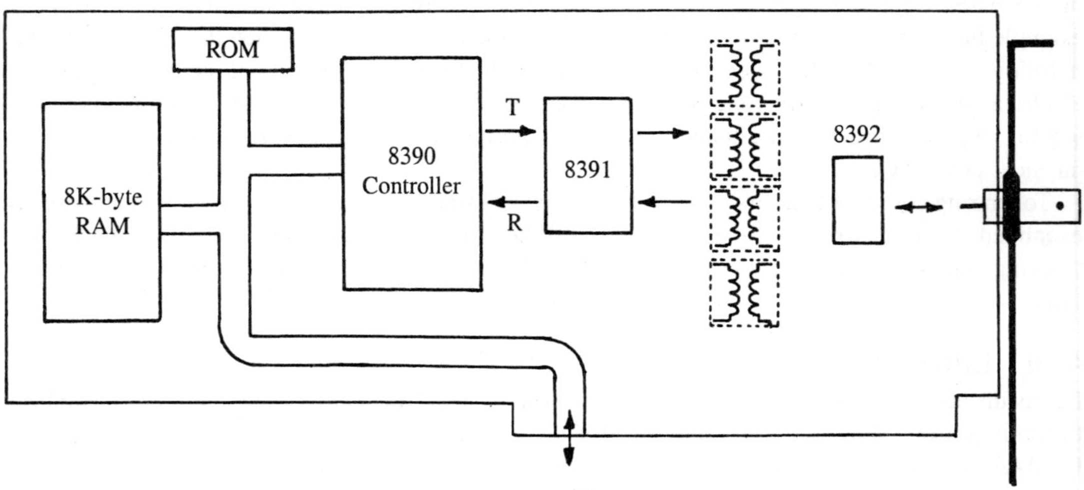

Fig. 4.12 Ethernet add-on card (Hardy, 1999)

decoding, and the actual interface to the cable. The only other major chip needed is an 8K, or greater, RAM chip used as a packet buffer. A small ROM is also used to hold the card's network address. The 8390 Network Interface Controller (NIC) is the heart of this chip's features. Let us briefly discuss each of these chips.

8390 Network Interface Controller (NIC): This 48-pin chip is responsible for the Ethernet protocol and for movement of information to and fro from the buffer RAM. Its operations include:

- moving data between NIC and the adapter's RAM buffer
- moving data between the host and the adapter buffer
- re-transmitting packets if collisions occur
- interpreting commands from the host
- notifying the work station of incoming packets or error
- serializing and deserializing (transmit and receive)
- generating and checking CRCs (cycling redundancy checks)
- recognizing station, multicast, and broadcast addresses
- recognizing and tallying (counting) various errors
- performing diagnostic tests including loop back.

8391 Serial Network Interface (SNI): The SNI chip is smaller with only 24 pins. Its functions are:

- adding the 10-MHz clock to the transmitted data using Manchester encoding
- extracting a 10-MHz receiver clock from the incoming waveform using a phase locked loop
- turning the 10-MHz "collision signal" from the cable interface into a High/Low logic signal for the NIC.

8392 Coaxial transceiver Interface (CTI): The CTI chip is used only in coaxial versions of Ethernet. It is the smallest of the three chips with only 16 pins. It must be isolated from the rest of the card with small transformers and must receive its power through an isolated supply. The CTI requires supply voltages of 0 and –9 volts. Its functions are:

- converting transmit voltages from the SNI chip into current levels on the coaxial cable
- squelching and receiving incoming transmission from the cable.
- detecting collisions on the cable and signaling the network controller; also periodically testing with a direct 'heart beat' circuit
- monitoring product lengths and turning local transmitter off if lengths become excessive.

Packet transmit: A packet to be transmitted is assembled in some place in the adapter buffer area. The total number of bytes including addresses, data, and any necessary padding is loaded into the 16-bit Transmit Byte Count register, and the transmit command is given with the destination address (6 bytes); Source address (6 bytes); Length (2 bytes); Data (46–1500 bytes) and padding if needed.

The NIC starts transmitting the 64-bit synchronizing preamble while it moves the data, using its local DMA channel, in bursts of several bytes at a time, into its internal 16-byte FIFO (First In First Out) memory. When the preamble is completed, the addresses and data provided are transmitted in turn. The bits within each byte are sent least significant first. When the FIFO needs more bytes, it will automaticaly get them from the buffer using NIC's local DMA. As these bytes are being sent, the NIC continuously calculates the 32-bit CRC. When the last bit of data is sent, the CRC "remainder" is added to the end of the data stream, and the transmission ends.

4.11 Example of a Distributed Controller

Let us discuss in detail a typical industrial Distributed computer control system SCAP EUROPA, S.A (Fig. 4.13). The distributed structure is composed of a global optimization scheme (GOS). The GOS is composed of a high-speed local area network of computers, (e.g., ARCNET), which meets the international specifications of MAP. Each of these computers are associated to a global operation centre (GOC). From each GOC the user may monitor, using the system for the acquisition and graphic analysis of data, the computer process and decide the corresponding actions to take. Likewise the global optimization system (GOS) has available the master system, which actuates on a global level, but does not execute any specific operation of logic nor any control algorithm. These capabilities of the systems are performed in the local optimization system (LOS).

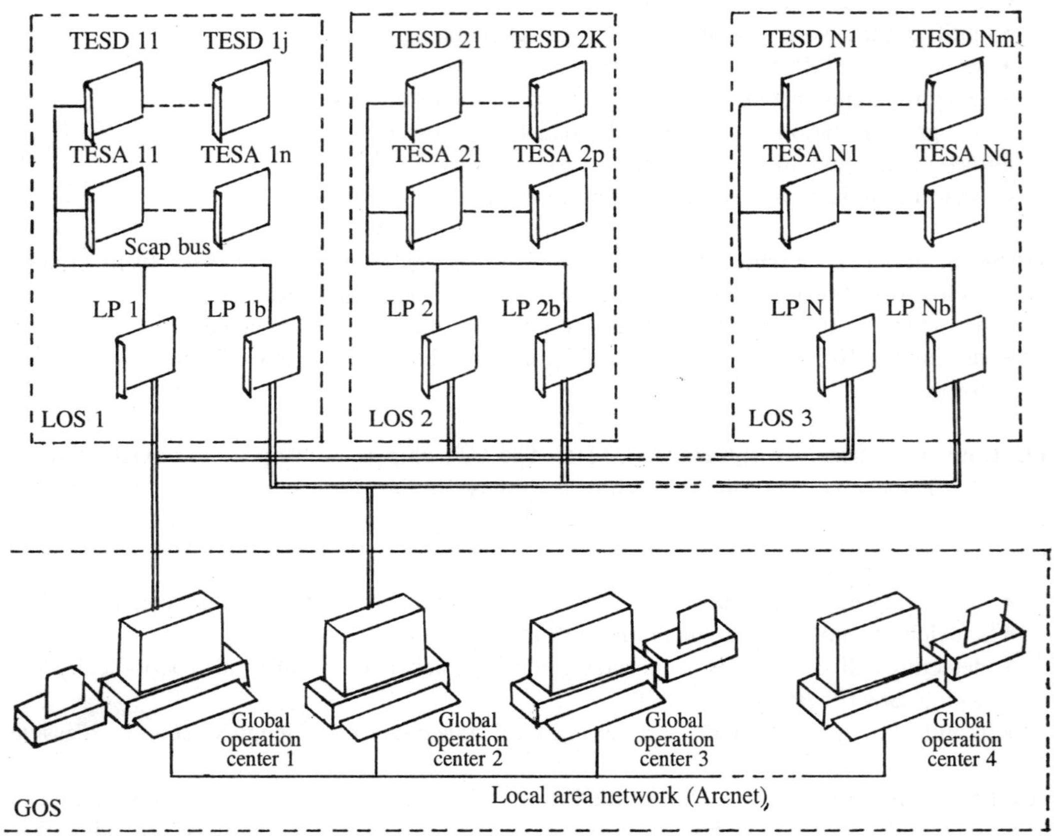

Fig. 4.13 Typical block diagram of a distributed computer control (SCAP EUROPA, S.A)

Local optimization system (LOS): Each global operation centre (GOC) maybe connected through a serial communication with a LOS network. Each LOS includes a Local Processor (LP), wherein its intelligence resides, and a set of analog and digital I/O boards. The LPs have available several control algorithms like adaptive control, predictive control, *fuzzy* control etc.

The LOS's group together given data sets of process variables, which are functionally interconnected, and allow optimization of local operations. These local optimizations and the global one are, respectively, coordinated and carried out from the GOS. Any variation in the LOS's is communicated to the GOS, where the common database resides. The LOS, might be conveniently distributed throughout the plant in order to reduce cabling cost and increase service security. The structure allows redundancy to the levels of GOC and local processor, which might be extended to the I/O signal level, although in the figure this is not represented.

Networks: The features of different networks and buses used in previous structures are:

(i) **ARCNET:** It intinterconnects the different computers using coaxial cable connection. Speed of transmission is 2.5 MBauds with maximum cable length of 600 m.
(ii) **RS422/RS485:** It connects the computers with the I/O stations, PLCs, or in the case of distributed system it connects the global optimization system (GOS) with the local optimization systems (LOS). The physical support is 2 shielded twisted pair; speed of transmission is 38400 Bauds and the maximum cable length is 1500 m.
(iii) **System bus:** In the distributed structure it connects 10 twisted pairs; has speed of transmission of 4 MBauds and the maximum cable length is 30 m.

The three buses in the architecture of the system allow the incorporation or removal of elements of the network without stopping its activity, that is to say, this system carries out dynamically its hardware/software reconfiguration in such a way that the element incorporated/removed is accounted for in a manner transparent to the user. For example, if it is required to extend the capabilities of the system, adding a GOC, a LOS or an I/O block, it may be done without interrupting its operation.

Components: The GOS's are connected to the local optimization system (LOS) by the intelligent serial communication board (RS-422 network as shown in Fig. 4.13). The communication board is based on a 16 bit microprocessor with its own data and program memory, and implements the lower level of communication protocol without disturbing the GOS main CPU.

LOS hardware: The LOS hardware, double Euroboard size printed circuit boards, have been implemented on 19″ modular architecture racks. One elementary LOS is composed of one main unit, a 19″ rack with its back-plane for the system bus; one plug-in power supply, one or two local processors and upto eight I/O boards. It is possible to add more I/O boards with expansion units, similar to main unit but without local processor. All the units of the same LOS must be tied together with a 20 wire flat cable.

I/O boards: There are one or more I/O boards with 16 digital inputs for DC or AC voltage. There are one or more I/O boards with 16 digital output for DC voltage. There are I/O boards with 8 analog inputs for current, voltage, or resistance and similarly I/O analog output boards with 8 analog outputs for current or voltage. Let us briefly describe each unit.

The local processing units have 19″ 6U rack with a back-panel for the system bus and 5 VDC distribution, one plug-in power supply and the connectors for the I/O cables. Main unit has a capacity for two LOCAL PROCESSOR boards and upto 8 I/O boards. The two local processors work in a backup configuration. There are expansion units which are similar to the main unit without local processors mounted. All the expansions unit must be tied together to the main unit with a 10 twisted pairs flat cable which interconnects the system bus. The maximum number of expansion units is 4 per main unit, and the length of interconnection flat cable must be less then 30 m. Maximum I/O configurations have 512 digital and 64 analog signals or 640 digital signals.

The power supply board: The different units in the main- and expansion-units require only 5 VDC, which are supplied with a high reliable plug-in power supply. Each system has one on-board DC-DC converter which provides the different internal voltage supplies. This distributed power supply technique features a high reliable system. There are two models of plug-in power supply: one for 220 VAC power distribution and one for uninterrupted power supply systems with 48 V batteries.

Local processor: The local processor is based on a 16 bit microprocessor with its own data and program memory, that works as a control unit for the system bus, executes commands received from the RS-422 communication line. Its architecture allows it to work in parallel with other local processor in such a way that in case of failure of one, the other assumes automatically the total job load. It is possible by jumper configuration in the main unit, to install the backup processor in a second RS-422 network to achieve communication redundancy. There may be upto 50 LPs installed in one RS-422 network in multidrop configuration with repeaters. The application program is loaded via the RS-422 serial channel from the GOS, for maximum flexibility and versatility.

Analog input board: The I/O boards have 8 analog inputs for current, voltage and resistance operation. Each signal has an isolation of 250 VDC from any other signal and all of them are opto-isolated from the system bus and individually protected against over voltage by a built-in mini fuse. There is an individual low-pass filter for each input to avoid high frequency noise in the measured signal.

Analog output board: This I/O board has analog outputs for current and voltage operation. Similar to the analog input boards, the analog output boards also have isolation and overload protection facility. When there are communication failures or the local processor is disconnected, the output boards are capable of maintaining the output signals for approximately 10 secs or permanently, depending on the board jumper setting.

Digital input board: The I/O board has 18 digital inputs for DC/AC voltage operation. Each signal has isolation of 250 VDC from any other signal and all of them opto-isolated from the system bus and individually protected against overvoltage by a built-in mini fuse.

Digital output boards: This I/O board has 16 digital outputs for DC voltage operation. The output driver is an open collector NPN transistor. The logical '1' corresponds to the ON state of the transistor (driving current) and the logical '0' to the OFF state. Similar to the digital input boards, the output boards also have isolation and overload protection facility. There are sets of active terminals, for DIN rail mounting, based on solid state relays to drive a wide range of different power signals. When there are communication failures or the local processor is disconnected, the output boards are capable of maintaining the output signals for approximately 10 secs or permanently, depending on the board jumper settings.

Intelligent communication board: This board is built around a microprocessor with one high speed serial communication channel. It is especially suited to carry lower level of the communication protocol without disturbing the main CPU. The board is based on INTEL's 8096 16 bit microprocessor running at 12 MHz with zero wait states. It has 8K EPROM, with the bootstrap program and some basic subroutines, and 40K of RAM, for protocol dependent program and data. The serial interface meets the standard RS-422/485, with all the signals opto-isolated and maximum baud rate of 250 KBaud. The interface with the main CPU is a dual ported RAM, allowing a peak data throughput of 16 Mbit/sec. This board uses a software selectable interrupt to indicate to the main CPU that the command is completed and the data is in the dual ported RAM. Upto 6 such boards can be mounted in one GOC.

4.12 Selection of Operator Interface

Input devices: Keyboards and mice do not hold up well in a dirty environment, and many factory-floor operator find them hard to use. Four industrial alternatives that need consideration are:

(i) Touch screens can be very intuitive for operators, but they are susceptible to abuse and they must also be kept clean and free from abrasives. Adhesive packed maylar overlays are readily available to help this. (ii) Function keys are the most rugged input devices, but care must be given to designing displays so the operator can readily associate the keys and their corresponding actions. Some units also offer renderable keys to facilitate this. (iii) Industrial scaled keyboards that are AT-compatible and designed for NEMA 4 are another choice to consider. (iv) Industrial mice specially designed for the factory floor may be an option as they are sealed and designed for NEMA 4 environments.

4.13 ERP and Process Control

ERP (Enterprise Resource Planning) is an industry term for the broad set of activities supported by multi-module application software that help a manufacturer manage the important parts of its business, including product planning, parts purchasing, maintaining inventories, interacting with suppliers, providing customer service, and tracking orders. Typically, an ERP system uses or it is integrated with a relational database system. The deployment of an ERP system can involve considerable business process analysis, employee retaining, and new work procedures. As of now, at least a dozen vendors including SAP, Oracle, QAD, Baan, Ramco, J.D. Edwards, People Soft, SSA Binary Semantics, Avalon etc., are offering ERP solutions.

Enterprise Resource Planning (ERP) helps us to automate each of the functional areas such as production, planning, control, inventory management, procurement, customer order management finance etc. Theses areas are integrated in an ERP system. Process control option is available with ERP systems that have been designed specially for process industry. One can connect field instruments-flow meters, pressure transducer, load cell etc., to an ERP system, through more sophisticated process control systems i.e., PLCs (programmable logic controllers) and/or DCS (distributed control system). These instruments collect detailed field level information at frequent intervals. Accurate resource consumption and production information is thus received by the ERP system through an interface. This information can be processed and used for addressing not only the production related issues but also certain business related issues.

Production planning and control (PP and C) module of an ERP system is connected to the DCS though an interface. The orders that are received from customers are entered into the COM (customer order management) module of an ERP. PP and C looks for pending orders for manufacturing and accordingly selects the orders that can be accommodated (based on certain parameters, viz., priority, delivery schedule, etc.) in a single production run. It then prepares the 'recipe' for production that includes the list of raw materials and other production resources alongwith the estimated quantity of each one of them required for producing the desired order quantity. PP and C also receives information regarding availability of these materials in stock from the inventory management module. (If the materials are not available in stock, they need to be procured and thus, the appropriate information is forwarded to the procurement module that initiates the purchase routine.)

The PP and C module receives inputs from the DCS regarding availability of production facility at a given point of time (e.g., say out of five production lines, one is under maintenance, one is processing batch, one has been allotted for the next batch, and two are available). The module accordingly selects the production line combination to schedule the batch and monitor the production run. Since dynamic link is established between the PP and C module and DCS, on-line updating of information takes place as per the inputs received through the DCS system and this information is categorized and shared by other modules of the ERP system.

Summary

Initially, control was performed local to the equipment control. The advantage was low wiring costs; disadvantages were not much control, monitoring, alarming, and history. With the advent of the microcomputer, sensors and actuators were wired into a central computer. This concept was called 'centralised' control. The advantages were centralized viewing and control of the process, plus much more flexible control, monitoring, alarming, and historic data capabilities. Disadvantages were the cost of massive wiring, the new risk of losing all control if the computer failed. With the advent of the microcomputer, distributed control units were located around the plant, closer to the process. They were connected to the control room via proprietary digital communications lines referred to as data highways. This concept was called 'distributed' control. The advantages are greatly reduced wiring costs, much more limited failure consequence, and less cost to add more points. Disadvantages are lack of interoperability among controllers of various manufacturers due to proprietary protocols. Hence, the user was locked into a single vendor. In the next chapter we will look at Fieldbus control. Here the digital communications link can interconnect sensors, actuators, and analyzers to the control room. The control can again be located to the devices and intelligent devices can communicate directly with each other and fully exploit each other capabilities. Its advantages include even lower wiring costs and failure consequences, smaller expansion costs (scalability). DCSs and PLCs can be more closely and more efficiently connected.

References

Ahson, A.I., Microprocessors with applications in process control, Tata McGraw Hill, 1986.

Boland, M., Boost plant performance through upgraded instrumentation, Chem. Engg. Sept., 1995, 80–87.

Doyle, L.F., Computer peripherals, Prentice-Hall Pvt. Ltd., Engelwood Cliffs, New Jersey, 1996.

Hardy, J.K., Inside Network, Prentice-Hall India, New Delhi, 1999.

Jhonson, C.D., Process control instrumentation technology, IInd Ed., Prentice-Hall of India Pvt. Ltd., New Delhi, 1998.

Mhashelkar, S., Effective means of controlling your process by using ERP, Chemical Industry Digest, Dec., 1998, pp 71–73.

Norton, P., Introduction to computers, Tata McGraw Hill, New Delhi, 1998.

Olsson G. and G. Piani, Computer system for automation and control, Prentice-Hall international, U.K., 1992.

Polke, M., (ed), Process control engineering, VCH, New York, 263–281, 1994.

Reiss, Introduction to local area networks with microcomputer experiments, Prentice-Hall, 1987.

Seal, A.M., Practical process control, Arnold Press, U.K., 1998.

Technical literature supplied by SCAP EUROPA, S.A, Espanna.

Weitzman, C., Distributed micro/mini computer systems, architecture, implementation, and application, Prentice-Hall, 1985.

Wheelis, J.D., Process control communications: token bus, CSMA/CD. token ring?, ISA Transactions, 32 (1993) 192–198.

Review Questions

1. Explain with block diagrams (i) supervisory computer control (ii) Direct digital control.
2. Explain with a block diagram the distributed computer control.
3. What are the advantages of distributed computer control over centralized computer control of processes?
4. What are the process control requirements of computers?
5. What are the basic multi minicomputer architectures?
6. What is meant by peer-to-peer communication and server-based communication? Compare these two methods.
7. What are the components in a network adapter card?
8. With a block diagram explain the components used in networking different topologies in a LAN.
9. Discuss briefly the different layers of OSI?
10. What is a CSMA/CD protocol?
11. Explain a token ring protocol?
12. Discuss the token bus protocol.
13. What are the advantages and limitations of different protocols?
14. With a block diagram explain an Ethernet card.
15. Explain the basic components of any commercial DCS.
16. What are the common input/output devices used for interacting with computers?

5

Examples of Experimental Computer Control of Processes

In this chapter, some examples of reported computer control of processes are reviewed. These includes computer control of liquid level, heat exchanger, trajectory tracking of an optimization path in a batch reactor, control of fedbatch fermenter, control of an injection molding process, and a distillation column. Details of the flow diagram and the interfacing circuits used are discussed.

5.1 Computer Control of Liquid Level System

Srinivas and Chidambaram (1996) have reported an experimental study on computer control of liquid level system. The schematic diagram of the liquid level control system is shown in Fig. 5.1. The tank can hold water upto a height of 100 cm. Water is pumped into the tank continuously by a small pump. The water height is measured by a differential pressure (DP) transducer attached to the bottom of the tank. The pneumatic signal from the DP transducer is converted into an electric current signal. This current signal, after appropriate signal conditioning, is converted to a digital signal by a 12-bit A/D converter. The computer used in the experiment is an IBM PC/XT having an Intel 80286 16-bit CPU with a 20 MHz clock speed. The real time control and interface routines are written using Turbo Pascal. Controller output is converted into an analog electrical signal by a 12-bit D/A converter. An electro-pneumatic converts the current signal into a pneumatic signal, which is sent to actuate a control value.

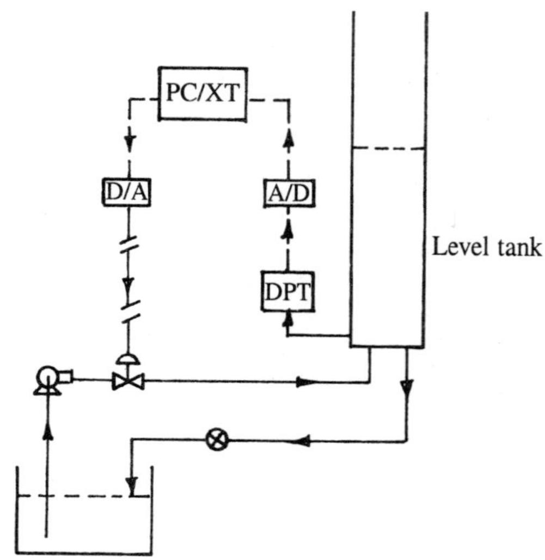

Fig. 5.1 Liquid level control system using a PC

5.2 Computer Control of a Heat Exchanger

Singh et al. (1996) have reported computer control of heat exchanger. Let us discuss briefly the reported setup. The process consists of 1-2 shell-and-heat exchanger (Bell and Gossett Co., USA). The carbon steel shell is 11.5 cm in diameter and 74 cm long, whereas the copper tubes are 1.27 cm in diameter and 122 cm long. The hot fluid, which is steam in the present case, passes through the shell side and cold process stream (water), passes through the tube side. The steam line consists of a steam regulator, a control valve, an absolute pressure transmitter, a pressure gauge, a temperature transmitter and a steam trap. The water line consists of a control valve, a rotameter, an orifice meter along with a differential pressure transmitter at the inlet of the heat exchanger. A schematic diagram of the process along with the control configuration is shown in Fig. 5.2. An I/O card (DT 2805, Data Translation, USA) is used for communication between the process (heat exchanger) and the PC/XT. It also provides a link between the computer and a microprocessor-based PID type hardware controller (SDC 350, Yamatake-Honeywell, Japan). The dynamic matrix control (DMC) and adaptive DMC computations are carried out in the PC/XT whereas the actual control is implemented through the hardware controller which is cascaded to minimize the fluctuations in the steam line pressure. While the DMC works with a sampling time of 10 s, the cascade control operates with a fast sampling period of 0.5 s.

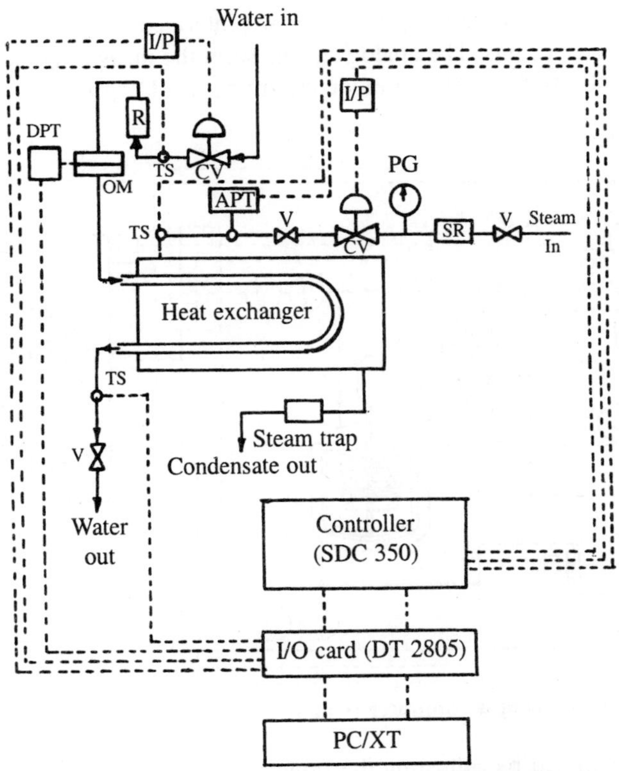

Abbreviations	
TS :	Temp. Sensor
R :	Rotameter
SR :	Steam Regulator
CV :	Control Valve
PG :	Pressure Gauge
I/P :	Current to Pressure Converter
OM :	Orifice Meter
APT :	Absolute Pressure Transmitter
DPT :	Differential Pressure Transmitter
V :	Valve

Fig. 5.2 Heat-exchanger control using a PC (Singh et al., 1996)

5.3 Trajectory Tracking of an Optimizing Path in a Batch Reactor

Chang et al. (1996) have reported an experimental study on an integrated method for optimization and control of a batch reactor. A schematic diagram of the experimental apparatus is provided in

Fig. 5.3. The reactor is a 10 litres stainless-steel cylindrical vessel with a 1 inch drain centered at the bottom. The vessel is composed of an annular jacket with baffling and internal helical coils. The reacting mixture is blended with an agitator (diameter = 0.22 m) at one end of which is a pitched flat-blade paddle and at the other end is a pitched anchor-type paddle. Adoption of an anchor-type paddle at the lower part of the agitator is to maintain perfect mixing. The rotational speeds of the agitator are 0-120 rpm.

The pyrex graham condenser with a drip tip is attached to a port of the reactor head. The condenser is used to condense any vapor escaping from the reactor. Tap water (roughly 25°C) is used as the coolant on the shell side of the condenser. A check valve is installed at the outlet of the condenser to prevent any back-flow of water from filling in the vapor-trap bottle. Two vessels, V1 and V3, are used to provide two temperature levels of the coolant. Another vessel, V2, is installed to serve as the heating medium.

The three vessels are about 0.08 m^3 (65 litres). An overflow pipe is inserted to maintain a constant head for each vessel. During the reaction, the temperature of water in the vessel V1 could be adjusted with an on/off solenoid valve. Meanwhile, an electric heater installed in the vessel was adjusted with a solid state relay (SSR) to maintain the desired temperature of the heating medium.

Computer control and the installed instrumentations for the experimental apparatus are shown in Fig. 5.3. The computer (IBM PC/AT type) is connected to a PCL-718 A/D interface card and a PCL-726 D/A interface card (both from Advanced Corp.). These two interface cards used 12-bit converter; therefore, the digital signal was 12 bits. The analog signals from the measuring elements are amplified and conditioned by AT740 (4-20 mA/0-5V) modules (from Acel Corp.).

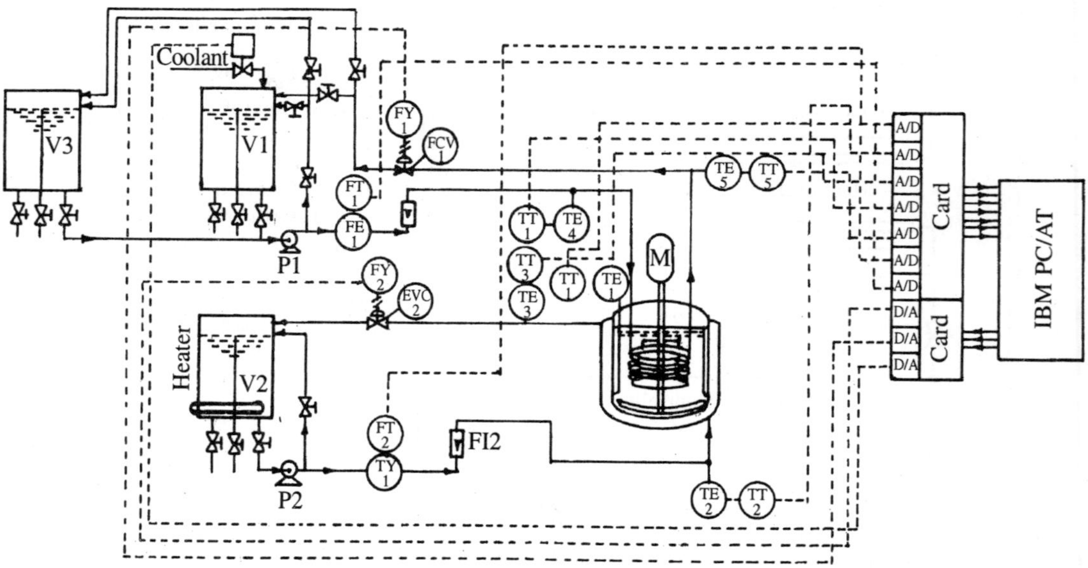

Fig. 5.3 Reactor control using a computer (Chang et al. 1996)

Heating of the reaction mixture is achieved by adjusting the flow rate of hot water F_{wj}, through the jacket side of the reactor with a 3.72 kW centrifugal pump. Heat removal is carried out by manipulating the flow rate of coolant, F_{wc}, through the internal helical coil with another 3.72-kW centrifugal pump. Two pneumatic control valves, two flow meters, and two rotameters are installed in the flow control loops. At the inlets and outlets of the jacket and the coil, four temperature sensors are located (0-120°C resistance temperature detector (RTD); accuracy ±0.1 to 0.3°C). The reactor temperature is measured by a RTD of the same type (accuracy ±0.1°C). The volumetric flow rates inside the jacket and the coil are measured by two flow transmitters (MK508 and MK2350 from Signet Industrial Inc.) which output a 4-20 mA signal proportional to the flow rate.

5.4 Computer Control of a Fedbatch Fermentor

Queinne and Dahhou (1994) have reported control of a fed batch fermenter. A schematic diagram of the experimental fed batch alcoholic fermentation plant is shown in Fig. 5.4. The growth medium, containing a carbon-rich substrate, flows into a bioreactor, inside which biological activity is sustained in order to accomplish the conversion of substrate to product alcohol. The environment variables, derived from standard sensors, correspond to the operating conditions. Stirrer speed, temperature and pH are maintained by a local analog controller. Typical operating variables and parameters of the experimental plant are: temperature 30°C, pH 3.8, stirrer speed 200 rpm, aeration 25 liter/h, active volume 6-16 liter, flow rate 0.0 to 2 liter/h.

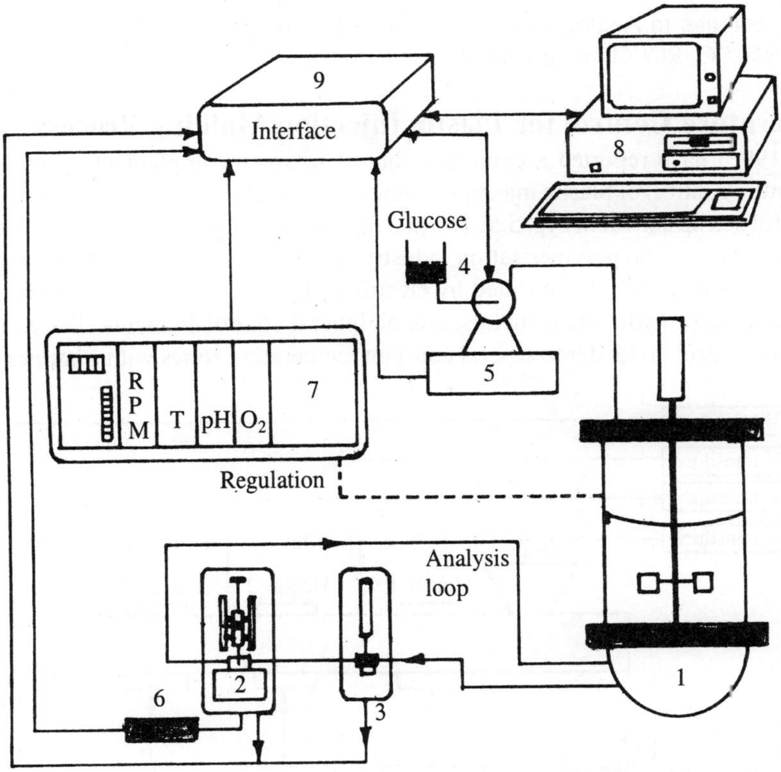

(1) reactor (2) glucose analysis (3) Do analysis (4) feeding (5) balance
(6) programmable logic controller (7) local analog controller (8) microcomputer

Fig. 5.4 Control system of a fed-batch process (Quennec and Dahhou, 1994)

The state variables indicate the state of the system at each sampling time. However, they are difficult to measure compared with the environmental variables because of the lack of specific sensors. To obtain these state variables measurements, system of measurement corresponding to the dimension of the reactor used were designed, especially to determine the substrate and biomass concentrations in the reactor.

To measure the glucose concentration, an automated injection system associated with laboratory equipment is used. The latter consists of a YSI 27 industrial analyzer which delivers an analog output voltage proportional to the glucose concentration. A set of pneumatic jacks around a calibrated syringe ensure sampling, injection into the glucose analyzer and rinsing of the syringe before the next measurement. By first selecting the syringe travel, it is possible to set the working scale for the substrate measurement. Substrate up to 200 g/litre can be measured with an accurancy of approximately 5%. The minimal sampling period is 2.5 min.

Biomass concentration is determined from the optical density of the culture medium. It is derived from the absorption of a laser beam in a photovoltaic cell with linear response. This equipment furnishes a measurement of the concentration of microorganisms up to 8 g/litre with good accuracy. The plant is linked to a PC/AT-compatible micro computer supporting a multitasking environment developed by Queinnec and Dahhou (1994) that allows real time identification, monitoring and control of the process. A RTI 815 board is used for communications between the micro computer and the process. It belongs to Analog Devices family of real-time interface input/output board and is designed for IBM PC, PC/XT or compatible.

5.5 Temperature Control for Plastic Injection Molding Process

Tsai and Lu (1998) have reported a computer control setup (for implementing multivariable self-tuning temperature control) of plastic injection molding process. Let us discuss briefly the experimental computer control set-up shown in Fig. 5.5. The plastic injection molding process is a well-established technique widely used in the polymerization industry. Such a process usually consists of an extruder, one (or two) powerful screw (s) controlled by electric or hydraulic actuators, a molding mechanism controlled by electric or hydraulic actuators, and peripheral control elements. The extruder typically consists of a large barrel divided into several constant temperature zones with a hopper at one end and

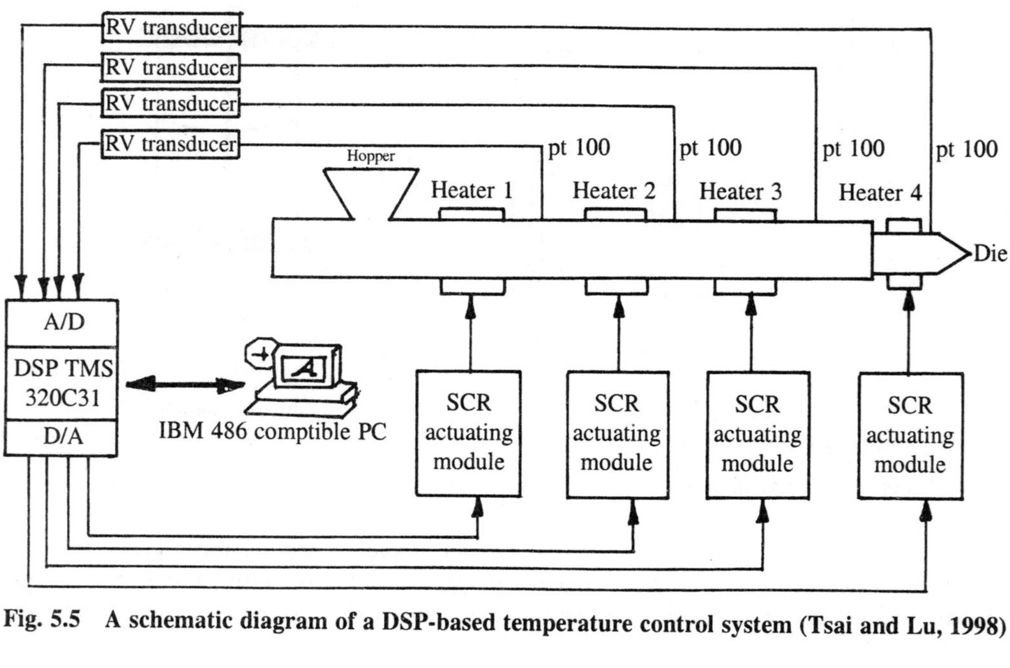

Fig. 5.5 A schematic diagram of a DSP-based temperature control system (Tsai and Lu, 1998)

a die at the other. Polymer is fed into the barrel in raw and solid particle from the hopper and pushed forward by a powerful screw. While passing through the temperature zone with gradually increasing temperature, this raw polymer is gradually heated. The heat produced by the heaters in the barrel, together with the heat released from friction between raw polymer and the surfaces of the barrel and screw, causes melting of the feed polymer, which is then pushed by the screw into the molding mechanism from the die. Generally speaking, the quality of extruder depends upon the uniformity of temperature distribution, magnitude of temperature in the barrel, back pressure, and the homogeneity of the physical mixing. To produce a good quality of extrudates, temperature in each zone must be appropriately set and precisely controlled.

To achieve high-precision and high-performance goals, the designed temperature control system of the extruder barrel is composed of four channel steel heated barrel with four heaters and a digital signal processor (DSP)-based controller with sensing and actuating modules. The heater barrel is a prototype for use in industry, where the temperatures to be controlled in the four channels are in ranges of 100-400°C. The heater's capacities are 500 W for the temperature zone, close to the die, which is called channel 4, and 1.5 kW for the other three temperature zones, channels 1, 2 and 3. All heating channels can be controlled simultaneously. The DSP-based controller consists of plantinum (the resistance of which is 100 ohms at 0°C, Pt 100) temperature sensing modules with accuracy of ±0.4°C (1.1% accuracy) of resistance-to-voltage (R-V) transducers, a DSP (TMS320C31) control board inserted in the slot of the industrial 80486 personal computer (PC), four channel 16 bits A/D and D/A converters, and four silicon controlled rectifier (SCR) actuating modules. Figure 5.5 shows a schematic diagram of the temperature control system. In the temperature sensing modules, four Pt 100 sensors are used to measure the temperature of four temperature zones in the heater barrel, and four R-V converters are employed to precisely convert the resistance output of Pt 100 sensors into the corresponding voltages ranging 0-3V. The DSP processor is adopted for reading the input data via the four 16-bits A/D converters, sensing out the control output via the four 16-bits D/A converters ranging 0-3V, and performing high-performance, but sophisticated, real time estimation and control algorithms. The industrial PC is responsible for editing, compiling, and simulating the C program codes for both the PC and the DSP. The PC also communicates with the DSP via the EISA bus during the operating period. The SCR actuating module (Fig. 5.6), consists of a V-I (voltage-current) converter which converts voltage output of the D/A converted into the corresponding 4-20 mA, an SCR actuating circuit which controls the trigger angle of the SCR and then switches on and off its corresponding heater, and a limiter which uses an adjustable resistor to limit maximum energy rate into the heated barrel. The sampling period is 20 s.

5.6 On-Line Optimizing Control of a Distillation Column

Ganguly and Saraf (1992) have reported on-line optimization and control of distillation column. The pilot plant facility for testing on-line optimization includes a 15 sieve tray distillation column, 22.8 cm in diameter and 10 m high with a vertical thermosiphon type reboiler and a water cooler 1-2 shell and tube total condenser. The column is heated by steam from an oil fired steam boiler. Typically it processes 150 kg/h of ethanol-water feed with 20% by weight of ethanol and produces 81% by weight ethanol in the top product and 6% by weight ethanol in the bottom with feed entering at the 11th stage from the top and reflux is around 3. The top few stages are not sensitivie enough to changes in manipulated variable (reflux flow) and hence the sixth stage temperature is used as the control variable. The reboiler steam pressure is kept constant. Secondary variables which include reflux accumulator level, bottom accumulator level, feed and reflux temperatures (using pre-heaters) and feed flow rate must be controlled for steady state operation of the column and can be accomplished

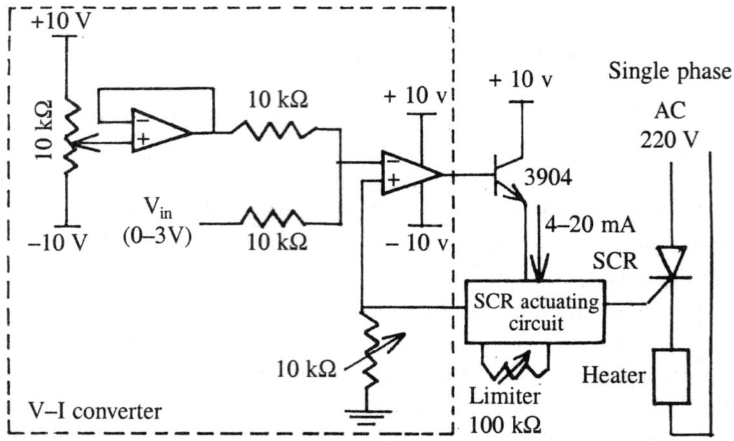

Fig. 5.6 A schematic circuit diagram of a SCR actuating module (Tsai and Lu, 1998)

using single loop controllers. All the single loop controllers are connected through a communication controller (Fig. 5.7) to the serial port of a PC/XT. The primary loop can be connected either to front end single loop controller or to a PC/XT through an I/O card. The nominal optimal trajectory is generated using the semi rigorous model residing in the PC/AT 80386 system and the bidirectional data communication between the supervisory and the regulatory levels is through the serial interface of the computer.

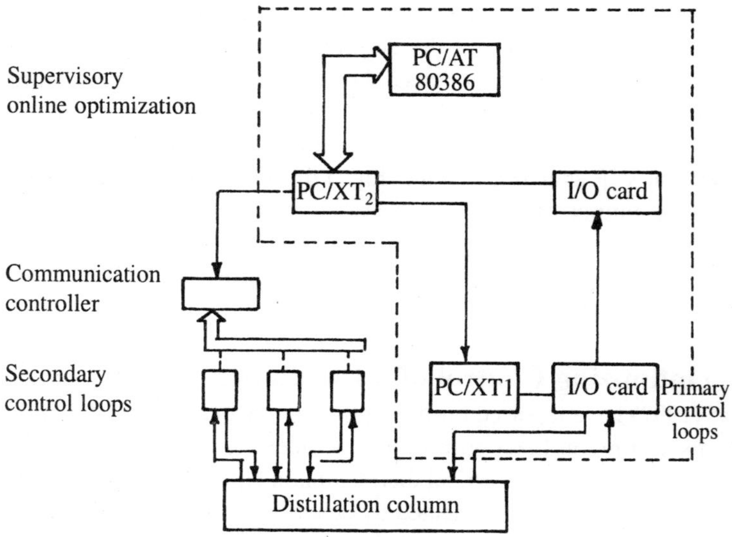

Fig. 5.7 Hierarchical control structure for distillation column (Ganguly and Saraf, 1992)

A process can be consisdered at pseudo steady-state when the persistent disturbances acting on the process are much slower than the dynamics of the process itself. The first layer takes care of regulation and allows the process to be considered at pseudo steady-state and the supervisory layer is responsible for determining the optimal set points under the influences of disturbances with a larger time scale.

It should be noticed that the time horizon for regulatory control is significantly shorter than for optimization.

In the study reported by Ganguly and Saraf (1992), the supervisory on-line optimizer receives the process output data at a slower sampling rate (30-120 secs interval) and estimates parameters for the semi-rigorous plant model using a nonlinear optimization scheme. The identified model is then used to generate nominal control trajectory which is communicated to the regulatory level. The regulatory level samples the plant output at a faster rate (10-15 sec interval) and uses a linearized process model alongwith the nominal control trajectory to generate a neighbouring optimal control action. This control action is implemented through a cascade controller which works at an even faster rate (1-2 secs interval) to minimize auxiliary effects (such as boiler output pressure fluctuations, piston pump pluses and valve non idealities etc.) The stationary disturbances get eliminated at the regulatory level and non stationary ones reach the on-line optimizer. At the supervisory level, the continuous time model parameters take most of the non stationary disturbances into account.

Summary

In this chapter, we have seen typical examples of computer control of processes. The flow sheets and interface circuits used in the reported studies were briefly discussed.

References

Chang, J.-S., J.-S. Hsu and Y.-T. Sung, Trajectory tracking of an optimizing path in a batch reactor: experimental study, Ind. Eng. Chem. Res., 35 (1996) 2247–2260.

Ganguly, S. and D.N. Saraf, On-line optimization: a hierarchical scheme for distillation column control, Chem. Engg. & Processing, 31 (1992) 337–347.

Queinnec, I. and B. Dahhou, Optimization and control of a fedbatch fermentation process, Optimal Contr. Appln. & Methods, 15 (1994) 175–191.

Singh, K., S.N. Maiti and D.N. Saraf, Adaptive dynamic matrix control of a heat exchanger: an experimental study, Indian Chem. Engr., 38 (1996) 8–17.

Srinivas, N. and M. Chidambaram, Experimental comparison of on-line identification methods for controller tuning, Process Contr. & Quali., 8 (1996) 85–89.

Tsai, C.-C., and C.-H. Lu, Multivariable self-tuning temperature control for plastic injection molding process, IEEE Trans. Indut. Appln., IA-34, No. 2, (1998) 310–318.

Review Questions

1. Explain with a block diagram, the computer control of a fedbatch fermenter.
2. Give a flow sheet of a plastic injection molding process and associated interfacing circuit for computer control.
3. Explain the multilayer decomposition of control task in an optimizing control scheme of a distillation column.
4. Rohani et al. (Canadian J. Chem. Engg., 1990, V68, 260–268) have presented a computer control of size distribution in a batch cooling crystallizer. Study the flow sheet and give a summary of the sensors and actuators used and the interfacing details.
5. Renard et al. (1988) have presented a computer control study on an anaerobic digestion process (Biotech. & Bioeng., 1988, V31, 287–294). Study this work and give the details of flow sheet, sensors, actuators and the associated interfacing circuits.

6. Herbst et al. (Powder Tech., 1992, V69, 21–32) have presented a computer control of ball mill circuit used in mineral processing operations. Study the work and give a report on the sensors, actuator and the interfacing circuits used.
7. Machado and Bolzan (The Chem. Engg. J., 1998, V70, 1–8) have presented a computer control study on batch polymerization reactor. Give a write up on the experimental facility and the sensors, actuators and interfacing circuit used.

6

Smart Sensors and Fieldbus

This chapter give details on the smart sensors (differential pressure transmitter, temperature transmitter) and smart valve positioners. The need for a field bus and details on commercial field buses such as foundation field bus and Profibus are also discussed.

6.1 Smart Sensors

Smart sensors/transmitters have an integrated single chip microprocessor for signal conditioning and transmission. They were developed to deal with non-ideal behaviour of sensor mechanisms, accommodate communications systems and improve the performance and reliability of instruments. Smart sensors can compensate and correct periodically various error modes including ambient temperature variations, noise, non-linearities signal interfaces/pickup, drift, mounting angle effects etc. The built-in microprocessor stores correction functions of factory calibrated data in its memory so that the above information is accessed for a continuous calibration. Smart instruments can identify mal-operating modes within itself which greatly facilitates trouble-shooting functions. On-line rangeability or span adjustment is significantly increased from the conventional average of 6:1 to 400:1. Calibration, re-arranging and self-diagnostics can be performed without the service interruption necessary in conventional instrument systems.

Smart instruments have linear/square-root curve fitted output, selection of any engineering unit and direct or reverse electrical output. Such instruments provide two-way digital/analog communication links and can easily interface with modern digital computer control networks and distributed control systems. Smart sensors can also have in-built PI/PID controller (i.e., software in the microprocessor) and alarm also. That is, these sensors can calculate the manipulation action and send signal directly to the final control element. In large plants, remote diagnostics, self calibration, maintenance free operation and improved reliability of the smart transmitter can be a justifiable investment. Todays smart instruments are part of a continuing evolutionary trend, from pneumatic instruments (3-15 psi), through analog electronic (4-20 mA) and simultaneous analog/digital communication (HART) to full digital communications (Fieldbus).

6.2 Smart Differential Pressure Transmitter

Differential pressure is applied via the seal diaphragm and the filling liquid to the silicon pressure sensor (Fig. 6.1). If the measuring ranges are exceeded, the overload diaphragm is flexed until one

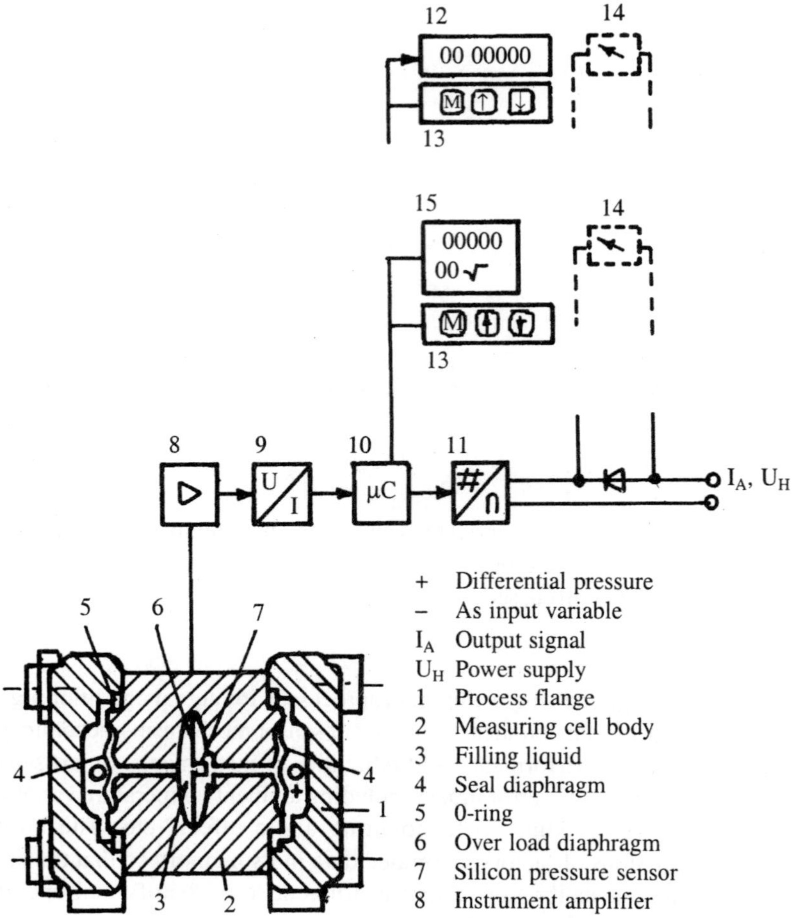

+ Differential pressure
– As input variable
I_A Output signal
U_H Power supply
1 Process flange
2 Measuring cell body
3 Filling liquid
4 Seal diaphragm
5 0-ring
6 Over load diaphragm
7 Silicon pressure sensor
8 Instrument amplifier
9 Voltage-to-frequency converter
10 Microcontroller
11 Digital-to-analog converter or PROFIBUS-PA interface
12 Digital display for parameterization
13 Keys for parameterization
14 Analog indicator (option)
15 Digital display for Parameterization and measured value

Fig. 6.1 Functional diagram of a smart DP transmitter (Siemens Ltd., 1998)

of the seal diaphragms rests on the measuring cell body, thus protecting the silicon pressure sensor from overloading. The measuring diaphragm is flexed by the applied differential pressure. The resistance of four piezo-resistors fitted in the diaphragm in a bridge thus changes. This change in resistance results in a bridge output voltage proportional to the differential pressure. The voltage is amplified and converted into a periodic signal by means of a voltage-to-frequency converter. This signal is evaluated by a microcontroller, and its linearity and temperature response corrected. The processed signal is converted by a digital-to-analog converter into an output current of 4-20 mA, or a digital bus signal. The data specific to the measuring cell as well as the parameters of the transmitter are stored in a nonvolatile memory (EPROM). The parameterization of transducer can be done using 3 external keys or built-in-digital display.

Parameterization: Depending on the version, there are different possibilities for parameterization of the transmitter and for setting or scanning the parameters. When parameterizing with a laptop PC, the connection to two-wire systems is via the HART modem. When parameterizing with a HART communicator, the connection is made directly to the two-wire system. The signals required for the communication according to the HART protocol are superimposed on the output current according to frequency shift keying (FSK). When parameterizing with a PROFIBUS interface, the connection occurs starting from a master, using signals transmitted via PROFIBUS and converted by a DP/PA interface with power supply into a signal for PROFIBUS. A bus transmitter is required for cable lengths greater than 2 m.

6.3 Smart Temperature Transmitter

The measured signal supplied by a resistance-based sensor (2, 3 or 4 wired connection) or by a thermocouple is amplified in the input stage (Fig. 6.2). The voltage, which is proportional to the input variable, is then converted into digital signals by an analog/digital converter. These signals are forwarded electrically isolated to the microprocessor. They are converted there in accordance with the sensor characteristic and further parameters (damping, ambient temperature etc.). The signal is then converted in a digital/analog converter into a load independent direct current of 4-20 mA. The power supply is located in the output signal circuit. The smart transmitter is parameterized and operated using a PC connected to the two-wire line via the HART modem. A handheld communicator can also be used for this purpose.

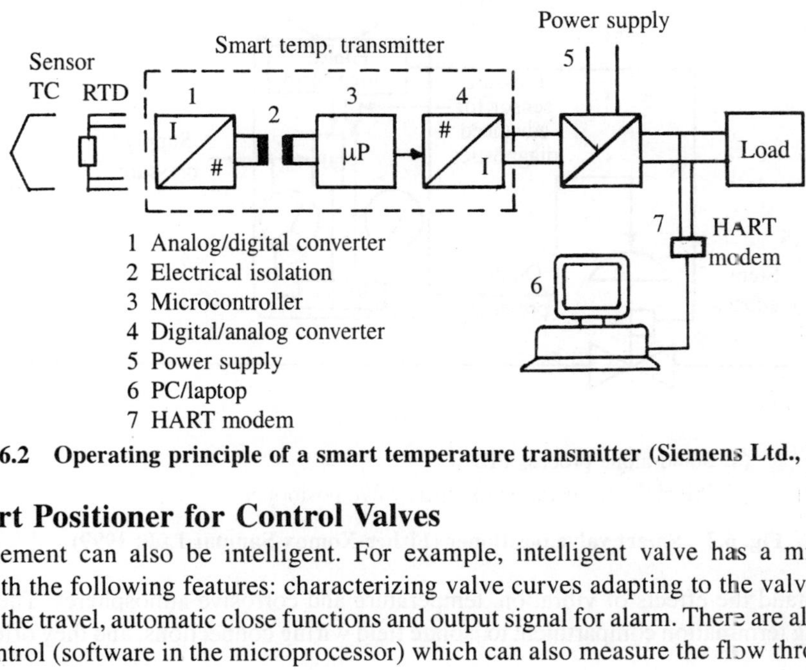

Fig. 6.2 Operating principle of a smart temperature transmitter (Siemens Ltd., 1998)

6.4 Smart Positioner for Control Valves

The final element can also be intelligent. For example, intelligent valve has a microcontroller (Fig. 6.3) with the following features: characterizing valve curves adapting to the valve specialities, limitation of the travel, automatic close functions and output signal for alarm. There are also positioners with PID control (software in the microprocessor) which can also measure the flow through its valve and control the flow.

A typical digital valve controller has a single module base which may be easily replaced in the field without disconnecting field wiring or tubing. The module base contains the following subunits: miniaturized I/P converter, printed wiring board PWB assembly, pneumatic relay and gauges (refer to (Fig. 6.3a). The module base can be rebuilt by replacing the subunit. The PWB is fully encapsulated

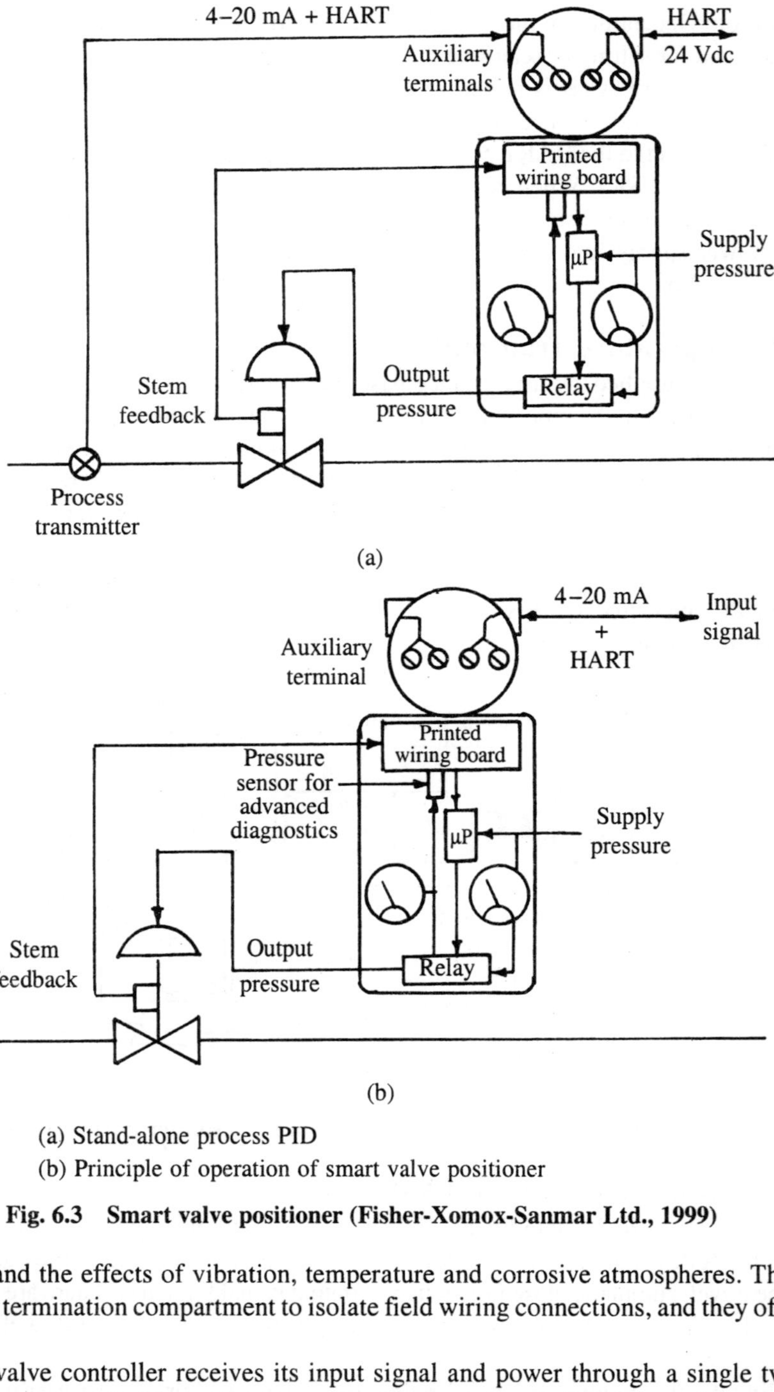

(a) Stand-alone process PID
(b) Principle of operation of smart valve positioner

Fig. 6.3 Smart valve positioner (Fisher-Xomox-Sanmar Ltd., 1999)

and can withstand the effects of vibration, temperature and corrosive atmospheres. The valve has a separate wiring termination compartment to isolate field wiring connections, and they offer a weather-tight housing.

The digital valve controller receives its input signal and power through a single twisted pair of wires brought in the terminal box. The input signal is routed to the PWB assembly subunit, where it can have many parameters applied such as: characterization, limits and others. The PWB assembly subunit sends a signal to the miniaturized I/P converter subunit which transform the input signal to a pressure signal. The signal is then amplified in a pneumatic relay which delivers the output signal to an actuator. The output signal can also be sensed by a pressure sensor located on the PWB assembly subunit and used for valve/actuator diagnostics. Stem position of the valve and actuator is brought as an input to the PWB assembly module and used as a feedback for the digital valve controller. Gauges indicated the inlet and outlet pressures.

The digital valve controller has also the following options for communications: (i) handheld communicator (ii) stand alone communicators and (iii) integrated to DCS (distributed control systems).

The digital valve controller can be powered independently by a 24 V dc power supply. A process transmitter can then be wired directly into the instrument auxiliary terminals. The process PID within the instrument (microprocessor) can be used to control the process. A handheld communicator can be used to change set point and set various control parameters or settings.

The HART protocol gives the capability of multidroping where you can network several devices to a single communications bus (Fig. 6.4). This process is well suited for remote applications such as pipelines, custody transfer sites etc.

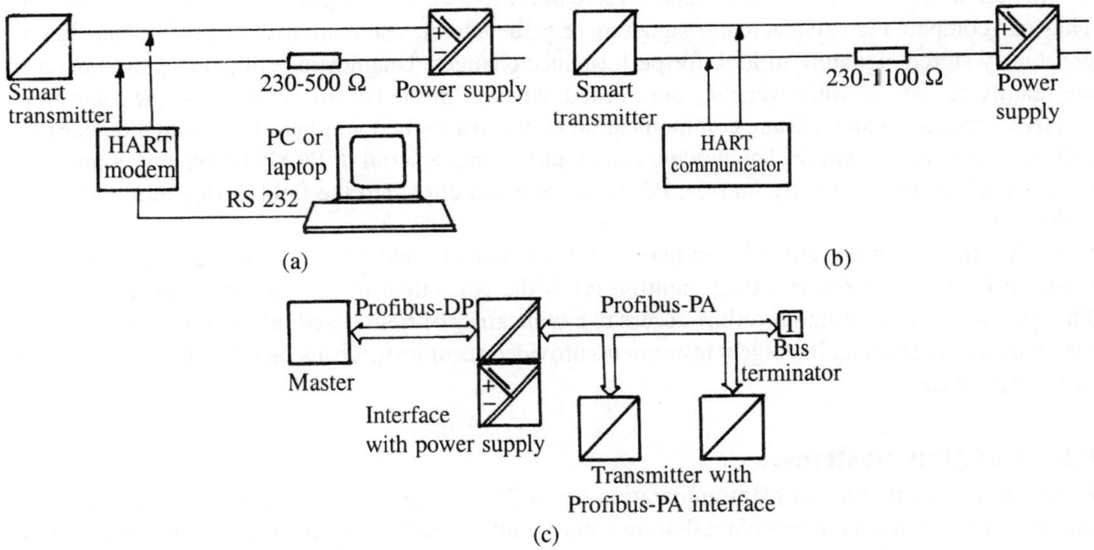

(a) Communication between PC or Laptop computer and transmitter
(b) Communication between HART comunicator and transmitter
(c) Communication via PROFIBUS-PA interface

Fig. 6.4 Communications in a smart transmitter

6.5 Advantages of Smart Sensors

Installation of small transmitters reduces unplanned maintenance calls. This is a considerable reduction in downtime required for calibration checks. Use of HART and eventually, fieldbus protocols is highly desirable because it brings the diagnostics into control room and makes the instrumentation data part of a system that can document and analyze failures.

Smart sensors (known as multivariable devices) also offer the chance to sense, use and transient auxiliary process variables in addition to the targeted measurement. Such multivariable transmission can provide more-accurate signals, by actuating alerts or correcting for significant environmental effects or changes in process conditions. For example, certain smart pressure transmitters offer measurement of both gauge and differential pressure by utilizing only one resonant sensor; others require a secondary, typically a piezo sensor. The multivariable transmitter is therefore able to perform more advanced functions such as pressure and temperature compensated level measurement and, reducing the need for additional devices/cabling as well as offering significant cost savings on equipment, installation and commissioning.

The use of digital signals by HART and fieldbus increases instrument availability in itself. Digital outputs, such as frequency and pulse count data from flow meters, can increase the accuracy of the measurement by more than a factor or two by eliminating D/A converter in the transmitter and A/D converter in the distributed control system or the programmable logic controller. However, the use of HART frequency outputs for a process variable add dead time to the control loop. Fieldbus reduces this delay by supporting much higher information communication rates. Many plants are equipped with HART based field devices. In such cases, an engineer or operator can use a handheld communicator to obtain information from these devices for eventual uploading onto a personal computer (PC) running assert management systems.

Maintenance personnel can know about valve performance without pulling the valve from the line. They can compare the valve/actuator signature (e.g., bench set, seat load, friction, and others) against previously stored signature to look for performance changes. Diagnostics critical to plant operators are readily accessible whenever they are needed, whether in the control room or on the plant floor.

HART modems can facilitate communication between a limited number of individual devices and a PC running such software. The modem can employ either a serial or PCMCIA type of connection: either of which plugs directly into the PC at one end and clips into the field device terminals at the other end.

Such offline solutions are only temporary; for use with all field devices and it can be permanently wired to HART multiplexers. Each multiplexer is the equivalent of 32 individual device modems. There exists direct continuous online access to a wide array of field based information by a PC in the maintenance department. Intelligent instruments provided accurate feedback on the status and conditions of the equipment.

6.6 Fieldbus Systems

Conventional techniques in plant automation use 4-20 mA standard for transmission of measured process variables from process plant floor to control unit. Though sophisticated Distributed Computer Control Systems (DCCS) are setting the pace in plant automation, the one-to-one connectivity of field devices like sensors and actuators to process controllers remain the same. However, during recent years, the concept of incorporating local intelligence in field devices and providing peer-to-peer communication is widely getting accepted. In order to gain the advantages of digital technology, there is a growing demand in the industrial world for a new digital standard for low level communication; this standard is known as Fieldbus. The fieldbus interconnects field devices such as sensors, actuators, I/O racks and local controllers at the field level. This nework provides new functions such as automatic calibration of field devices, self test, download of parameters, identification of devices, configuration, on-line diagnostics, calibration, pre-computing of measured values, network monitoring etc. Such networks reduce cabling costs and improve availability.

An obvious advantage of digital vs analog technique is the saving in the amount of cabling; a single digital loop can replace a large number of 4-20 mA conductors. The possibilities opened by fieldbuses are notable. A large share of the intelligence required for process control is moved out to the field. The maintenance of sensors becomes much easier because operations such as test and calibration can be remotely controlled and require less direct invention by maintenance personnel.

The phase-out of all the old instrumentation may take some years. During this period, one will need more than one communications technology. This is where the HART (Highway Addressable Remote Transducers) and Fieldbus communication protocols came in. The primary strength of HART is that it allows augmentation of any existing DCS, because it has co-resident digital and 4-20 mA analog communications capability. This permits the installation of advanced digital devices as needed while the majority of process variables continue to be communicating in the analog mode and controlled by the installed DCS. Thus existing field instrumentation, software, and input-output controllers do not become instantly obsolete.

However, the fieldbus signifies total conversion to the digital mode of communications. Technology can be applied in new process units, unit expansions, and new plants where the legacy of old equipment does not impede the total adoption of digital technology. One of the major hurdles to wider acceptance of digital technology for field devices has been the lack of interoperability between many different existing devices made by various manufacturers. Interoperability was a problem with all the first generation protocols. HART is the first protocol that effectively solved this problem. As a result, fieldbus is the direct beneficiary of the development process, and thus offer interoperability. In evolving from pairs of cable connecting field devices to a communications system, all the existing field units need to be able to communicate over the new network in order to operate together efficiently. Theoretically, this can be done by agreement of the different equipment manufacturers to build products that reflect certain specified parameters. In practice, it is not that simple because not everyone is bound by the agreed-upon specifications of protocols.

Fieldbus is an all-digital, serial, two-way communication system that interconnects process instrumentation and actuators to controllers. Field devices in use today may be grouped into three categories: units with traditional analog and discrete I/O; hybrid analog and digital devices; and purely digital devices (Fig. 6.5). Devices of the first type connect to an I/O system or controller by

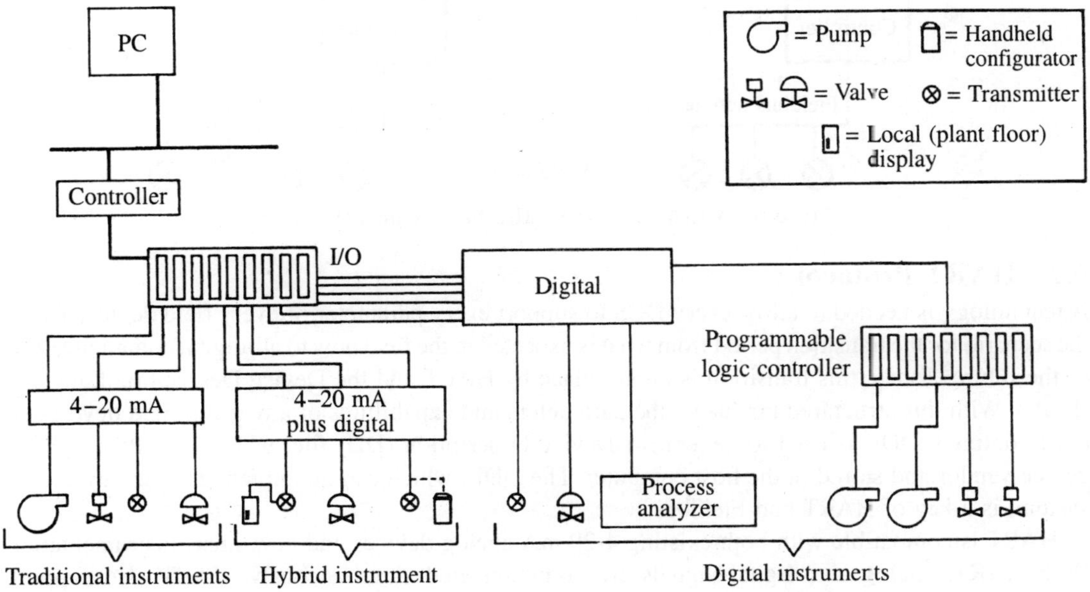

Fig. 6.5 Mix of traditional analog control loops, mix of analog and digital interfaces and a proprietary digital scheme (Santori and Zench, 1996)

means of dedicated 4 to 20 mA analog current loops. These are strictly point to point systems in which each device has its own connection to the host controller. The second category, hybrid analog and digital devices, can be used with both analog and digital communication systems. For example, the well established HART concept super imposes a digital communication signal on top of a standard 4 to 20 mA analog signal. The digital signal is made to have a zero average value so as not to affect suitably averaged readings of the analog current. If few HART devices are integrated into traditional point to point systems, each will require a dedicated line, just like the traditional instruments. Purely digital devices, the third category, often have open interfaces. However, they usually require custom interfacing hardware and necessitates the use of custom software drives in the control system. Devices in this category include weighing scales, process analyzers (gas chromatography, for example), supervisory and data acquisition systems, programmable logic controllers and even some intelligent valves. Even when connected via standard ports, for example, RS-232 or RS-485, they still require the custom software.

Fieldbus segments may interface with a variety of host equipments, ranging from simple PC-based operators interface (OIs) to small, scalable controllers/systems to full-fledged DCSs. Communications above host level will typically be on Ethernet (Fig. 6.6).

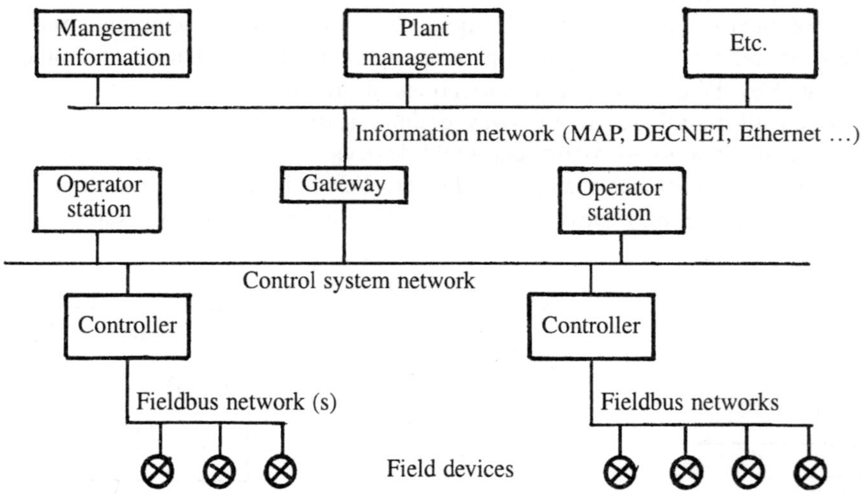

Fig. 6.6 A fieldbus system with hierarchical structure

6.7 HART Protocol

A technology is needed to allow every DCS to support every parameter of every field device. This is the scenario for the transition period from what is expected in the field now to all-digital communications of the future. Today, this transition is taking place by HART and the Device Description Language (DDL). With this structured language, the parameters and capabilities of any given field devices are easily defined. DDL is used to generate a Device Description (DD) file, which is written by each device vendor and stored in the host computer. The DDL, which can be implemented in any control system, is a key of HART and Fieldbus.

HART is compatible with both existing 4-20 mA analog devices and new smart instrumentation. With HART, analog and digital signals are communicated simultaneously. HART also supports multivariable instruments and provides access to the online status of smart field devices.

Physically, HART runs on a protocol based on the Bell 202 telephone standard, using the forward channel of that standard, which is 1,200 baud. As shown in Fig. 6.7, it modulates on two frequencies, namely 2,200 and 1,200 Hz (with an amplitude of –0.5 to +0.5 mA) that provides an effective data rate of about two poll responses per second for sensing a single process variable. A high speed option increases the ratio to 3.7 poll responses per second. A HART network supports upto 256 variables. Transmitters can be networked together in a multi-drop (that is multiple devices connected through one bus) configuration to power and communicates with upto 15 devices, so a single network can accommodate a large number of input-output points. The average value of HART signal is zero and hence the analog signal is unaffected. The HART signal is removed from the standard analog signal by standard analog input circuit filtering. Finally the HART is a master-slave protocol, i.e. a device replies only when it is spoken to. A HART device can have more than one master at a time, for example handheld terminal, PC or a control system. HART is a field-proven technology that is up and running today communicating both analog and digital signals concurrently. A large selection of products are available with HART in process control, maintenance, and data acquisition applications.

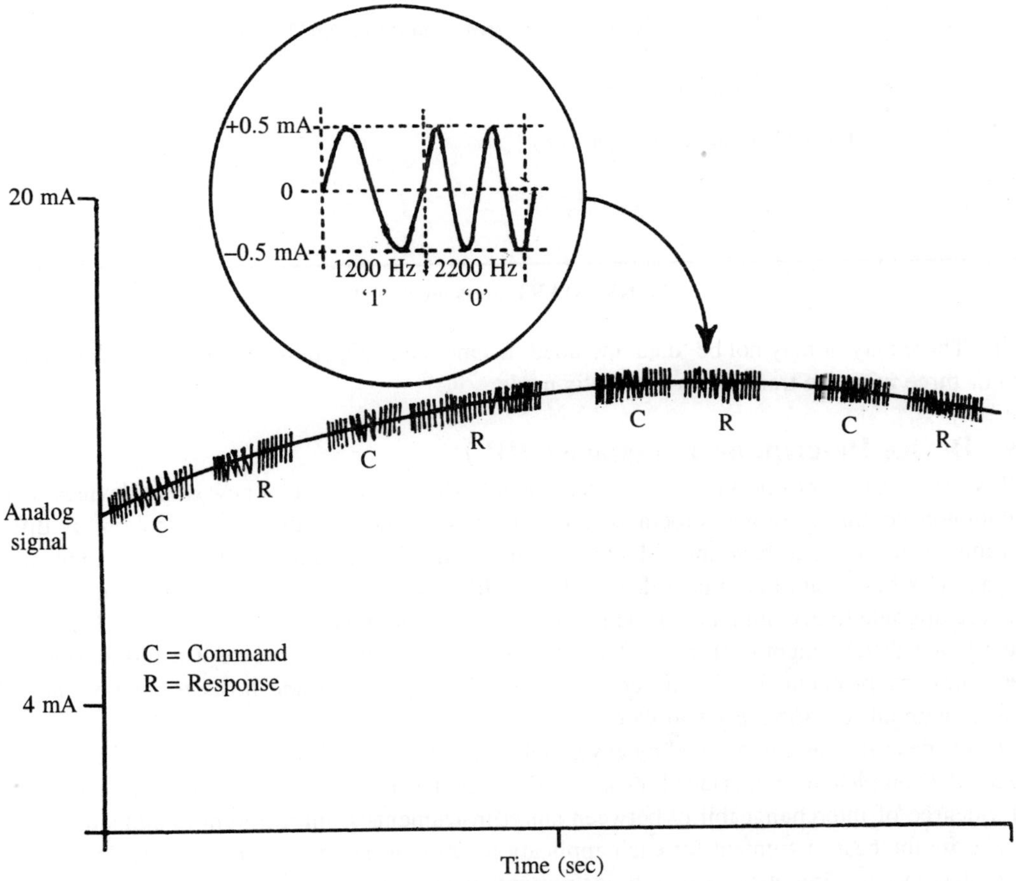

Fig. 6.7 The HART signal

Each message (Fig. 6.8) includes the addresses of its source and destination, to ensure that it is received by the correct device, and has a 'checksum' to allow detection of any corruption of the message. The field device's status is included in every reply message, indicating its continued good

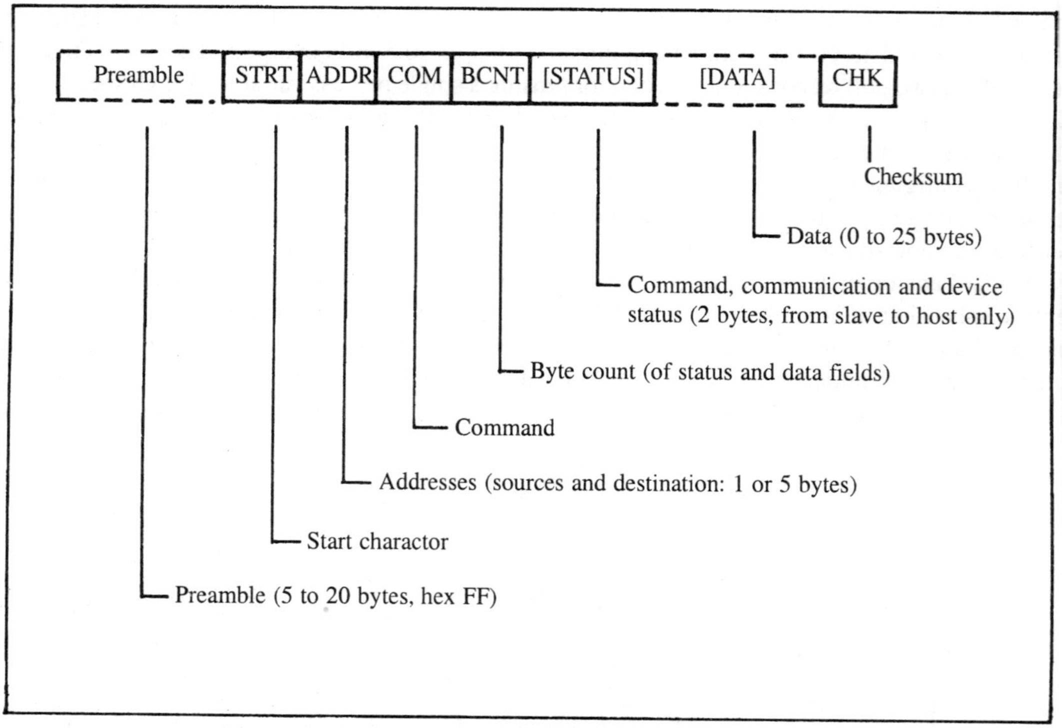

Fig. 6.8 HART message structure

health. There may or may not be 'data' included in a message, depending on the particular command. Two or three message transaction per sec is made .

6.8 Device Description Language (DDL)

DDL solves a number of problems relating to the introduction and use of new smart devices. The use of common communication protocol is not enough to ensure useful communication, as different instruments are likely to have individual variations on the data available, its use and its meaning. In the past, this has meant that a host device (handheld communicator, control system etc.) requires a software upgrade to accommodate product-specific features of each new field device. This involved a significant development cost, and often there would be a time lag before it could be done. DDL overcomes the problem. Device description make it easy to upgrade hosts to support new field devices, without rewriting any software.

Device descriptions can be used by any suitably designed host devices to automatically provide a correct and complete user interface for each field device. Device description allow full interoperability and a degree of interchangeability between smart instruments from different manufacturers. Users can choose the best instrument for each application, without being locked into a single supplier for a complete system. Innovations in new instruments is encouraged.

DDL is a formal language, used to describe completely and unambiguously, what a field instrument looks like when you talk to it through the "window" of its digital communication link. It form an additional "user layer" on top of the OSI protocol reference model. DDL includes description of accessible variables, the instrument's command set and operating procedures such as calibration. It also includes a description of a menu structure which a host device can use for a human operator. The device description (DD), written in a readable text format, consists of a list of items ("objects") with a description of the features ("attributes" or "properties") of each.

The principle constructs or object type of DDL are:

Variable: Refers to any item of data contained in the field device such as measurements, operating parameters, device information. Among the attributes of a variable are a label for display and the specific data type.

Commands: Refer to HART commands which the device will accept. Defined by the content of their request and reply data fields and the response codes implemented.

Menus: Menus for presentation to an operator (specially for a handheld communicator) is defined as a list of items (variables, displays, methods, further menus).

Edit displays: Displays for presentation to the operator (specifically for a handheld communicator) are defined as a list of variables for display and/or editing, and procedures ('methods') to be executed before or after editing.

Methods: Defined sequences of interactions with the field devices and with the operator executed by a host to achieve specified operations on the field device (e.g., calibration or re-ranging) using a subset of the ANSI C programming language. A library of built-in functions is available for use within the methods, including sending commands to the field device, inspecting responses, displaying messages to the operator, and accepting keyboard input from the operator. These allow a method to deal correctly with error and failure conditions, as well as normal operation. A method can also be used to warn the operator of the implications before proceeding with an action which might adversely affect system behaviour.

Arithmetic, logical or conditional expressions are permitted in the definition of many attributes. In this way the device can be treated differently depending on present circumstances (e.g., configuration parameters or operating modes).

Fieldbus provides the environment in which each smart field device is an integrated component of the overall control network, permitting unlimited access to the data stored in each device by all other devices on the network. Fieldbus does for control systems what Local Area Networks (LANs) have done for PCs in offices where a desktop PC becomes an autonomous node on the LAN, allows the user to work interactively with other computers on the network.

Fieldbus software resides in all the connected field devices making use of parameter, trend, display, alarm, function block and scheduling. These are used in building up a field-based control system, one device at a time, beginning with something as simple as a transmitter and a valve, and allowing more devices to be added at will. A system can grow incrementally from the bottom up, because it is inherently scalable. This is the key attribute of field-centered architecture.

6.9 Topology of Fieldbus

Physically, fieldbus consists of two types of communications buses (Fig. 6.9). The H1 low-speed fieldbus operates at a speed of 31.25 kilobaud, and H2 high-speed fieldbus operates at either 1 or 2.5 megabaud. Both are capable of handling multi-drops, each accommodating a maximum of 32 nodes. The H1 bus goes out to the field, and field devices are connected to it. It can accept upto 32 self powered, four wire devices, 12 devices that require power from the bus, or just six devices if intrinsic safety is required.

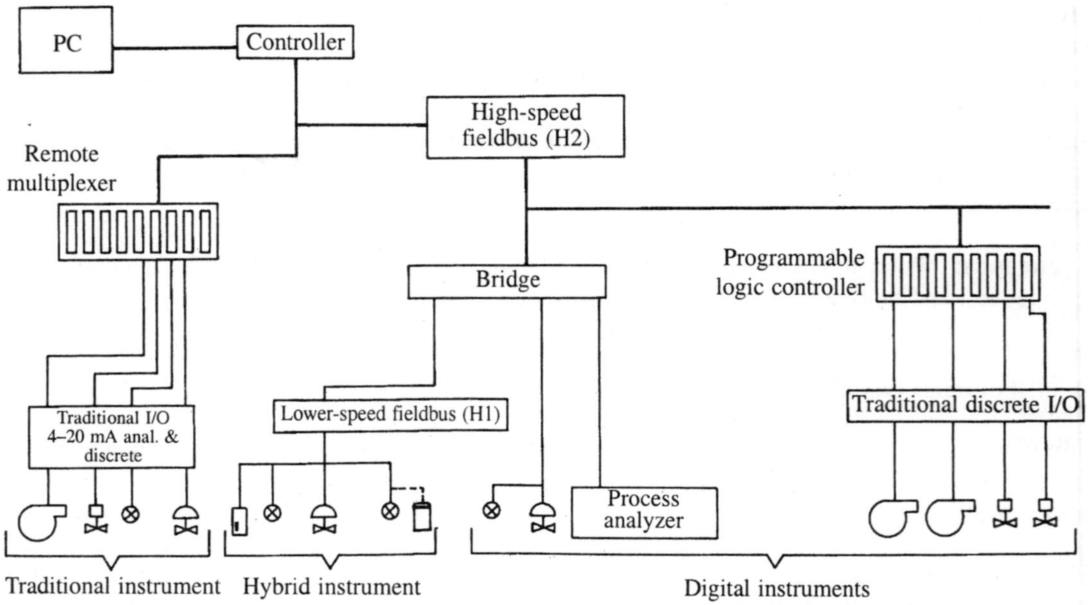

Fig. 6.9 Field bus structure (Santori and Zech, 1998)

The H1 fieldbus can be used for control applications such as temperature, level and flow control. The DC supply voltage can range from 9-32 VDC.

H2 is the unit wide network, and H1 buses are connected to it by means of a bridge, which is a protocol converter from the low-speed buses to the high speed H2 buses. All the nodes on H2 are self powered, which is equivalent to today's proprietary unit-wide networks, and this is where the actual controllers reside. Each controller can handle hundreds of loops. With each type, once the 32-node limit is reached, a bridge can be installed to another bus. Fieldbus software provides for interoperability as well as secure and scalable network architecture. Further, its object-based modular software makes it easy to build a whole plant one loop at a time. The combination of low-and high-speed bus supports a highly efficient digital communications network capable of accommodating all the field devices and other nodes required in an extensive network. The H2 fieldbuses will typically be used for advance process control, remote input/output, and high speed factory automation applications.

6.10 Industrial Field Buses

Let us briefly examine the main features of the fieldbuses FIP from France and PROFIBUS from Germany, the industrial BITBUS developed by Intel and the Foundation Fieldbus from U.S.A.

6.10.1 FIP (Factory Instrumentation Protocol)

FIP uses a twisted pair conductor and the transmission speeds are 31.25 kbits/s, 1 Mbit/s and 2.5 Mbit/s. The cable length for a segment is 500 m for operations at 1 Mbit/s and varies depending on the actual speed. Several segments can be connected together with repeaters, the maximum number of stations connected to a segment is 32 and for a whole system is 256. All the interface functions are contained in a special integrated circuit called FULLFIP. This integrated circuit is connected with the CPU of the communicating device (PLC, intelligent sensor or other) and on the other side with the FIP bus.

6.10.2 PROFIBUS (Process Fieldbus)

PROFIBUS combines specifications at OSI layers 1, 2 and 7 (physical medium, data link, and application). It follows existing standards at the different layers and attention has been paid to compatibility with the hardware already in use. The most important bus operations do not require specialized processors but can be carried out with program running on common CPUs. The PROFIBUS physical medium is a screened twisted pair cable according to RS-485 specifications, with a maximum length of 1200 m (upto 4800 m may be reached with the help of repeaters). Data transmission speeds are 9.6, 19.2, 187 and 500 kbit/s. PROFIBUS uses a hybrid version of the IEEE 802.4 Token Bus model with master and slave stations. Upto 127 stations, divided into active and passive, can be connected to the bus. Active stations circulate the token and may transmit when they hold it. Passive stations do not circulate the token and must be addressed by an active station to be able to reply with a message. An innovative feature of PROFIBUS is the definition of management functions, e.g., to change the bus configuration, add or take away station as well as detect and isolate faulty units. Figure 6.10 shows details of the PROFIBUS.

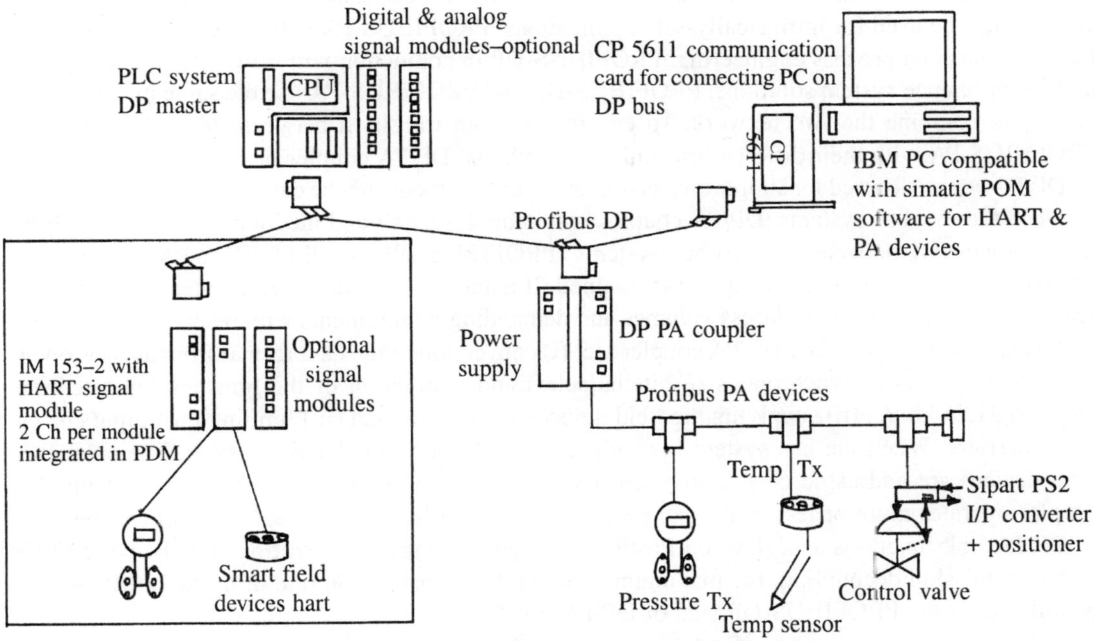

Fig. 6.10 PROFIBUS architecture (Siemens Ltd., 1998)

PROFIBUS has been designed to support the connection on the same bus of intelligent sensors together with more complex units such as PLCs, regulators and small process computers.

6.10.3 Details on PROFIBUS

Practical industrial installations require multiple cell-level controllers and programming and configuration devices to be able to communicate both with each other and with devices in the field. PROFIBUS therefore provides a mechanism for multiple master stations each to gain access to the shared bus and to communicate entirely with other devices.

The hybrid PROFIBUS medium access protocol establishes a 'Logical token ring' between masters, each master being given exclusive access to the bus for a precise time interval while it holds the token, during which it can communicate with other masters and with 'slaves' such as intelligent actuators, sensors and drives. In principle, only master stations can initiate communicatons, although special facilities allow slaves to send unsolicited messages such as alarms. Additional facilities ensure a high level of data integrity and provide for the detection of defects in both the transmission medium and the transceivers.

For applications requiring real time deterministic performance PROFIBUS-DP (H2 Fieldbus) trades the amount of information which can be transferred for higher performance. Traditionally, PROFIBUS-DP has been used for fast cycle data transfer between PCs/PLCs and decentralized I/O. However, with the growing use of intelligent drives, and positioners, many more parameters have to be changed during set-up and operation. Profibus-DP has been enhanced to allow acyclic read or write of any type of data, confirmation of diagnostic and alarm messages and transmission of device status information.

PROFIBUS-PA, a variant for the applications in Process and Power Plant Automation, has been developed. In PROFIBUS, the RS-485 transmission protocol that is used in Profibus-DP was replaced with the transmission method to IEC 1158-2. This method is in accordance with international standards and can also be used for intrinsically-safe applications. PROFIBUS-PA is the open fieldbus solution for chemicals and process engineering. PROFIBUS-DP in conjuction with PROFIBUS-PA supports high-performance system solutions. PROFIBUS-DP and PROFIBUS-PA use the same protocol, so it is easy to combine the two network. To ensure a smooth transition between PROFIBUS-DP and PROFIBUS-PA with their different transmission methods, DP/PA coupler or DP/PA link is used.

DP/PA coupler is used for simple conversion of the bus protocol and the data transmission procedure between the two bus systems. DP/PA coupler is suitable for small and medium-scale plants. DP/PA link is a gateway between the two bus systems (PROFIBUS-DP and PROFIBUS-PA) in a manner whereby data transmission decoupling is retained. It is ideally suitable for implementation in large-scale plants with extensive address volumes and demanding requirements with respect to cycle time.

The main functions of the DP/PA coupler are: (i) conversion of the data format from asynchronous (11 bits/character) to synchronous (8 bit/character) and conversion of the transmission rate from 45.55 to 31.25 kbit/s; (ii) supplying the field devices with power and (iii) limiting the output current using barriers. When the bus system is configured, the DP/PA coupler is invisible. The connected field devices are addressed or accessed directly from the programmable controller or system. The coupler operates more or less in the same way as a wire and is not detected by the bus nodes.

DP/PA link functions as a slave on Profibus-DP and as a master on Profibus-PA. Because of the data transmission decoupling, the maximum Profibus-DP transmission rate of up to 12 Mbits/s can be utilized on the PROFIBUS-DP side of DP/PA link.

FMS (Fieldbus message specifications) is designed for general purpose communication primarily between programmable controllers, such as PLCs (programmable logic controller) and PCs (personal computers). FMS contains an application layer with communication services available to the user. These services make it possible to access variables, transmit program, and control program execution as well as transmit events. RS-485 and fiber-optic physical layers are available for PROFIBUS-FMS.

RS-485 physical layer is most frequently used in PROFIBUS applications. Baud rates of 9.6 kb/s to 12 Mb/s can be used.

6.10.4 Bitbus

Bitbus consists of the two first OSI layers: (1) physical and (2) data link. At the physical level, Bitbus uses twisted pair cable according to the RS-485 standard. A maximum of 28 devices can be connected to a single bus and several buses can be joined together with the help of repeaters. Transmission speeds are 62.5 kbit/s, 375 kbit/s and 2.4 Mbit/s. At the lowest speed, the distance between repeaters can be up to 1200 m.

Bitbus is structured hierarchially. One of the connected devices acts as a master, all other devices are slaves. The master is thus always in control of the communication process. The communication protocol is strictly causal: the master sends a message to a slave device and the slave must reply; a slave cannot initiate communication on its own, but must wait for a request from a master. A slave must reply to a master request within a specified time.

Bitbus is apt for use in smaller production cells or processing units with relatively limited data traffic, where there is no need for several masters to control communication. In large scale applications, the master units are connected together with high speed local area networks and transfer the concentrated data to a central process computer at higher levels.

Real-time oriented fieldbuses like FIP or bitbus can be used at the lowest to collect sensor data in a cell database. PROFIBUS may connect together several independence devices and intelligent sensors in a hybrid configuration.

6.10.5 Foundation Fieldbus

Foundation fieldbus is an all digital, serial, two-way communication protocol that interconnects field equipment such as transmitters, valves and controllers. Fieldbus is a Local Area network (LAN) for instruments used in process control with built-in capability to distribute the control application across the network. This technology is owned and maintained by the Fieldbus foundation, a non-profit organization that consists of more than 100 of the worlds leading control and instrumentation suppliers and end users. Figure 6.9 shows the functional description of foundation filedbus.

6.10.5.1 Controller and I/O interfaces

The compact, modular design allows you to build controller nodes that meet your process needs in a cost effective manner. The controller and I/O packaging design supports installation into field enclosures, near the field devices. The hardware design is based on DIN rail-mountable, modular carriers. The controller and I/O interfaces are designed to complement the installation thereby reducing wiring and marshaling panel costs and saves valuable control and instrument room space. All components are modularly designed, enabling plug and play. To speed up installation, the I/O interfaces are composed of two separate pieces: the I/O card and the I/O terminal block. The modular design allows the I/O terminal blocks to be installed prior to installing the I/O cards. We can pre-wire the field devices and carry out the wire continuity test directly from the I/O terminal block. The physical design ensures that the I/O card type always matches the installed I/O termination.

The controller offers an unique ability to automatically identify itself to the control network. When the controller is first linked to the control network, it automatically notifies the network of its connection. The workstation automatically assigns the controller a unique address. All controller information is now available to all nodes on the control network. The controller can provide a visual recognition of the controllers physical location, matching the controller with the software. This feature will be useful in distributed applications requiring multiple controllers when we need to quickly identify which physical location is sending data. The controller also automatically detects any connected I/O card. As soon as an I/O card is plugged in, the controller detects the new I/O functionally and notifies the system. Each I/O interface contains a pre-defined default settings. This minimizes the controller configuration effort and helps us get our process on-line faster.

The I/O interfaces support a broad range of analog and discrete signal types, including HART signals. All I/O cards are enclosed in an attractive form factor that plugs into the I/O interface carrier.

Each I/O card is clearly labelled. LEDs indicate power availability, I/O card condition, and status of the individual eight channels. Each output channel maintains its own fail-safe value. If an I/O card detects a communication time out with the controller, it will force the outputs to their fail-safe values.

The fieldbus H1 card allows adding fieldbus devices to the system. This card slips into any vacant slot in the I/O carrier, alongside HART and conventional I/O. Pulse applications can use discrete input I/O cards. The most commonly used cards for such applications are the discrete input 24 VDC cards. The pulse counter can manage constant pulse signals upto 100 Hz. Serial interface cards allow easy integration with popular industrial serial devices, such as weighing scales and PLCs. I/O termination options provide local wiring, channel isolation, and fusing.

6.11 OLE for Process Control (OPC)

Object Linking and Embedding (OLE) for Process Control (OPC) is an emerging software technology standard that connects window based process control systems to hardware devices on the plant floor. OPC based on Microsoft's COM/OLE technologies, a specification that defines how individual software components can interact and share data under windows. There are two sets of interfaces, OPC Custom interfaces for high performance applications written in a compiled languages such as C++, and the OPC Automation interface for higher level applications development in, for example, Microsoft Excel or Visual Basic.

6.11.1 Benefits of OPC

OPC promises to provide plug and play connectivity between software and hardware components from different automation suppliers. Manufacturing companies can easily integrate these components into corporate-wide automation and business systems, sometimes that has been virtually unachievable in the past. In addition, integration costs can be reduced because all software and hardware components can adhere to a single industry standard. OPC servers can replace proprietary I/O device driver software. OPC client applications, such as HMI/SCADA software can communicate and interoperate with any OPC server. The OPC can provide a single, consistent industry standard interface so that software suppliers can focus on adding new features to their software instead of developing a long list of proprietary hardware device drivers. OPC technology can extend beyond hardware I/O to more complex control and business systems. DCS, SCADA, HMI, batch control, maintenance, and other manufacturing applications can act as OPC clients and servers to permit real-time data exchange between cooperating applications; the end user can focus on value added business activities versus systems integration problems.

Components to support a simple bus system that have received attention in component development are cable assemblies and distribution (splitter) boxes. Taking assemblies first, it is worth noting that some suppliers have ensured that the same design may be deployed with a number of different standards. Standard cables may be used to interface with conventional sensors in both electro-mechanical or field-bus systems. These cable assemblies are also acceptable for use with the emerging range of intelligent sensors or recently announced components using analog inputs.

To simplify wiring, distribution (splitter) boxes were introduced into equipment where several sensors/actuators were concentrated in a relatively small area. This concept has now been extended to fieldbus systems. Special splitter boxes have been developed with active circuitry that can handle slave components in a fieldbus circuit, whilst retaining the rugged standards offered with the electro-mechanical design.

The fieldbus network can also be connected to a PLC. A foundation fieldbus module (hereafter named FB card) installed in the PLC is responsible for communication between the PLC and the fieldbus network. The module allows for the transfer of discrete and analog information and can communicate the status of fieldbus devices, which improves failure detection and maintenance. The foundation fieldbus protocol and a PLC allow available resources to be applied to sophisticated control strategies with minimal human interference.

A casual observer would find Profibus-PA and its competitor, Foundation fieldbus H1, virtually the same. The key differences are in the protocol implementation. As low speed networks PA and H1 comply with the same standard. They do the same job; auxiliary power to the application, with a data rate of 31.25 kbits/sec. Similarities include function-block-based architecture and a device description language (DDL). They use the same physical layer for digital data transfer. PA has another key difference: its protocol stack can operate devices on the same RS-485 and fiber-optics-based, high speed network wiring used for the other version of Profibus family, at data rates upto 12 Mbits/sec.

Summary

Fieldbus systems are all about the economical transfer of data and instruction codes from any kind of controller to sensors and actuators in a plant. Aside from the economic benefits of faster design and commissioning times and reduced cabling costs, fieldbus systems also allow a level of diagnostic control. Device level buses require intelligence to communicate and, using microprocessors, a few lines of diagnostic code can be written for almost no extra cost. The nature and location of existing or potential system faults can be transmitted by the fieldbus. A third benefit of the fieldbus is its flexibility. Buses are designed to be changed quickly, and most of their components can be changed or reused.

The fieldbus will be to the DCS as the PC has been to the mainframe computer. The large DCS marshaling cabinets will shrink as sets of fieldbus network bring in thousands of field variables (in place of thousands of pair of wires). Equipment rooms need not be so large. The lowest level control migrates to the field network. Control that remains in the DCS will be a higher-level area, inter-area, and optimization control. The DCS will be more of an information manager and engineering tool providing consolidated reporting, computer-assisted configuration, archiving, and maintenance facilities. High-speed fieldbus networks will eventually allow a mixture of speciality companies to operate together. Both the fieldbus user layer and the object linking and embedding (OLE) for process control (OPC) specification will be enabled for this. Just as the minicomputer enabled centralized control and the microprocessor enabled distributed control, advent of fieldbus specifications has enabled the onset of fieldbus control systems. Fieldbus is not a device communication standard. It is the next process control architecture.

References

Boland, M., Boost plant performance through upgraded instrumentation, Chem. Engg. Sept., 1995, 80–87.
Bowden, R., HART, Field communication protocol, IInd (Ed.), Fisher-Rosemount, May 1999.
Olsson, G. and G. Piani, Computer System for Automation and Control, Prentice-Hall International, U.K. 1992.
Polke, M. (Ed.), Process Control Engineering, VCH, New York, 263–281, 1994.
Santori, M. and K. Zench, Field bus brings protocol to process control, IEEE Spectrum. March, 60–64 (1996).
Seal, A.M., Practical Process Control, Arnold Press, U.K., 1998.
Technical literature supplied by Fisher-Rosemount (India) Ltd., 1998.
Technical literature supplied by Siemens Ltd., 1998.

Review Questions

1. What is a smart sensor? Give the advantages of using smart sensors.
2. Discuss with a diagram the details of a smart temperature transmitter.
3. Discuss with a diagram the details of a smart differential pressure transmitter.
4. Discuss with a diagram, the details of a smart control valve positioner.
5. What is a fieldbus? What are the advantages of using fieldbus than using 4–20 mA current signal.
6. What is a HART protocol?
7. Explain what is a device description language.
8. Explain the details on a PROFIBUS.
9. Give the details of a foundation fieldbus.

7

Programmable Logic Controllers (PLCs)

7.1 Sequential Control

In the process industry and in industrial automation there is a wealth of application of switching circuits. During 1970s programmable logic controllers (PLCs) became more common, and today sequencing is implemented in software. PLCs were originally developed for use in discrete manufacturing applications. The batch process control requires sequencing and interlocking schemes similar to discrete manufacturing, and a PLC can easily replace the relay's and tenor drum control common to many batch/filtration systems. A PLC can also substitute for process operators in certain areas. For example, a PLC can automatically start a sequence when material is present, shut down when it is not, and initiate special cycles such as cleaning on pre-programmed cycle counts or other measurement criteria. PLCs are often thought of as being for batching operations, packaging, and material handling. While DCSs are associated with continuous flow systems or complicated loop configurations, traditionally many PLCs have been used in plant control applications where they provide interlock logic and alarm monitoring of purely digital signals. In such applications, inputs would typically come from digital sensors such as micro switches, and outputs would drive digital devices such as relays and indicators.

The typical decisions and calculations may be simple in detail, but the decision chains in large plants are very complex. Sequencing networks operate asynchronously i.e., the execution is not directly controlled by a clock. The chain of execution may branch for different conditions and concurrent operations are common.

Both an electric switch or relay contact and a valve intended for use in logic circuits are binary and are designed to operate in the on/off mode. A binary variable is represented as a variable having values 0 or 1. For a small contact, relay circuit (labelled X) the statement $X = 0$ means that the element is open and $X = 1$ means closed. Many switches are realized by solid state gates, but electro-mechanical relays are still in use in many applications. Relays remain, a necessary interface between the control electronics and powered devices.

Relay circuits are either normally open (n. o.) or normally closed (n.c.), where 'normally' refers to the state in which the coil is not energized. Relays can implement elementary circuits such as AND and OR as well as SR flip flops. The relay symbols are shown in Fig. 7.1

Advantages to relay output switching: contacts for giving to a temporary overload; immune to false trips from electrical noise; little voltage drop across contacts; no restrictions when connecting

Fig. 7.1 Relay symbols for n.o., n.c. contacts and relay coil

in serial or parallel configurations; definite ON or OFF state; with contacts physically open; cheap price; contacts generate little heat; and no leakages.

Limitations of relay output switching: mechanical switching is slower; mechanical life is limited by demands of the load and the contacts; require 50 milliamps or more to energize; subject to contact arcing or welding; subject to contact bounce and cannot be completely sealed.

Advantages of solid state switching: fast switching speed; high reliability and almost infinite life; low power required to energize; no contact arcing; little or no switch noise; positive switching and no contact bounce; can be hermetically sealed-good for hostile environments.

Limitations of solid-state switching: switch may be destroyed by an overload; tends to fail in the ON-state; heat dissipation; expensive and possibility of false trips from electrical noise.

7.2 Programmable Controllers

Programmable logic controllers (PLCs) are microcomputers developed to handle boolean operations. A PLC produces on/off voltage outputs and can actuate such elements as electric motors, solenoides, fans, heaters and light switches. The basic operation of a PLC corresponds to a software-based equivalent of a relay panel. A PLC can also execute operations such as counting, delays and timers. A single PLC can replace hundreds of relays. Many modern PLCs also accept proportional signals and they can perform simple arithmetic calculations and handle analog input and output signals and replace PID controllers. Programmable logic controllers imitate relay acion through software. PLCs have the following features: (i) they are rugged and designed to withstand vibrations, temperature, humidity and noise. (ii) the interfacing for inputs and outputs is inside the controller and (iii) they are easily programmed and have an easily understood programming language. The programming is concerned with logic and switching operations.

One of the advantages of PLC is that they are modular. PLCs being event-driven makes them different from general purpose computers, which are driven largely by stored information. The original PLCs are nothing more than mimic relay logic. But as solid state and microprocessor technique became more advanced, the PLCs become more powerful. The original languages for PLCs served only to replace relay logic. Later languages were enhanced with such added refinements as arithmetic, branching, data manipulation and data transfer.

The availability of smaller and more powerful processors, and microprocessors allowed size reduction and increased capability of all PLCs. The smaller units still remain basically sequence controllers, but the larger units are soon capable of handling reasonable level of analog control, along with many other facilities. Two types of PLCs are available, the small, stand alone limited functions, low cost PLCs, and the powerful, networked, multifunction PLCs.

The PLC first came into the markets at the beginning of the 1970s as a replacement for relay circuits. Since then, these systems have been continuously developed, supported by the progress in microelectronics. For that reason, the present systems are much more powerful than those which were originally named PLC. Therefore, the terms PLC, process computer, microcomputer, etc., often overlap one another. Technically, these systems are often indistinguishable. In addition, there are often many similarities with regards to their functions. With most PLCs it is possible to process simultaneously a number of bits, e.g., 8 bits (rather than just 1 bit such as a switch signal). This is known as *word* processing. Thus it is possible to connect an A/D converter to the PLC and thus evaluate analog signals supplied by the process. PLCs possess convenient interfaces by means of which they are able to communicate with overriding systems (e.g., control computers) or link up with systems of equal status (e.g., other PLCs).

There are PLC systems available for small, medium, and large control tasks. The required number of inputs and outputs, i.e., the sum of all sensors and actuators, and the necessary program memory capacity generally determine whether a control task is to be designated as large or small. To facilitate matching PLC systems to requirements they are generally offered as modular systems. Input and output modules are obtainable as compact external modules or as plug-in-electronic cards, generally with 8 or 16 inputs/outputs. The number of such modules to be coupled depends on the size of the control task.

PLCs are at home in a real-time environment. A real-time environment is one which decision and actions must be performed fast enough to respond to changes in the real world. Many standard programmable logic controllers do not have permanent operator console, but are designed to be connected to portable keyboards for programming, and then immediately disconnected. Once they are programmed, these controllers are allowed to run alone. The user interactive will have I/O modules designed for operator interface and durable enough for their environment.

A fundamental difference between PLC and a personal computer (PC) is that the PLC is specifically suited, not adapted, to and for factory use. The PLC is not a disk based system like a PC is. With a disk based system, there is constant checking of what to do next. With a PLC the answer of what to do next is inherent; there is no consulting. The PLC program is stored in battery-packed RAM or EPROM. The electrical fields generated in factories from induction motors or heaters do not affect it. There is a custom silicon in PLCs-counters and timers, for example, which is not found in a PC.

The important difference between a PLC and a PC is that the PLC contains the operating system programs and application programs in ROM memory. The operating system does not have to load an application program, as it too already is in ROM. It would be difficult to differentiate between PLC's 'BIOS', 'operating system', and 'application programing'. Another difference between PLC and PC is that I/O devices for PLC, called I/O modules, plug into a rack alongside the CPU, and are available only from the PLC manufacturers. Screw terminals are provided to connect sensors, actuators, or communication wiring. While typically much more expensive than the equivalent interface card for a PC, they are often cheaper in the long term because the user does not have to learn a new set of interface protocols for each module.

7.3 Advantages of PLCs

Flexibility: The PLC is a general purpose machine that is usually programmed from scratch for process control. It is not preloaded with a dedicated program of loops and screens, then configured to an application. Thus a PLC-based process control system allows a more customized final product. Systems Engineers often use packaged Operator Interface (OI) software to prepare standard PLC process control loops and screens. This shortens programming time while retaining end-user flexibility. Standard libraries contain ratio/cascade and single loop controls, scaling and linearization, recipe handling, and records storage/retrieval. These are inserted into PLC programs using cut/paste functions of word processor-like programming tools. Standard OI templates provide single loop, group, and overview displays, alarming, and trend displays.

Maintainability: PLCs are typically maintained by electricians while computers are maintained by engineers and programmers. With PLCs, plant operating personnel seldom require engineering assistance after startup.

Connectivity: PLC-based process control systems offer superior communications with external devices such as driver, temperature controls, islands of automation, and so on. A PLC has numerous ports that can be dedicated to third-party devices; these ports provide an easy path for information flow in and out of the control system.

Ruggedness and reliability: PLCs don't need to be housed in air-conditioning environments. They have battery-backed memory for data storage, and offer hot-standby configurations of redundant CPUs and I/O cabling.

7.4 Basic Structure of PLC

Figure 7.2 shows the basic structure of a PLC. The inputs are read into the input memory register. This function is already included in the system software in the PLC. An input/output register is often not only a bit but byte. Consequently one input instruction is capable of governing the status of 8 different input ports.

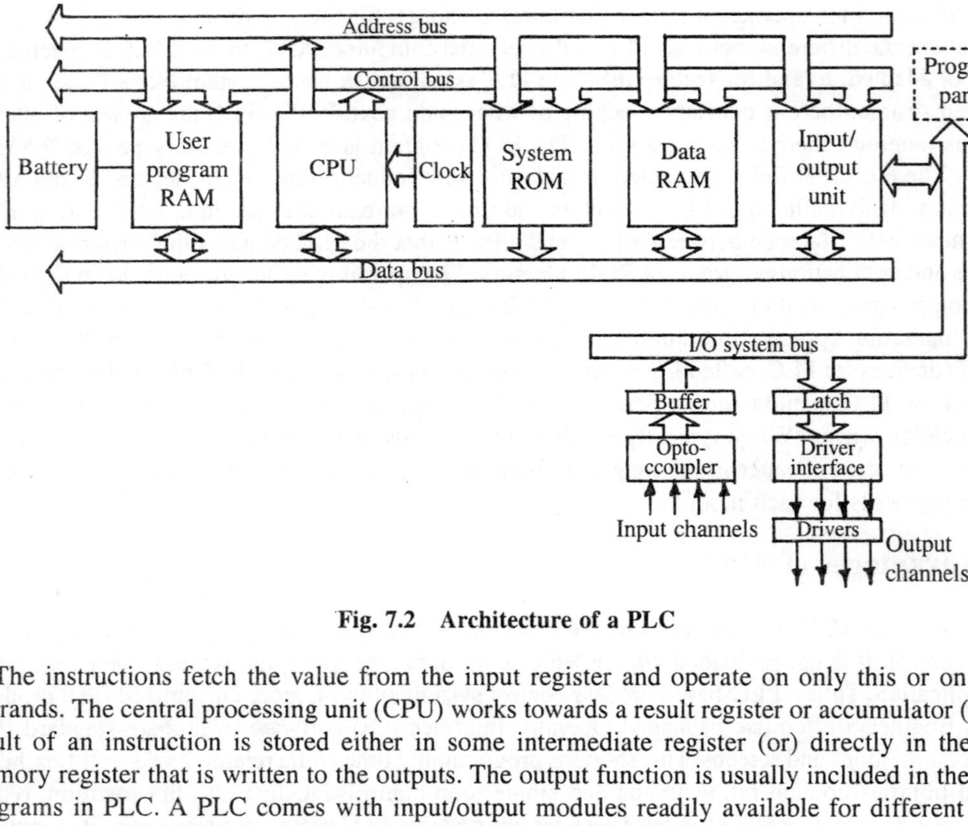

Fig. 7.2 Architecture of a PLC

The instructions fetch the value from the input register and operate on only this or on several operands. The central processing unit (CPU) works towards a result register or accumulator (A). The result of an instruction is stored either in some intermediate register (or) directly in the output memory register that is written to the outputs. The output function is usually included in the system programs in PLC. A PLC comes with input/output modules readily available for different voltage levels and can be easily interfaced to hundreds of input and output lines.

The input/output unit provides signal conditioning and isolation functions so that sensors and actuators can be generally directly connected to them without the need for other circuitry. The input voltages are 5 and 24 V. Common output voltages are 24 and 240 V. Outputs are often specified as being of relay-, transistor- or triac-type. The relay isolates the PLC from the external circuit and can be used for both AC and DC switching. The transistor type of output uses a transient to switch current through the external circuit. Optoisolators are used with transistor switches to provide isolation between external circuit and the PLC. The transistor output is only for DC switching. Triac outputs can be used to control external loads which are connected to ac power supply.

Speciality modules: Speciality modules fall into the category of smart modules. A smart module contains its own microprocessor. Speciality modules are designed to provide a specific function. Some of the speciality modules are BASIC modules, communication modules, remote I/O sub scanners, RTD input modules, Stepper motor control modules, and Thermocouple/millivolt input modules.

Communication modules: Here we have two classes of modules: ASCII I/O module and RS-232C interface modules. ASCII I/O modules allow the interfacing of bar code readers, meters, printers, and data terminals to a PLC. ASCII modules, which accept only valid ASCII data, are not used as extensively as they once were. RS-232C interface communication modules are available that reside in a PLC chassis and enable us to connect a PLC to telephone lines using a modem. PLCs connected to a phone line allow central control room operators to examine ladder programs to modify or edit program operation at remote PLCs sites. Today many remote oil, gas and waste water treatment units applications are unmanned. Remote access by way of phone lines saves maintenance people from driving to remote sites each time a PLC encounters a problem or a program change is necessary.

Relay circuits are usually drawn in the form of ladder diagrams. They also serve as a basis for writing programs for programmable controllers. A ladder diagram reflects a conventional wiring diagram. A wiring diagram shows the physical arrangement of the various components (switches, relays, motors, etc.) and their interconnections, and is used by electricians to do the actual wiring of a control panel. Ladder diagrams are more schematic and show each branch of the control circuit on a separate horizontal row (the rungs of the ladder). They are meant to emphasize the function of each branch and the resulting sequences of operations. The base of the diagram shows two vertical lines, one connected to voltage source and the other to ground (Figs. 7.3 and 7.4).

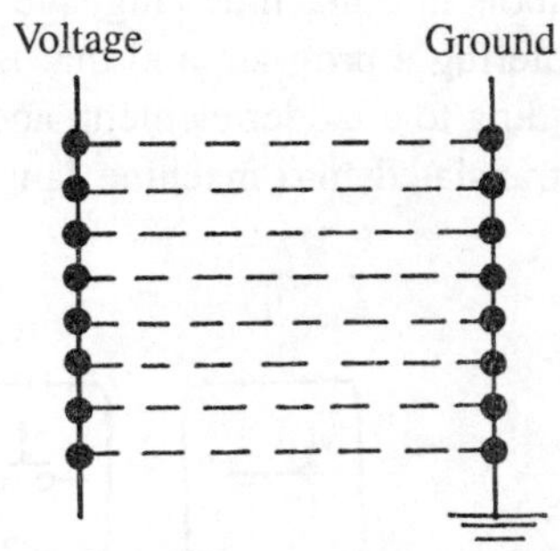

Fig. 7.3 Framework of a ladder diagram

7.5 Basic Instruction and Execution

The programming of a PLC consists of mainly the definition of sequences. The input and output functions are already prepared. The instructions from a ladder diagram, a logical gate program or boolean expressions are translated to machine code. At execution, the program memory is run through in a cyclic manner in an infinite loop. Every scan might take some 15–30 ms in a small PLC and the scanning time is approximately proportional to the memory size. In some PLCs the whole memory is always scanned even if the code is shorter. In other systems, the execution stops at the end statement that concludes the code; thus the loop time can be made shorter for short programs. The response time of the PLC depends on the processing time of the code. While the instruction and the

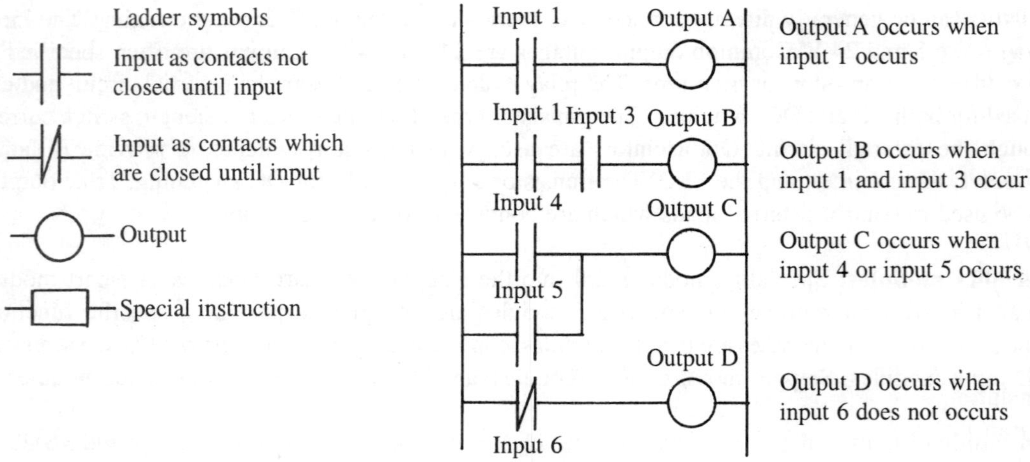

Fig. 7.4 Ladder logic diagram

output execution are executed, the computer system cannot read any new input signals. The ladder diagram can be considered as if every rung of the ladder were executed at the scan time. Thus it is not possible to visualize the ladder being executed sequentially on a row-by-row basis. The execution has to be very fast compared to the time scale of the process under control.

Each horizontal rung on the ladder in a ladder program represents a line in the program and the entire ladder gives complete program in 'ladder language'. The programmer can enter the program into the PLC using a keyboard with the graphic symbols for the ladder elements (Fig. 7.5), or using a computer screen and a mouse to select symbols, and the program panel or computer then translate these symbols into machine language that can be stored in the PLC memory. There is an alternative way of entering a program and that is to translate the ladder program into mnemomics, each code corresponding to a ladder element, and then enter these into programming panel or computer. These are then translated into machine language. The mnemonics used by different PLC manufacturers differ.

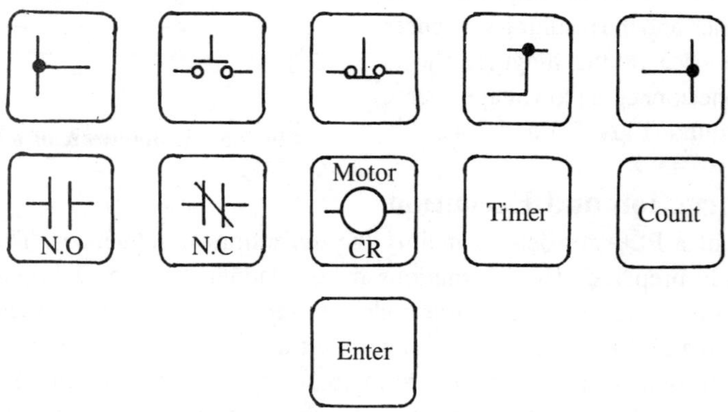

Fig. 7.5 Key board of a typical PLC depicting the components used in constructing ladder diagrams

There are four fundamental instructions:

LD, LDI: a number from the computer input memory is loaded (LD) or inverted (LDI) before it is read into the accumulator (A).

AND, ANI: an AND or AND inverse instruction executes an AND logical operation between A and an input channel, and stores the result in A.

OR, ORI: an OR or OR inverse instruction executes and OR logical operation between A and an input channel, and stores the result in A.

OUT: the instruction outputs A to the output memory register. The value remains in A, so the same value can be sent to several output relays.

Note that the logical operations may be performed on bits as well as on bytes.

Translation from a ladder diagram to machine code

The translation from ladder diagram to machine code is illustrated by Fig. 7.6. A logical sequence or ladder diagram is often branched. Then there is a need to store the intermediate signals for use later. This can be done with special help relays, but in a PLC it is better to use two instructions *orb* (OR block) and *anb* (AND block). They use a memory stack area (Last in First Out) in the PLC to temporarily store the output.

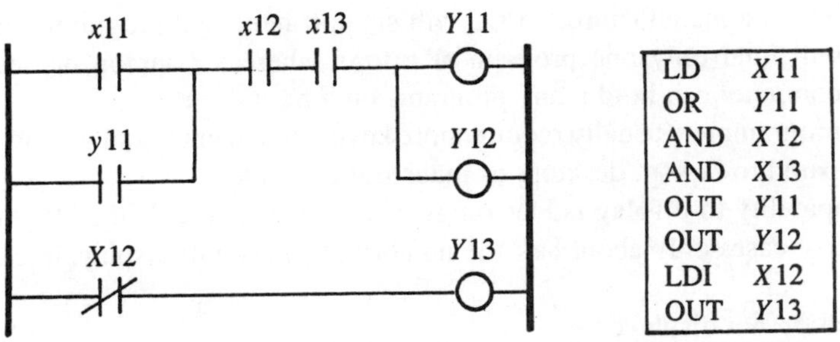

Fig. 7.6 Translation of a ladder diagram to a machine code

7.6 Programming a PLC

Programming devices: The program device is a device that is used to load the program into the processor. It can be a personal computer, or a dedicated piece of equipment made by the controller manufacturer. It is the program that makes the PLC flexible. Normally the program loader is not needed during normal operation once the system has been started up and debugged.

A PLC is programmed via an external unit; this unit is not needed for the PLC operation and may be removed when the PLC is in operation. Programming units range from small handheld portable units, sometimes called 'manual programmers' to personal computers. A manual PLC programmer looks like a large packet calculator, having a number of keys and a simple display. Each logic element of the ladder diagram is executed separately, one at a time, with series or parallel connections achieved by using keys for AND, OR, and NOT. Although typically their editing and display capabilities are limited, their main advantages is their capability. Then can be taken into industrial environments, and programs can be changed on-site. Included in their keyboards you will always find the standard symbols of relay logic. A sophisticated programmer consists of a personal computer with a graphic display. The display shows several ladder diagram lines at a time and also indicate the power flow within each line during operation to make debugging and testing simpler. Other units are reprogrammed using logical gates instead of a ladder diagram.

Most desktop units have a full typewriter keyboard, as well as built-in-monitor. Desktop unit might also have a row of functions keys, called soft keys, that enable you to select among several display, and programming modes. For example, one of the keys calls up a ladder grid on the CRT on which you can develop your program graphically.

7.7 Advantages of a Smart, Handheld Programming Terminal (HHT or HHP)

1. Easy transfer of the PLC program to the HHT for editing or trouble shooting
2. Esay transport of a program to the field to update a current machine's program
3. Rugged and industrially hardened for the factory environment
4. Cheaper than a laptop computer
5. Easy to use and easy to learn; no software required
6. Compact size; some will fit in a packet
7. Program storage
8. Monitor resident PLC program for trobleshooting

Limitations of HHT

1. Not all CPUs in a manufacturer's PLC will support handheld programming.
2. An HHP will hold only one program at a time whereas a laptop or desktop personal or industrial computer can hold many programs on a hard drive.
3. Handheld programmers usually require more keyboards to enter and get the same information when compared to laptop, desktop, or industrial computer.
4. Limited capability to display ladder rungs due to screen size. If the HHP displays rungs of logic, in most cases only about half the number of rungs will display, in contrast to a laptop computer.
5. Documentation not displayed.
6. If there are different manufacturers' PLCs in the field, more than one HHP will be needed.
7. If the battery life is over, a program stored in memory will be lost.

Advantages of software programming

1. Personal computer monitor can display multiple rungs of logic.
2. Personal computer monitor allows a maintenance individual to monitor multiple program rungs for trouble shooting purposes. Scrolling through a program rung by rung is as simple as pressing the up and down key.
3. Program can be stored on the computer's hard drive.
4. Programs can be transfered from the hard drive to a floppy drive and then it can be transported.
5. Rung comments, instruction comments, or symbols can easily be added to any rung and they can be displayed with the related text.
6. Personal computer software offers cut-and-paste features for easy editing.
7. Data table can be easily monitered.

The CPUs all have operational switches, some of which require a key to prevent unauthorized personnel from running a turned-off processes. The key-typed switch also can prevent unauthorized alterations to the operating program of the system. Typical switch positions are:

- off—system cannot be run or programmed
- run—allows the system to run but no program alterations can be made

- disable—turns all outputs off or sets them to the inoperable state
- monitor—turns on screen that displays operating information
- run/program—system can be run, and program modifications can be made while it is running
- off/program or program—system cannot run but can be programmed or reprogrammed.

The programmer/monitor: There are larger programmer/monitors with large cathodic screens. The large monitor screen shows as entire circuit of one to five rungs. There are small hand held programmers/monitors with small display windows. The difference in display size is directly related to cost. The screen shows only one part of the circuit at a time.

The programmer/monitor (PM) is connected to the CPU by a cable. After the CPU has been programmed, the PM is no longer required for CPU and process operation and can be disconnected and removed. We may need only one PM for a number of optional CPUs.

7.8 Input/Output Modules

We get the information in and out of the PLC through the use of input and output modules. The input module terminals (Fig. 7.7) receive signals from wires connected to switches, indicators, and other input information devices. The output module terminals provide output voltages to energize motors and valves, operate indicating devices and so on. Each module can contain typically 4, 8, 12 or 16 terminals.

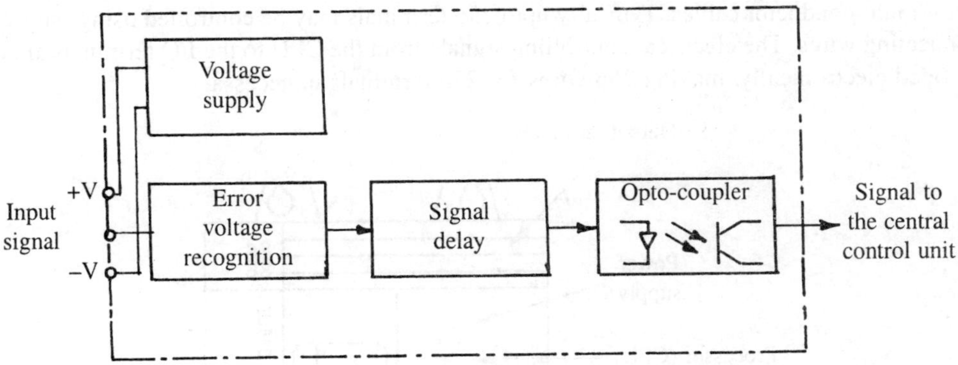

Fig. 7.7 Elements of an input module

The module contains an error voltage recognition facility, which is activated when the input voltage exceeds specified tolerance limits. A trigger circuit with signal delay (added in sequence) ensures that momentary interference peaks are suppressed. Positive and negative supply voltages (+V, –V) are provided at the input terminals for connection of various sensors. Opto-couplers separate internal and external circuits so that interference is not able to penetrate the PLC via conductive lines. The input module contains visual indicators (mainly LEDs) which indicates whether a 1 or 0 is present at the input.

The *requirements at the output side* are amplification of the output signals, and protection against short circuiting. Figure 7.8 shows the basic elements of an output moldule. An opto-coupler is used to separate the internal from the external circuit to prevent interference via conductive connections. The output signals need to be amplified so that final control elements and actuators which require additional current can be directly connected to the PLC. There are three methods of safeguarding the

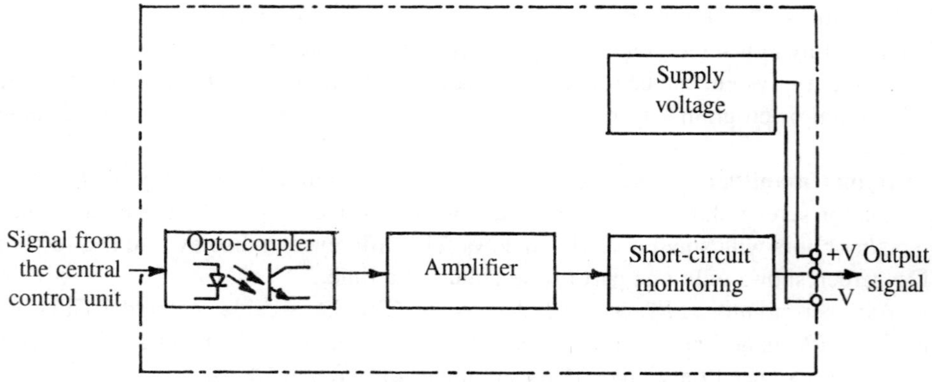

Fig. 7.8 Elements of an output module

outputs against short circuiting: protection by a safety fuse, conditional short-proof outputs (momentary protection) and sustained short-circuit-proof outputs. The signal status of the outputs can be read from the indicators.

In the small systems, the input and output terminals may be included on the same frame as the CPU (Fig. 7.9). In large PLC systems, the input and output modules are separate units (Fig. 7.10). Here the modules are placed in groups on racks. The racks are connected to the CPU via appropriate connector multi conductor cables. Typically upto 256 terminals may be controlled using only 9 to 24 interconnecting wires. The electrical controlling signals from the CPU to the I/O terminals are coded and decoded electronically, making 256 wires for 256 terminals unnecessary.

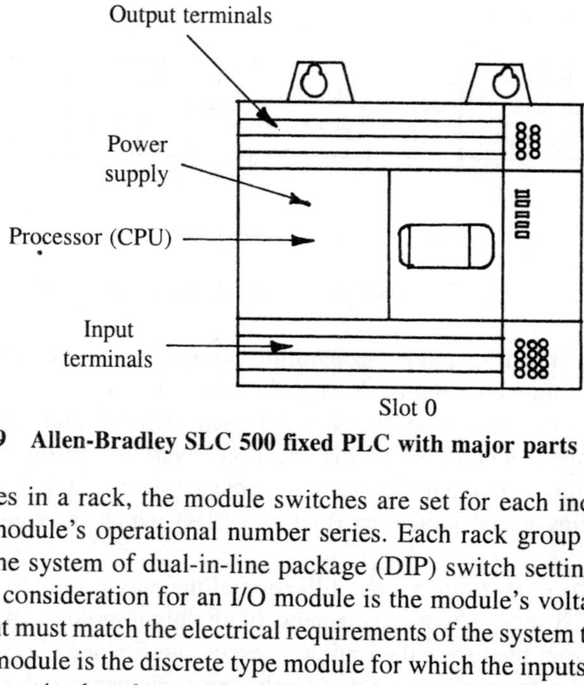

Fig. 7.9 Allen-Bradley SLC 500 fixed PLC with major parts labeled

For multiple modules in a rack, the module switches are set for each individual module. These settings specify each module's operational number series. Each rack group knows what number it should respond to by the system of dual-in-line package (DIP) switch settings.

The most important consideration for an I/O module is the module's voltage and current ratings. Both voltage and current must match the electrical requirements of the system to which it is connected. Most common type of module is the discrete type module for which the inputs are either on or off and the outputs are either energized or de-energized. There are analog type modules which take variable signals with varying values.

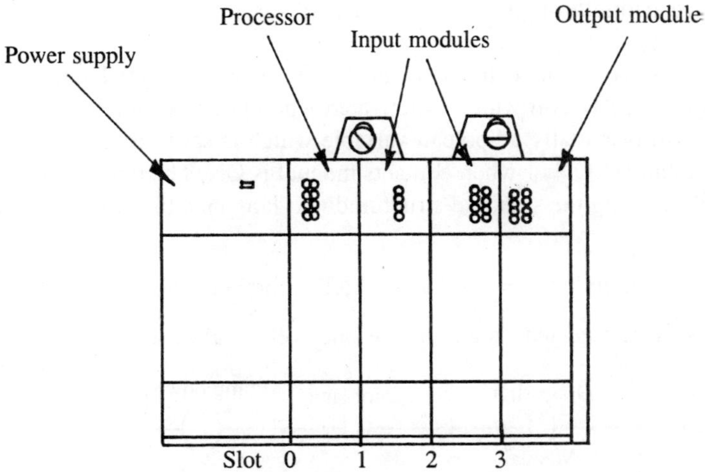

Fig. 7.10 Allen-Bradley SLC 500 modular PLC

Most PLCs continuously execute a cyclic scan. The cycle starts with a hardware I/O system gathering the latest values of all input signals and storing the values in fixed region in memory. The rungs of the ladder program are then executed in a fixed order starting at the first rung. During the program scan, new values of physical outputs, as determined from the logic of the various ladder rungs, are initially written to output memory region. Finally, when the ladder program is completed, all the output values held in memory are written out to the physical output by the PLC hardware in one operation. Generally, the frequency of the ladder program execution depends to a large extent on the length and complexity of the ladder program. Adding more rungs increases the scan time. Usually the critical section of the ladder program will be scanned frequently and switching in different sections of the less critical ladder rungs in the background.

For programming, we use a language, called a source language, that is easily understandable. Like all computers, PLCs have a number of different source languages; these include Ladder logic, Boolean, GRAFCET, BASIC and others. The actual vocabulary used in the programming is called the instruction set.

Ladder logic: Ladder logic is a language that differs from most other computer languages in that it is a graphic language. The programmer simply draws a schematic diagram of the logic on the screen. It resembles the elementary diagrams used in the hard-wired relay logic systems. The instruction set consists of coils, normally open contacts, normally closed contacts, and other familiar sysmbols. The symbols represent the function of the component.

7.9 Ladder Diagram

Ladder diagram is based on a technique used to design logic using relays. A ladder diagram always has left hand vertical power rail that notionally supplies power through contacts spread out along horizontal rungs. Each contact represents the state of a boolean variable. For example, the contact may depict the state of a limit switch of heat sensors. When all contacts in a horizontal rung are made, i.e., in the 'true' state, power can flow along the rail and operate a coil on the right of the rung. Contacts which are normally open present a 'true' state when the contacts are closed. Conversely contacts that are normally held closed, present a 'true' state when the contacts are opened. By arranging alternative paths through a network of contacts it is possible to describe various logic connections.

Figure 7.11 depicts the main features of the ladder diagram (LD) language. In this example, pumpdown represents a relay coil which is activated when power is able to flow from the left hand power rail. Pumpdown is actually a boolean variable which is set to true state either when contacts Drainshut and Doorshut are ON or when contacts manual is ON. The contacts represent the status of the boolean variables. Using the standard structured text language the ladder diagram in Fig. 7.11 is equivalent to:

Pumpdown: = (drainshut AND doorshut) OR manual;

The network that is associated with the logic for one coil is called a 'ladder rung'.

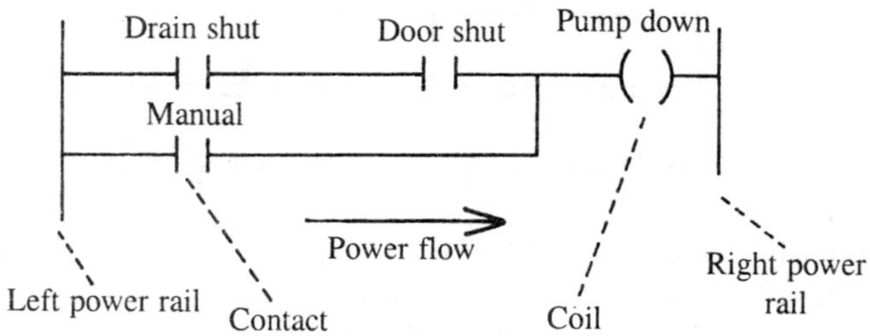

Fig. 7.11 Ladder diagram concept

In ladder diagram a contact is assumed to have ON or OFF (for a normally open contact these states are sometimes referred as OPEN and CLOSED) which are equivalent to boolean states of 1 (true) and 0 (false). A contact always represents the status of the associated boolean variable. A contact provides a 'read only', access to the state of the boolean variable; it cannot be used to change the value of the variable. On the other hand, a coil provides 'write only' access and can only be used to update the states of the associated boolean variable when notional power flow can occur via contacts from the left power rail. Ladder diagrams are quite understandable even if the right power rail is not shown.

7.10 Relay Logic

From the nature of electrical circuits, contacts in a horizontal rung represent logical AND operations. Alternative paths from the main rung provided by vertical paths represent logical OR operations.

In Fig. 7.12, the ladder diagram logic is equivalent to the following structured text:

X1: = B1 AND B2 AND B3

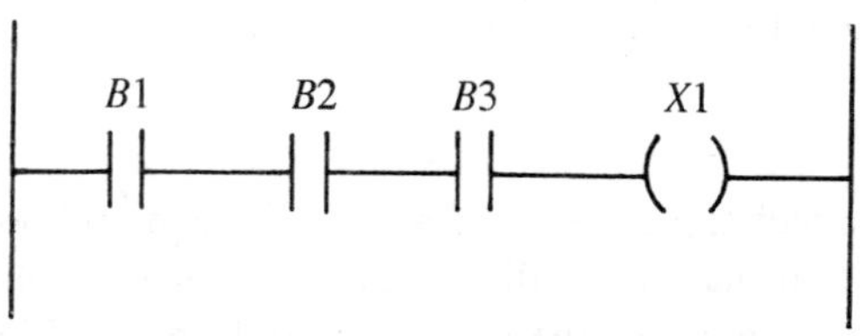

Fig. 7.12 Ladder diagram AND logic example

The use of alternative power flow paths via other horizontal rungs connected by vertical paths provides logical OR operations.

In Fig. 7.13, the addition of alternative power paths modifies the logic. We have:

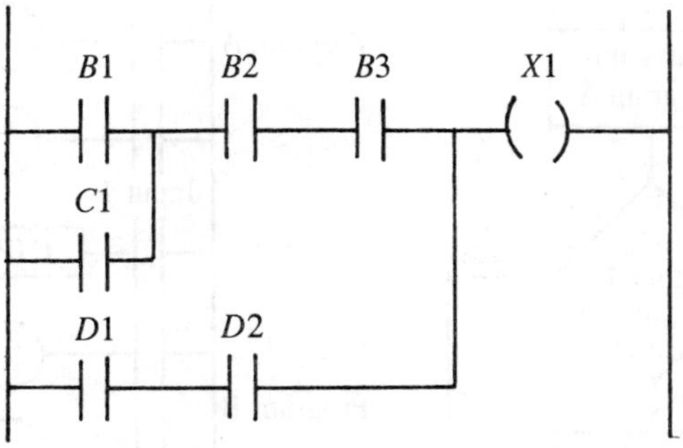

Fig. 7.13 Ladder diagram OR logic example

X1: = (B1 OR C1 and B2 and B3) OR (D1 and D2)

Timer: The timer circuit is specified by starting the interval to be timed and the conditions or events that are to start and/or stop the timer. They are commonly regarded as relays with coils which, when energised result in the closing or opening of input contacts after some preset time. Figure 7.14 shows part of a program involving a delay-on timer. When there is an input, the timer is energised and starts timing. After some preset time the contacts associated with the timer close and the output occurs. Timers can be linked together to give larger delay times than is possible with just one timer. Figure 7.15 shows part of a program involving a jump.

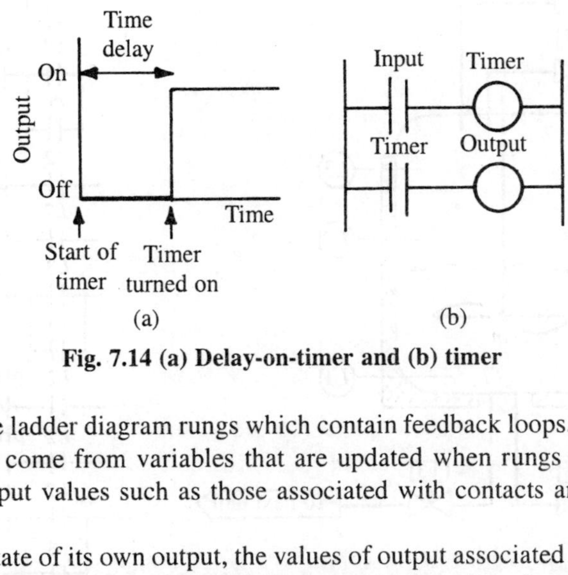

Fig. 7.14 (a) Delay-on-timer and (b) timer

It is possible to create ladder diagram rungs which contain feedback loops, i.e., where one or more values used in contacts come from variables that are updated when rungs are executed. In ladder diagram, all external input values such as those associated with contacts are gathered before each rung is evaluated.

If the rung uses the state of its own output, the values of output associated with a contact is always the value resulting from the previous evaluation. However, if the value of output is used in any of the following rungs, the latest evaluated state is used.

7.11 Example 1

Consider a neutralization system with the piping and instrumentation diagram (P and ID) in Fig. 7.16a. Where a certain amount of solution is added to a tank, heated and chemically treated, and then sent out to the next tank. The sensors T1 and pH1 indicate whether or not the solution has the correct

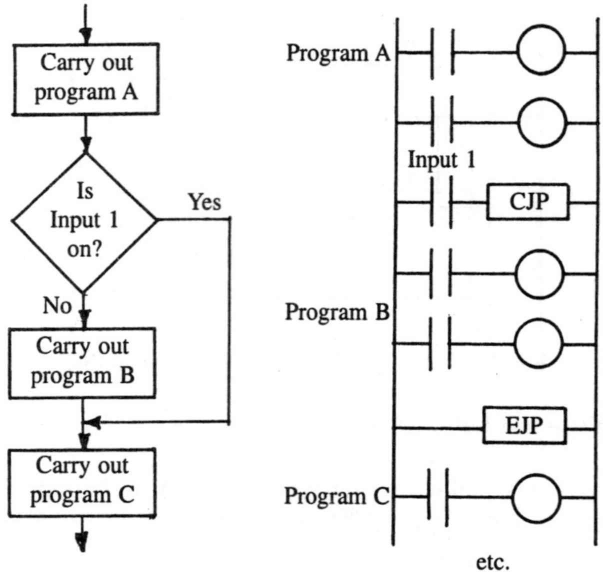

Fig. 7.15 Jump operation in a PLC

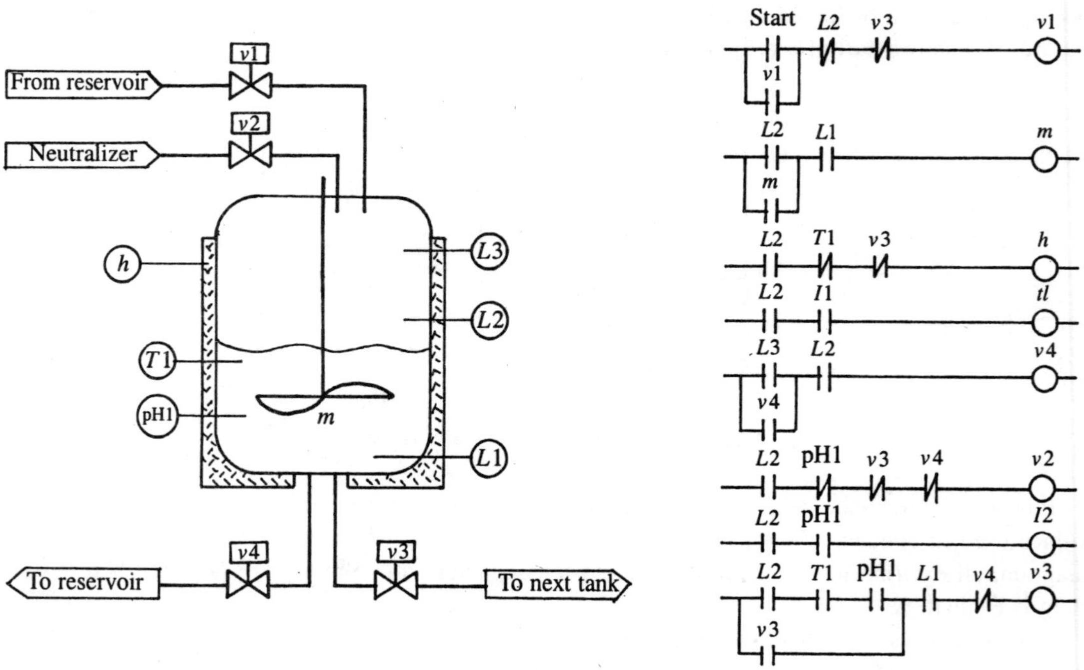

(a) (b)

Fig. 7.16 (a) Block diagram of a neutralization system (Falcione and Krogh, 1993); (b) Ladder logic used to control the example system in Fig. 7.16a (Falcione and Krogh, 1993)

temperature and pH respectively. When T1 and pH1 are both activated, the neutralization is complete. The level switches L1, L2 and L3 are activated whenever the level in the tank is at or above a given level. The neutralization process is to proceed as follows:

1. Initially, all the valves are closed, the mixer and heater are off, and the reaction tank is empty.
2. When the start button (not shown in Figure) is pressed, open V1 until L2 is activated. This fills the tank with the solution to be neutralized.

3(a). When the solution level rises above L2, start the mixer m. When the level drops below L1, stop the mixer.

3(b). Whenever the temperature of the solution is below a preset point, energize the heater h.

3(c). Whenever the pH of the solution is unbalanced, open V2 to add neutralizer.
(It should be noted that steps 3(a) to 3(c) are to occur simultaneously)
If the tank becomes full, indicated by the activation of L3, close V2 to stop the inflow of neutralizer. Next, open V4 to reduce the level of the solution to the point indicated by L2. Then close V4 and proceed with step 3(c).

4. When both the temperature and pH of the solution are correct, de-energize ther heater and close V2. Then open V3 to drain the tank. When the tank is empty, indicated by the deactivation of L1, close V3 and proceed with step 1.

In addition to the above control sequence, two indicator lights, I1 and I2 (not shown in figure), are regulated. Light I1 (I2) should turn on whenever the solution level is above L2 and the temperature (T1) has reached the preset value.

The desired sequence of operations is activated by the relay ladder diagram (Fig. 7.16b). The logic is more compactly represented in Boolean equation form (We use · for AND logical operation; + for OR; ! for NOT):

V1 = (Start + V1) · !L2 · !V3
m = (L2 + m) · L1
H = L2 · !T1 · !V3
I1 = L2 · T1
V4 = (L3 + V4) · L2
V2 = L2 · !pH1 · !V3 · !V4
I2 = L2 · pH1
V3 = (L2 · T1 · pH1 + V3) · L1 · !L4

As discussed in previous chapter that in a decentralized control system the sensing, actuating and processing sections of subsystems are connected to a master computer which evaluates signals applied by each sub process and sends back control signals. These subsystems can be realized in different ways: e.g., a PLC or a single-board computer could be used as the control system for the sub process. A process computer, personal computer, or mainframe can be used as a master computer.

7.12 Example 2

Figure 7.17 shows a simple control task from chemical process technology as an example of the networking of various systems. In a liquid mixing plant, there different liquids, A, B and C, are mixed in a specified proportion. Filling is affected by three valves, V_A, V_B and V_C, and the mixture is extracted via one valve, V_D. Before the mixture is removed, it is heated to pre-specified reference temperature T. A PLC controls the valves and the mixer. The temperature of the mixer is adjusted to the reference temperature by a two-step controller. An overriding computer prepares the reference values for the mixture addition, the computer stores operational data and downloads the data to the printer as documentation.

The reference values 1 and 2, which determine the mixture ratio, are passed on by the computer to the PLC. The PLC controls the remainder of the sequence. First valve V_A is opened and liquid A is poured into the reservoir. The liquid level rises until the level specified by reference value 1 has been reached. The liquid level in reservoir is determined by a suitable level sensor. If reference value and actual value of level agree, valve V_A is closed and valve V_B is opened. Next, liquid B is poured in until reference value 2 has been reached. Then valve V_B is closed and valve V_C is opened. Liquid C is poured in until the maximum value (reservoir full) is reached. Reference values 1 and 2 determine the mixture ratio A : B : C.

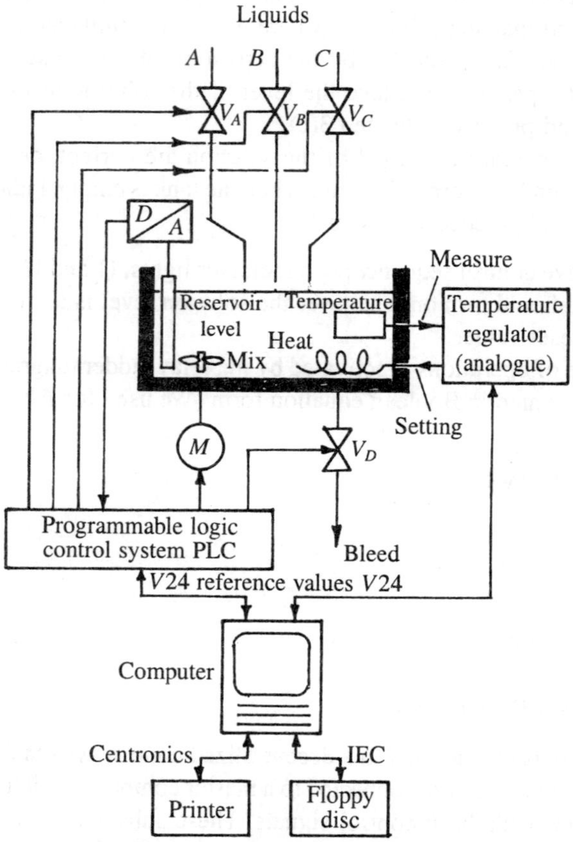

Fig. 7.17 Networking of a PLC, a local controller with a computer (Soloman, 1994)

On reaching the maximum level of the liquid, the mixing motor switches ON and sends on a signal to the computer, which, in turn, switches ON and sends on a signal to the regulator. The regulator adjusts the temperature to the reference value and informs the computer when the reference value has been reached. Next, the computer requires the PLC to switch OFF the mixing motor and open the bleed valve V_D. Once the reservoir has been emptied, the cycle is completed. A new cycle is now started up with the same or modified reference values.

7.13 Boolean Programming

It is written in the form of a listing. The series logic is represented by the Boolean AND function; parallel logic is represented by the Boolean OR function. Normally closed contacts are referred with the term NOT. All logic is in some combinations of these three logic operations: AND, OR, NOT. Unlike Ladder logic, Boolean logic places no limit on the number of contacts that can be put in series or parallel. Concept of OR statement rule needs to be understood first for Boolean programming. The rule explains how the OR statements works with LOAD statement. The OR statement rule: Each time an OR statement is used, that statement is combined with the previous LOAD statement.

Stack register: The stack register is a software device that aids in the assembly of the programming logic. The more complex the logic, the more levels are used. Let us introduce a stack rule: each time a LOAD instruction is used, the logic is pushed down one level into the stack. Each time an AND LOAD or OR load instruction is entered, the logic is pulled up one level in the stack. AND LOAD and OR LOAD instructions are used to combine groups of logic AND fashion and OR fashion respectively. Each manufacturer of PLCs using a Boolean language has a stack register with a given number of levels [4 or 6 or 8 or 16 levels]. For the ladder diagram of Fig. 7.18, we can write the following Boolean programming:

```
LOAD        X1
AND         X2
LOAD        X7
AND NOT     X2
LOAD NOT    X12
AND         X13
OR LOAD
AND         X10
OR LOAD
LOAD        X3
AND NOT     X4
AND         X5
LOAD        X11
AND         X4
AND NOT     X6
OR LOAD
AND LOAD
OUT         X30
```

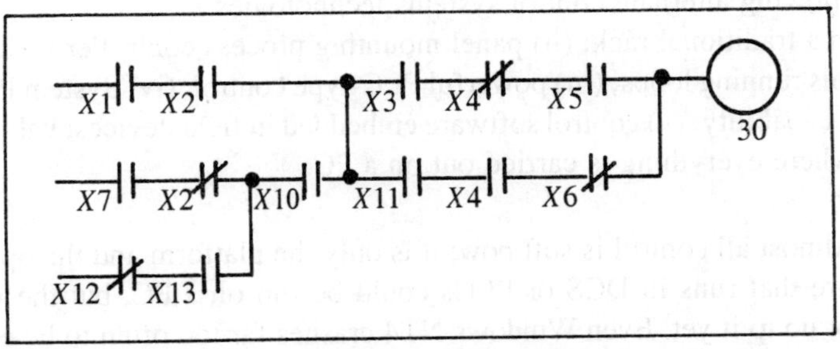

Fig. 7.18 Multi-level logic

7.14 Example 3

Let us discuss an example to illustrate how timers can be used in process control systems to energize and de-energize output devices. The mixing circuit (Fig. 7.19) shows the control system that controls three different steps in the process: fill, agitate, and drain, with a single timing function. When the

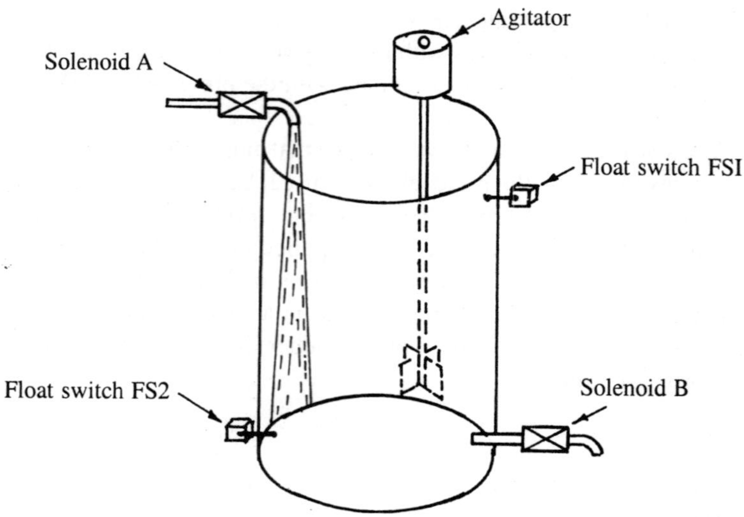

Fig. 7.19 Example for use of timers in liquid level control (Simpson, 1995)

start button is pressed, solenoid A is energized and a batch of liquid enters the mixing tank. Float switch FS1 detects the upper limit of the liquid level and FS2 detects the lower limit. As the tank begins to fill, FS2 closes. When the tank is full, switch FS1 shuts off solenoid A and starts the agitator to mix the liquid. The agitator mixes the liquid for 30 secs and shuts off. Then solenoid B is energized to drain the liquid. After the tank has been emptied, float FS2 opens off solenoid B shuts off. The ladder logic diagram for this problem is shown in Fig. 7.20. In this circuit, both the normally-open and normally-closed sets of contacts for the float switch FS1 are used. When the start button is pressed, control relay 020 00 is energized and solenoid A turns on. As the tank fills, FS2 (110 03) closes. When the tank is full, FS1 (110 02) changes state. This shuts off solenoid A, starts timer 050, and energizes the agitator motor (010 01). After 30 secs, the contacts associated with the timer (050 15) change state, causing the agitate motor to shut off and solenoid B to energize. As the liquid begins to drain, FS1 returns to its normal resting state. When the tank is completely empty, FS2 re-opens and the drain valve (010 02) is shut off.

We have the following alternate control systems technologies:

(i) Mini-DCS in a traditional rack; (ii) panel mounting process controller boxes, with front panel displays and controls running loops; (iii) powerful PLC-type control; (iv) System PLCs with enhanced continuous control capability; (v) control software embedded in field devices; valves and transmitters; (vi) soft control, where everything is carried out on a PC.

Soft contnrol: Almost all control is soft now; it is only the platform and the operating system that varies. The software that runs in DCS or PLCs could be run on a PC, but the windows operating system is really not up to it yet. Even Windows NT4 crashes far too often to be economic in critical applications. But, the trend is certainly for using softlogic. Hopefully, new versions of operating system will be more stable.

7.15 Connecting a Personal Computer Serially to a PLC

The personal computer can be connected to a PLC for software upload, download, or monitoring through a serial cable connection. The serial cable is connected to the personal computer's

COM1 port and the PLC's CPU communication port. The proper serial cable must be used for this purpose.

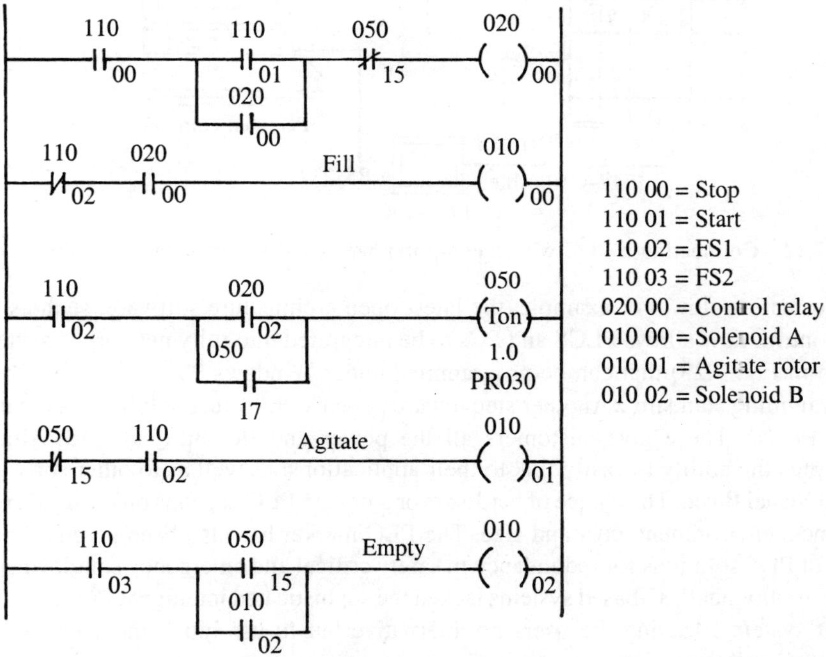

Fig. 7.20 Ladder logic diagram for example of Fig. 7.19 (Simpson, 1995)

In order to communicate to some PLC processors, a special interface converter box needs to be connected between the personal computer's COM port and the PLC controller (Fig. 7.21). The interface converter changes the serial RS-232C signal from the personal computer's serial port to a signal (data highway 485 protocol) understood by the PLC.

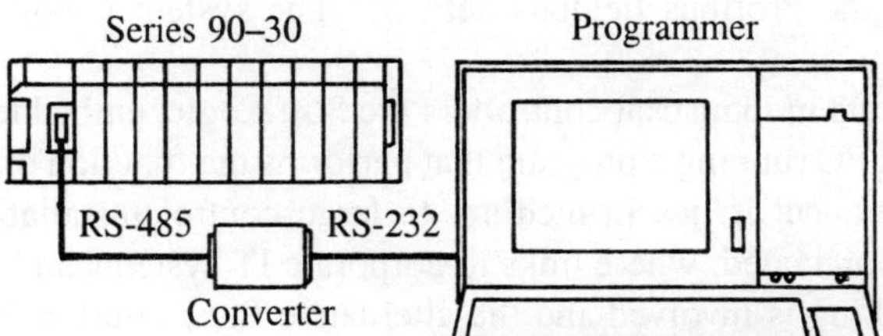

Fig. 7.21 Connecting a PC with a PLC using a RS-422 to RS 232 converter

In some cases, the PLC can be connected to a personal computer through an internal hardware card (I/O interface card) kept in an ISA or EISA expansion slot of the personal computer (Fig. 7.22). The proper interface card meant for the particaular PLC should be used.

The continuing growth in the PLC market is being fueled by developments (refer to Table 7.1 for the architecture evolution of PLCs), such as open architectures, which have resulted in a major step forward in integrating both PLCs and into networked industrial environments, with programs on industry standard personal computers (Fig. 7.10). The result for users is a significant improvement in reducing cycle times and training requirements, while providing far better communications and

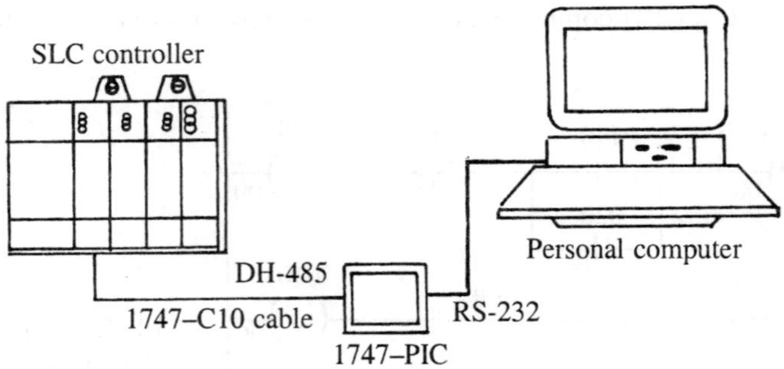

Fig. 7.22 Connecting a PLC with a computer using a PIC converter (Dunning, 1998)

increased information flow. For example, the latest open architecture software, such as GE Fanuc's Complicity control, now allows PLCS and PCs to be integrated into fully networked systems (similar to that shown in Fig. 7.23) that can be programmed under Windows 95 or NT using the latest IEC 1131–3 programming standards. Another step towards open architecture solutions is the development of 'software PLCs'. These give customers all the power and flexibility of MS Windows 32-bit architecture, plus the ability to easily link to their applications; as well as to other software products like Excel and Visual Basic. The choice of hardware or software PLC depends on the user's requirements for performance, environment, cost and size. The PLC market has also been given a boost with the development of PLC solutions for redundancy in safety critical and emergency shutdown applications. Until recently, traditional PLC-based systems lacked the sophisticated intelligent I/O capability required for redundant systems, leaving the users no alternative but to use much more expensive and less flexible non PLC solutions.

Simatic PCS7, the latest programmable system from Siemens uses standard automation components, including networks and controllers, configured via Windows. It makes use of a common database, saving on engineering and commissioning time and costs. Various levels of control management are possible with PCS7, from SCADA up to the top-end DCS. The graphical front end is based on Simatic WinCC software, and the system incorporates the S7–300 and S7–400 families of controllers. Core networking is via the Profibus fieldbus network. The system is compatible with IEC 61131 software.

The latest developments in industrial controllers are Soft Logic, embedded PLCs, and Java-based embedded controllers. A PC running a program that performs the function of PLC is called soft logic. Soft logic has its market, but it lies in medium to large control installations where the operator interface of the PC is appreciated, where links to corporate IT systems and archiving to hard disc is a benefit, where networking is involved and the alternative PLC solution would be quite costly.

An embedded PLC is a PLC stripped down to its essential components and packaged suitably for building into a product needing control facilities. At the extreme it could be a PLC on a chip. More usually it would be packaged on a small PCB sub-assembly, probably with a plastic case, that can easily be built into an end product needing control. Micro PLCs are PLCs with small number of I/O channels designed to control, say, a pump topping up a tank.

Along with self contained PLCs with inherent PID and analogue routines, there is also an emerging trend towards development of coprocessor for PLCs, offering functions such as temperature and PID control. In general, these coprocessors will be industry standard version of high specification personal computers able to run any of the standard operating systems (Windows NT, QNS, or UNIX) or simply running specialized embedded code. This powerful combination of soft PLC and PC topologies will provide an opportunity to run the most advanced forms of model based control, fuzzy logic control,

Table 7.1 Architecture evolution of PLCs (Rohen, 1999)

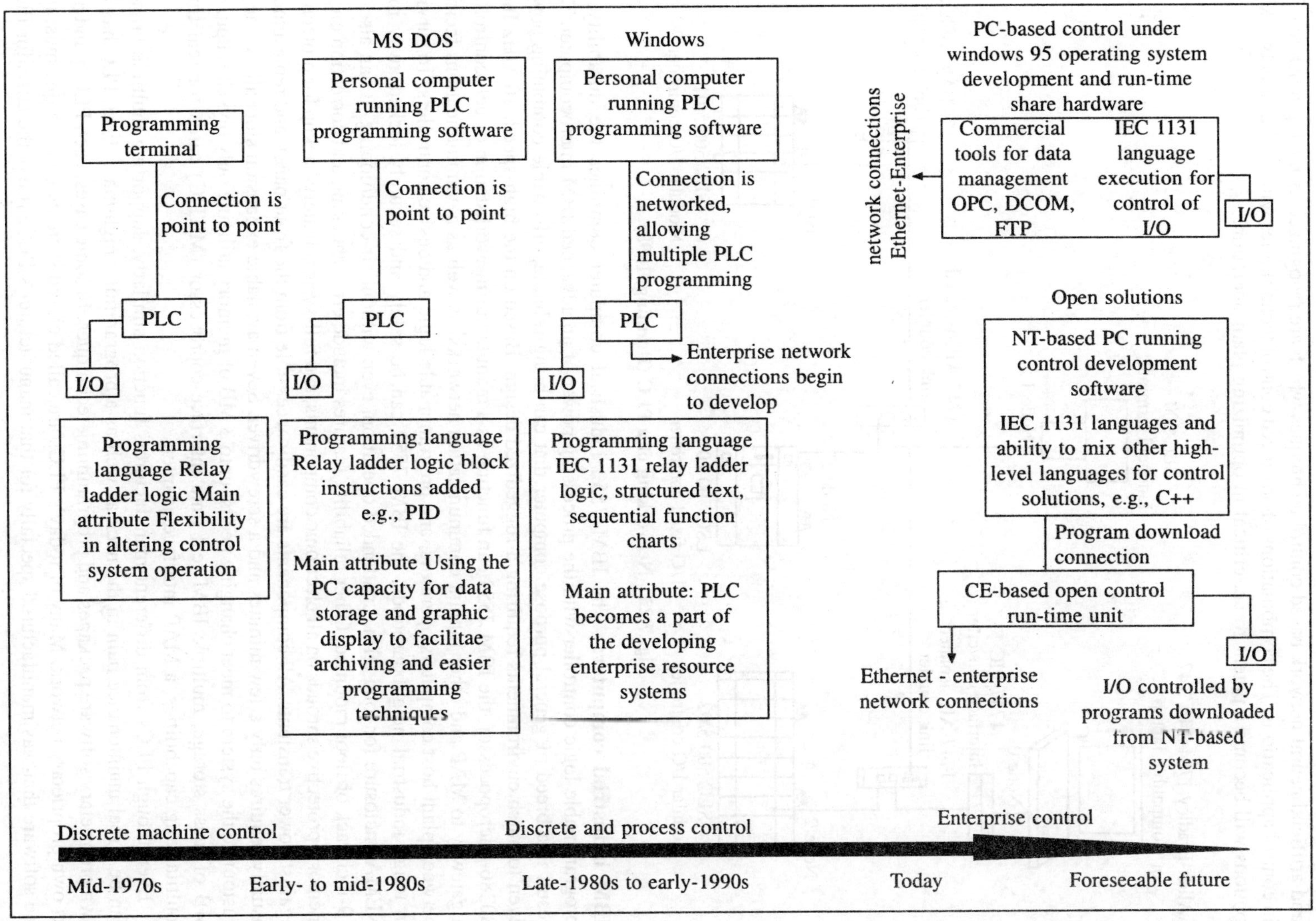

and artificial neural network based control techniques which are recognized as having much to offer in control applications. The combinations of advanced control technologies with the latest networking options will become a potent force, critical in optimizing plant operation.

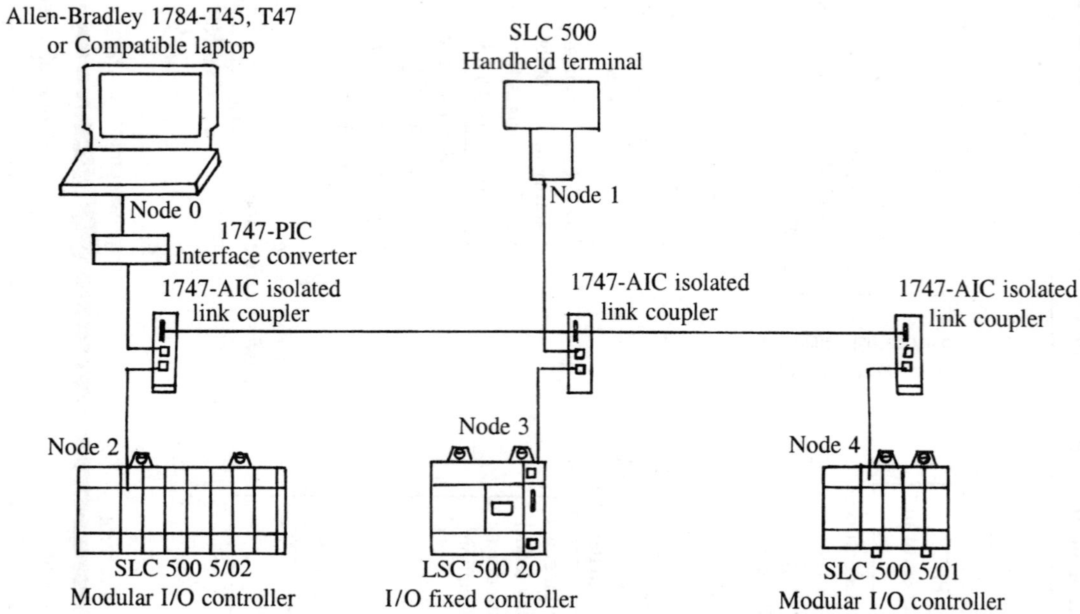

Fig. 7.23 Networking of a PLC (Dunning, 1998)

IBM industrial computer: The IBM 7552 industrial computer combines the durability of programmable logic controller with the processing power of an advanced IBM microcomputer. The result is a breed of general purpose computer that can bring reliable, affordable computing power even to harsh environments requiring a ruggedized design. Based on the high-speed, 10 MHz Intel 80286 micropocessor, the IBM 7552 can function as a manufacturing workstation or cell controller, a gateway to MAP and other inplant communication networks, as well as a communications facility between plant host computers operators, and programmable logic and device controllers. To survive in tough industrial neighbourhoods, the IBM 7552 can be safely and securely locked away in a NEMA enclosure for long term, unattended operation, even without air-conditioning. It can also be 19-inch rack or floor mounted. Other reliability features include error checking and correction on all memory, protective shrouds on all electronic components, and full-system battery backup for protection against power transients. All components are easily accessible from the front panel, and replacement usually requires only a few minutes and a screwdriver. Seven available expansion slots allow one to customize the system to meet changing needs: upto 3 MB of memory utilizing only one slot, upto 40 MB of mass storage, multiple IBM real time interface-coprocessor (MRTIC) cards for real time multitasking capabilities, a MAP interface adapter.

Even though PLCs from different manufacturers function similarly, supporting multiple brands can be a real maintenance pain in the neck as well as a programming nightmare. Each PLC has its own proprietary software packages and programming techniques. In some cases, each PLC brand has its own proprietary network. Many of today's PLCs are called closed systems because one must use the software that was manufactured specially for that manufacturer's PLC. Today the costs for PLC hardware are declining with the emergence of smaller, and faster 'micro' PLCs of less cost, while software is becoming more sophisticated as PLC functionality increases. Hence, the software cost is a larger cost item in a PLC system. The software development, debugging, installation, control system commissioning, trouble shooting and future system modification would be greatly simplified if there were standards governing the development, usability, and reusability of software. The IEC 1131 standard (released in 1982) standardizes PLC programming and consequently makes a program developed on one system usable on other PLC platforms with minimum modification.

Standard IEC 1131–3 defines a consistent set of programming languages for PLCs. The specification consists of four traditional languages and one higher-level programming language. The languages are broken down into two graphical languages, Ladder Diagram and Function block diagram, and two text-based languages. Instruction List and Structured Text, along with flow chart-type programming called Sequential Function Chart Programming.

Summary

PLCs are required for automating sequencing processes. To facilitate matching PLC systems to the requirements, PLCs are generally offered as modular systems. The input/output units provide signal conditioning and isolation function so that sensors and actuators can be directly connected to the I/O module without need for any other circuitry. The programming of a PLC consists mainly of the definitions of sequences. PLCs are programmed in any one of the languages: Ladder logic, Boolean, Grafcet and Basic. A PC running a program that performs the function of a PLC is called soft logic. Embedded PLC is a PLC stripped down to essential components and packaged suitably for building into a product required for control.

References

Bolton, W., Mechatronics, 2nd Ed., Addison Wesley Longman, Singapore, 1999.

Dunning, G., Introduction to Programmable controllers, Delmar Publishers, Singapore, 1998.

Falcione, A. and B.H. Krogh, Design recovery for relay ladder logic, IEEE Control Sys. Magazine, 1993, pp 90–99.

Lewis, R.W., Programming industrial control systems using IEC 1131–3 (revised Edition), IEE Control Engg. Series 50, The Institution of Electrical Engineers, London, 1997.

Loewenberg, M.J., How to choose a control package, Chemtech, 26 (1991) 686–690.

Olsson G. and G. Piani, computer system for automation and control, Prentice-Hall International, U.K. 1992.

Programmable Lolgic Controller, TPC training systems a division of Telemedia, Inc., USA, 1992.

Rohen, D., Tricky buying decision looms, To PC or not to PC?, Intech, Sept. 1999, 68–70.

Seal, A.M., Practical process control, Arnold Press, U.K., 1998.

Simpson, C.D., Programmable logic controllers, Prentice-Hall, New Jersey, 1995.

Soloman, S., Sensors and Control Systems in Manufacturing, McGraw-Hill, NewYork, 1994.

Webb, J.H., Programmable Controllers: Principles and Applications, Merril Pub. Co., 1988.

Review Questions

1. What are the advantages and limitations using relay circuits for sequencing operations?
2. What is a PLC? What are its advantages and limitations over that of a relay based controller?
3. How is a PLC different than a PC?
4. What are the advantages and limitations of handheld transmitters?

5. What are the advantages of software programming over that of HHT?
6. How is a jump operation executed in a PLC?
7. How is a timer used in a PLC?
8. Explain a ladder logic diagram.
9. Explain boolean programming of a PLC with an example?
10. What are the different control system technologies?
11. How do you connect a PLC with a PC?
12. Discuss on networking of PLCs.
13. How are the architectures of PLCs evolved?
14. What are the logic functions used for switches
    (a) in series and (ii) in parallel?
15. Draw the ladder rung to represent (a) two switches normally open and both have to be closed for a motor to operate; (b) either of two, normally open, switches have to be closed for a coil to be engaged and operate an actuator.
16. Design a timing circuit that will switch an output on for 10 sec.
17. Device a circuit that can be used to start a motor and then after a delay time of 80 secs to start a pump. When the motor is switched off there should be a delay of 8 secs before the pump is switched off.

Part 2

Design of Controllers

8

Discrete Time Systems and Control

Introduction to discrete time systems, block diagram representation of computer controlled systems, Z-transform and its inverse for solving discrete time systems, and design of controllers for discrete time systems are presented in this chapter.

8.1 Computer Control Loop

Figure 8.1 shows a block diagram representation of digital computer control loop. Sampling is a fundamental property of computer controlled systems because of the discrete-time nature of the digital computer. The computer receives the measurement from the process at discrete times and gives control signal value also at discrete times. For a digital computer both input (data) and output (data) are in digital forms and corresponds to discrete-time values. Hence, there is a need for hardware element between a digital computer and the process. A sampler is simply a switch closing at specified time intervals, which takes in values of a continuous signal and produces a sequence of sampled values at particular time instants. An analog to digital converter (ADC) converts the discrete value to a corresponding digital value. Most of the final control elements (pneumatic valves in particular) are actuated by continuous-time signals (e.g., compressed air). Therefore the control commands produced by the computer program should be converted from discrete time to continuous time signals. This is accomplished by zero-order hold element.

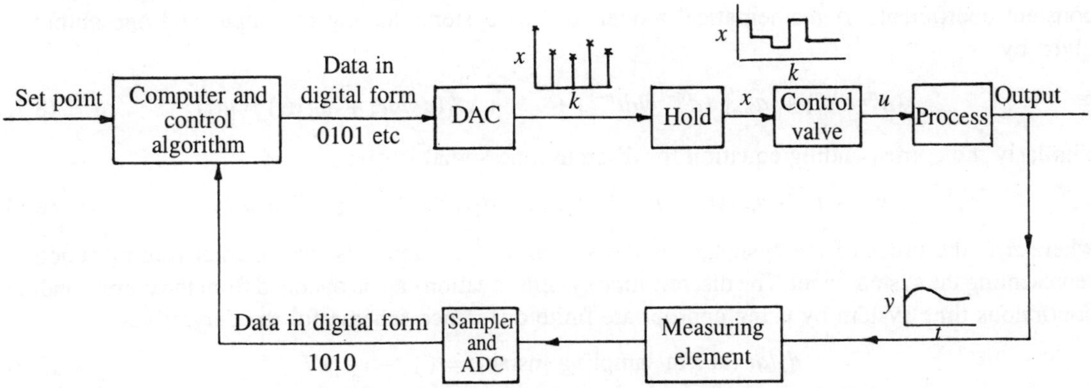

Fig. 8.1 Block diagram of a digital control system

The zero-order hold element keeps the value of a discrete time signal constant for the entire period until the next signal comes along. The model used for this purpose is to describe the change in signal from sample to sample. Time delays are associated with the measurements in the process. The analysis of continuous systems with time delay is complicated whereas the analysis is simple for discrete time systems with time delay. In order to calculate the sampled system from the continuous system, the value or the shape of the manipulated signal over the sampling instant must be known. These calculations are easy to do using zero-order hold because the control signal is piecewise constant. The zero-order hold sampling is also most common in practice because it is easily implemented using ordinary D-A converters.

8.2 Discrete Time Systems

We know that the continuous dynamic system is represented by a scalar differential equation

$$dx/dt = f(x(t)) \tag{8.1}$$

with an initial condition

at $$t = 0, \quad x(t_0) = x_0 \tag{8.2}$$

The corresponding dynamic system is represented in discrete time by a scalar difference equations as:

$$x(k+1) = f_d(x(k)) \tag{8.3}$$

The initial condition becomes

$$x(k_0) = x_0 \tag{8.4}$$

where k represents discrete-time, subscript d denotes discrete time function. By solving the system equation, the system response is evaluated. If the system is under the influence of some external forces or if we intend to change the system response intentionally by adding some external forces, the system is represented by the so called controlled difference equation, i.e.,

$$x(k+1) = f(x(k), u(k)) \tag{8.5}$$

$$x(k_0) = x_0 \tag{8.6}$$

We intend to vary the manipulated variable $u(t)$ such that the closed response has the desired specifications.

Continuous time invariant dynamic systems are designed by linear differential equations with constant coefficients. A mathematical model of such systems having one input and one output is given by

$$(d^n y/dt^n) + a'_{n-1}(d^{n-1}y/dt^{n-1}) + \ldots + a'_1(dy/dt) + a'_0 y(t) = u(t) \tag{8.7}$$

Similarly the corresponding equation for discrete time signal is given by

$$y(k+n) + a_{n-1}y(k+n-1) + \ldots + a_1 y(k+1) + a_0 y(k) = u(k) \tag{8.8}$$

where n is the order of the system, y is the system output, and u is the external forcing function representing the system input. The discrete time system equation can be obtained from the corresponding continuous time system by using appropriate finite difference scheme for the derivatives

$$dy/dt \text{ (at } k\text{th sampling instant)} = (y_k - y_{k-1})/T \tag{8.3}$$

$$d^2y/dt^2 \text{ (at } k\text{th sampling instant)} = (y_k - 2y_{k-1} + y_{k-2})/T^2 \tag{8.10}$$

Here T is the sampling period. The discrete time system model equations can also be obtained from the experimental data set.

For continuous time signal, the initial conditions are specified by known values of the system output derivatives upto the order of $(n - 1)$ at the initial time t_0, that is

$$y(t_0),\ dy(t_0)/dt,\ d^2y(t_0)/dt^2,\ \ldots,\ d^{n-1}y(t_0)/dt^{n-1}$$

In the discrete time domain, the initial conditions are specified by

$$y(k_0),\ y(k_0 + 1),\ \ldots\ y(k_0 + n - 1)$$

It is interesting to point out that in the discrete-time domain the initial conditions carry information about the evolution of the system output in time from k_0 to $k_0 + n - 1$.

For time invariant system the coefficients a_i, $i = 1, 2, \ldots, n - 1$ are constants. In general, the input function can be differentiated by the system so that the more general descriptions of time invariant continuous and discrete systems are respectively given by

$$[d^n y(t)/dt^n] + a'_{n-1}[d^{n-1}y(t)/dt^{n-1}] + \ldots + a'_1[dy(t)/dt] + a'\,y(t)$$

$$= b'_m[d^m u(t)/dt^m] + b'_{m-1}[d^{m-1}u(t)/dt^{m-1}] + \ldots + b'_1[du(t)/dt] + b'_0 u(t) \quad (8.11)$$

and

$$y(k + n) + a_{n-1}y(k + n - 1) + \ldots + a_0 y(k)$$

$$= b_m u(k + m) + b_{m-1}u(k + m - 1) + \ldots + b_1 u(k + 1) + b_0 u(k) \quad (8.12)$$

where for the coefficients a_i, $i = 0, 1, 2, \ldots (n - 1)$ and b_j, $j = 0, 1, 2, \ldots m$ are constants.

The input-output model is generally defined as

$$y(k) + a_1 y(k - 1) + \ldots + a_n y(k - n) = b_0 u(k) + \ldots + b_n u(k - n) \quad (8.13)$$

In most of the computer controlled system, we have $b_0 = 0$. There is no direct term in the discrete time model. $y(k)$ is measured first and then $u(k)$ is calculated. That is, $y(k)$ cannot be influenced by $u(k)$ even if continuous system has a direct form. There should be atleast one sampling time delay present (i.e., $y(k)$ is affected by $u(k - 1)$, $u(k - 2)$ etc.). In case the process itself has an additional delay, then sum of the sampling time and the process time dealy will be considered.

Consider the homogeneous first order difference equation

$$y_{k+1} - a y_k = 0 \quad (8.14)$$

with the given initial condition $y_0 = 1$. The characteristic equation $q - a = 0$ gives $q = a$.

The solution of the difference equation is then given by

$$y_k = y_0 q^k \quad (8.15)$$

i.e.,

$$y_k = a^k, \quad k = 0, 1, 2, \ldots \quad (8.16)$$

If the magnitude of a is less than 1, the natural solution returns to 0 and the system is stable. If the magnitude of a is greater than 1, then the system is unstable. When the value of a is negative, then the response is oscillatory between – and + values even for this first order system. For a first order continuous system, such a response is not possible and a is always positive.

8.3 Z-transform and Inverse of Z-transform

A discrete signal can be constructed by sampling a continuous signal, that is, every T seconds, a switch closes for an infinitesimally short time. The discrete time signal is given by

$$f^*(t) = f(kT) \quad \text{for } k = \ldots -3, -2, -1, 0, 1, 2, 3, \ldots \tag{8.17}$$

We can also mathematically model a discrete signal as the product of an impulse train and the original continuous signal (impulse modulation). That is,

$$f^*(t) = f(t) \sum_{k=-\infty}^{\infty} \delta(r - kT) \tag{8.18}$$

We find the Laplace transform of the sampled signal as:

$$F^*(s) = \int f^*(t) \exp(-st)dt \tag{8.19}$$

Substituting Eq (8.18) in the above equation we get,

$$F^*(s) = \sum_{k=0}^{\infty} \int f(kT) \exp(-st)\, \delta(t - kT)dt \tag{8.20}$$

$$= \sum_{k=0}^{\infty} f(kT) \exp(-skT) \tag{8.21}$$

It is to be noted that the Laplace transform of a sampled signal is not a rational function of s. We can define a new variable z, however, as

$$z = \exp(sT) \tag{8.22}$$

We have now defined the z-transform for a sampled signal:

$$F(z) = \sum_{k-0}^{\infty} f(kT)z^{-k} \tag{8.23}$$

Digital control textbooks give extended Tables of sampled signal, z-transform pairs. Table 8.1 gives the z-transform of simple systems. More important is the property of z-transform that allow us to transform a difference equation into a frequency domain (algebraic) equation:

$$f[(k+1)T] = zF(z) - zf(0) \tag{8.24}$$

$$f([(k+2)T] = z^2F(z) - z^2f(0) - zf(\mathrm{T}) \text{ etc.} \tag{8.25}$$

$$f[(k+n)T] = z^nF(z) - z^nf(0) - \ldots - zf(n-1)T] \tag{8.26}$$

Example: Given the zero state discrete difference equation:

$$y(k+3) + 0.2y(k+2) - y(k+1) - 0.1y(k) = 4x(k+1) + x(k),$$

the initial conditions $y(0)$, $y(1)$ and $y(2)$ each zero, find the relation between y and x in z domain. On Z-transformation, we get

$$(z^3 + 0.2z^2 - z - 0.1)\, Y(z) = (4z + 1)\, X(z) \tag{8.27}$$

that is

$$Y(z) = [(4z + 1)/(z^3 + 0.2z^2 - z - 0.1)]\, X(z) \tag{8.28}$$

Given the Z-transform of the input, we can find the output $y(k)$ by any number of inverse Z-transform methods. For instance, we can use the Partial fraction expansion method. It is common practice to convert $Y(z)$ in terms of polynomial in z^{-1} both in the denominator and in numerator (then

Table 8.1 Z-transform for various discrete functions

S. No.	$X(s)$	$x(t)$, for $x(kT)$ replace t by kT	$X(z)$
1.	$1/s$	1	$1/(1 - z^{-1})$
2.	$1/s^2$	t	$Tz^{-1}/(1 - z^{-1})^2$
3.	$1/(s + a)$	$\exp(- at)$	$1/[1 - exp(- aT) z^{-1}]$
4.	$a/[s(s + a)]$	$1 - \exp(- aT)$	$(1 - p)z^{-1}/[(1 - z^{-1})(1 - pz^{-1})$
5.	$(b - a)/(s + a)(s + b)$	$\exp(- aT) - \exp(- bT)$	$(p - p_1)z^{-1}/[(1 - p/z^{-1})(1 - p_1z^{-1})]$ $p = \exp(-aT)$; $p_1 = \exp(- bT)$
6.	$1/(s + a)^2$	$t \exp(- aT)$	$Tpz^{-1}/(1 - pz^{-1})^2$
7.	$\omega/(s^2 + \omega^2)$	$\sin(\omega t)$	$z^{-1} \sin(\omega T)/[1 - 2z^{-1} \cos(\omega T) + z^{-2}]$
8.	$s/(s^2 + \omega^2)$	$\cos(\omega)$	$[1 - z^{-1} \cos(\omega T)]/[1 - 2z^{-1} \cos(\omega T) + z^{-2}]$

Table 8.2 Important properties of and theorems of Z-transform

S. No.	$x(t)$	$Z[x(t)]$
1.	$ax(t)$	$aX(z)$
2.	$ax_1(t) + bx_2(t)$	$a_1X_1(z) + bX_2(z)$
3.	$x(t + T)$	$zX(z) - zx(t = 0)$
4.	$x(t - kT)$	$z^{-k}X(z)$
5.	$\exp(- aT)x(t)$	$X[z \exp(aT)]$
6.	$x(t = 0)$	$\lim X(z)$ if the limit exists $z \to \infty$
7.	$x(t = \infty)$	$\lim [1 - z^{-1})X(z)$ if $(1 - z^{-1})X(z)$ is analytic on $z \to 1$ and outside the unit circle.

only $y(k)$ is related to $y(k - 1)$ and $u(k - 1)$ etc.). Hence, we have to divide $Y(z)$ by the highest power of the denominator and then apply the partial fraction method. After the expansion coefficients are found, the inverse of z-transform for these simple terms are found out from look-up table.

Example: Find the inverse of $G(z) = (2 - z^{-1})/[(1 - z^{-1})(1 - 0.4z^{-1})]$
By partial fraction method. we get

$$G(z) = [A/(1 - z^{-1})] + [B/(1 - 0.04z^{-1})] \tag{8.29}$$

$$(2 - z^{-1}) = A(1 - 0.4z^{-1}) + B(1 - z^{-1}) \tag{8.30}$$

Let $z^{-1} = 1$, we get $A = 5/3$.
Let $z^{-1} = 1/0.4$, we get $B = 1/3$.

Hence,
$$G(z) = [(5/3)/(1 - z^{-1})] + [(1/3)/(1 - 0.4z^{-1})] \tag{8.31}$$

Referring to the Z-transform (Table 8.1) we can find the corresponding inverse of z-transform as

$$g(kt) = (5/3) + (1/3) \exp(- 0.92nT); \quad n = 0, 1, 2, \ldots \tag{8.32}$$

8.4 Transfer Functions

The system transfer function for a single-input single-output time invariant continuous system is defined as the ratio of the Laplace transform of the system output over the Laplace transform of the system input assuming that all the initial conditions are zero. That is, for the system Eq (8.11), we have

$$G(s) = (b'_m s^m + b'_{m-1} s^{m-1} + \ldots + b'_1 s + b'_0)/(s^n + a'_{n-1} s^{n-1} + \ldots + a'_1 s + a'_0) \quad (8.33)$$

For real system, always $n \geq m$.

In discrete time system, the Z-transform takes the role of Laplace transform so that the discrete-time transfer function is given by

$$G(z) = (b_m z^m + b_{m-1} z^{m-1} + \ldots + b_1 z + b_0)/(z^n + a_{n-1} z^{n-1} + \ldots + a_1 z + a_0) \quad (8.34)$$

A control system composed of blocks of z-domain transfer functions obeys the same principles of cascade parallel and feedback combinations as continuous systems. Mason's rule, likewise, can be applied to discrete-time systems.

Steady-state gain: In case of continuous-time systems, the steady-state gain is obtained by making $s = 0$ (zero frequency) in the transfer function. In the discrete case, $s = 0$ corresponds to

$$z = \exp(sT) = 1 \quad (8.35)$$

and thus the steady state gain $G(0)$ is obtained by making $z = 1$ in the pulse transfer function. Therefore, the steady-state gain for the discrete system of Eq. (8.34) is given by

$$G(0) = \left(\sum_{1}^{m} b_i\right) \Big/ \left[1 + \sum_{0}^{n=1} a_i\right] \quad (8.36)$$

We can easily check the following statements:

It is not possible for $G(z)$ to match the behaviour of the $G(s)$ from where it was derived. If the discrete model produces the sampled impulse response as the continuous model, it will not match the step response (and vice versa). The frequency response of the discrete model never exactly matches the frequency response of the continuous model.

Several techniques exist for discretizing continuous systems. Among them are time response matching (pulse or step response), and numerical integration methods such as Euler, backward, forward and trapezoidal. Some techniques preserve time response, whereas others preserve the frequency response better. From control point view, stability preservation is very important. Techniques that do not preserve system stability have to be used very cautiously otherwise misleading situation can result.

We know that $z = \exp(sT)$. Since it is a nonlinear relation, we need to approximate by a suitable linear relation for $\exp(sT)$. Any of the following linear relations between z and s can be used:

$z = 1 + sT$ (Forward Euler method) (8.37a)

$z = 1/\exp(-sT) = 1/(1 - sT)$ (Backward Euler method) (8.37b)

$z = (1 + 0.5sT)/(1 - 0.5sT)$ (Pade approximation or bilinear transformation or Trapezoidal method or Tustin approximation) (8.37c)

These equations can be rewritten in s as a function of z as

$$s = (z - 1)/T \quad (8.38a)$$

$$s = (z - 1)/Tz \tag{8.38b}$$

$$s = (2/T)(z - 1)/(z + 1) \tag{8.38c}$$

Forward Euler technique does not always preserve stability. For example, consider the stable continuous filter:

$$H(s) = 1/(s + 1) \quad \text{(pole at } s = -1) \tag{8.39}$$

The forward Euler's transformation gives

$$H(z) = T/(z - 1 + T) \quad \text{pole at } z = 1 - T \tag{8.40}$$

If $T > 2$, the discrete pole lie outside unit circle, and the filter response will be unstable.

Zero-order hold equivalent (or step invariant transformation)
It is, by far, the most common technique used by control engineers. We use this technique only to model a continuous plant, since we are interested in time response of the plant. We rarely use ZOH equivalent to convert a continuous filter to a digital filter, since the method does not faithfully reproduce frequency response.

8.5 Zero-Order Hold Equivalence

In practice, a discrete signal from the controller never directly drives a continuous plant. The discrete signal is first passed through a digital-to-analog converter (DAC). The DAC produces a continuous output that reflects its discrete input. The simplest and most commonly used DACs are devices that convert the binary computer output to a voltage level and then hold that level until the computer outputs the next data word T seconds later. This device is known as a zero order hold (ZOH). The output of the ZOH reconstruction of the signal is shown in Fig. 8.2. The zero order hold creates an output pulse for each input impulse. Because a step is derived from an impulse by integration, we get the ZOH transfer function (Laplace transform of a unit pulse of duration T)

$$H(s) = [1 - \exp(-sT)]/s \tag{8.41}$$

For smaller time delay, we can approximate the equation by using expansion for exp ($-Ts$) as

$$[1 - \exp(-Ts)]/sT = [1 - \{1 - Ts + 0.5(Ts)^2 + \ldots\}]/(Ts) \tag{8.42}$$

$$= [1 - 0.5(Ts) + \ldots] \tag{8.43}$$

$$\approx 1 - (Ts/2) \tag{8.44}$$

$$\approx \exp(-Ts/2) \tag{8.45}$$

Thus zero order hold effect is approximately delaying the input signal by half of the sampling time.

The transfer function combining hold element, final control element and the plant and the measuring element together is denoted by $HG_p(s)$ (Fig. 8.3). Whenever there is a ZOH preceding a plant, we take the z-transform of HG_p as follows:

$$\text{z-transform of } \{1 - \exp(-Ts)]/s\}\, G_p(s) = \text{z-transform of } \{G_p(s)/s\}$$
$$- \text{z-transform of } \{\exp(-Ts)\, G_p(s)/s\} \tag{8.46}$$

We use the notation that z-transform of $G_p(s)$ means the z-transform of a time function whose Laplace transform is $G_p(s)$. The second term in the preceding expression is the delayed version of the first term. Because the z-transform of a unit delay is z^{-1}, we get

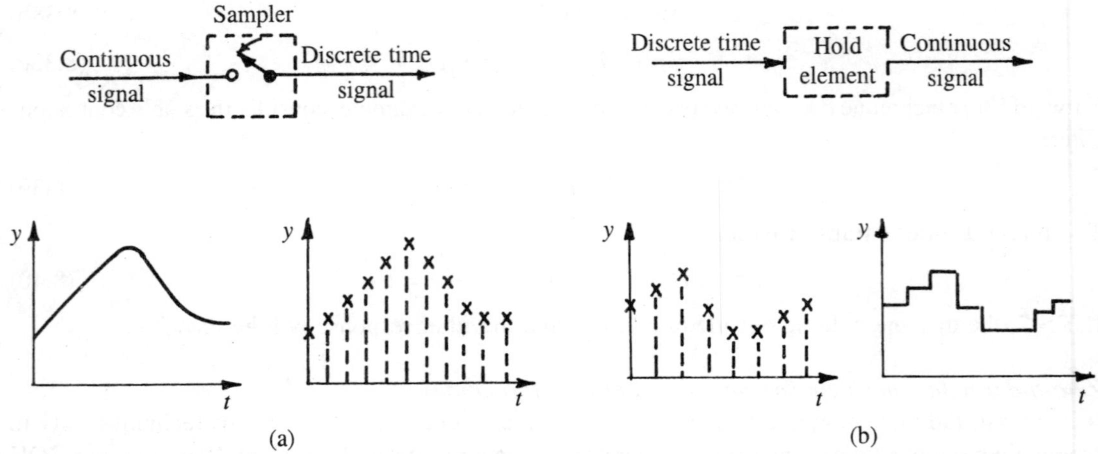

Fig. 8.2 (a) Conversion of continuous to discrete-time signal; (b) reconstructing of continuous signal from its discrete time equivalent

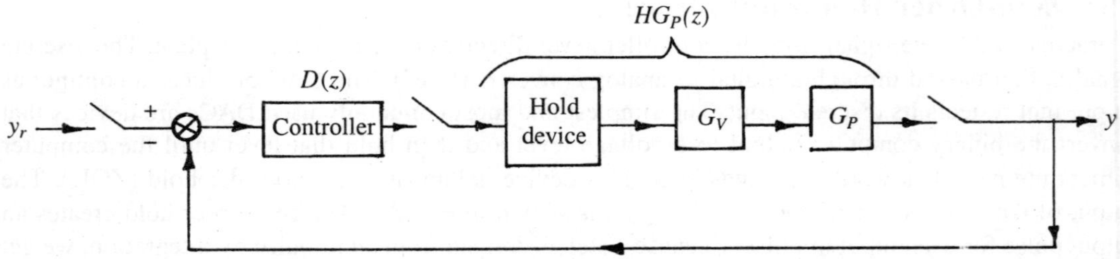

Fig. 8.3 Digital control system

$$HG_p(z) = (1 - z^{-1}) \, \{\text{z-transform of } \{G_p(s)/s\}\} \tag{8.47}$$

Hence, to find the z-transform of $HG_p(s)$, we first find the z-transform of a function whose Laplace transform is $HG_p(s)$. Usually the term $G_p(s)/s$ will be converted into simple terms by partial fraction method. Then for the resulting simple terms, the corresponding z-transform is found out from standard z-transform table.

Let us consider a first order transfer function model for $G_pG_v(s)$ together as $k_p/(\tau s + 1)$. Then $HG_p(z)$ can be obtained as follows:

$$HG_p(z) = k_p(1 - z^{-1}) \, \{\text{ z-transform of } 1/[s(\tau s + 1)]\} \tag{8.48}$$

$$= k_p(1 - z^{-1}) \, \{\text{z-transform of } [(1/s) - 1/[s + (1/\tau)]\} \tag{8.49}$$

$$= k_p(1 - z^{-1}) \, \{[1/(1 - z^{-1})] - 1/[1 - \exp(-T/\tau)z^{-1}]\} \tag{8.50}$$

$$= k_p[1 - \exp(-T/\tau)]z^{-1}/[1 - \exp(-T/\tau)z^{-1}] \tag{8.51}$$

The expression for $HG_p(z)$ for a given $G_p(s)$ and sampling time (T) is given for several $G_p(s)$ by Neuman and Baradello (1979). Table 8.2 gives few important pulse transfer functions of some analog transfer function models.

Table 8.2 The z-transfer function without and with zero-order hold

$G(s)$	$G(z) = \mathcal{Z}\{G(s)\}$	$HG(z) = \frac{z-1}{z}\mathcal{Z}\left\{\frac{G(s)}{s}\right\}$
$\frac{1}{s}$	$\frac{z}{z-1}$	$\frac{T_0}{z-1}$
$\frac{1}{s^2}$	$\frac{T_0 z}{(z-1)^2}$	$\frac{T_0^2(z+1)}{2(z-1)^2}$
$\frac{1}{s^3}$	$\frac{T_0^2 z(z+1)}{(z-1)^3}$	$\frac{T_0^3(z^2+4z+1)}{6(z-1)^2}$
$\frac{1}{s^4}$	$\frac{T_0^3 z(z^2+4z+1)}{6(z-1)^4}$	$\frac{T_0^4(z^3+11z^2+11z+1)}{24(z-1)^4}$
$\frac{1}{s+a}$	$\frac{z}{z-e^{-aT_0}}$	$\frac{(1-e^{-at_0})}{a(z-e^{-aT_0})}$
$\frac{1}{(s+a)(s+b)}$	$\frac{1}{(b-a)}\frac{(e^{-at_0}-e^{-bT_0})z}{(z-e^{-aT_0})(z-e^{-bT_0})}$	$\frac{1}{ab(a-b)}\frac{(Az+B)}{(z-e^{-aT_0})(z-e^{-bT_0})}$ $A = a - b - ae^{-aT_0} + be^{-aT_0}$ $B = (a-b)e^{-(a+b)T_0} - ae^{-aT_0} + be^{-bT_0}$

8.6 Study of the First Order Systems with Time Delay

The continuous-time model is characterized by the transfer function

$$H(s) = k_p \exp(-\tau_d s)/(1 + \tau s) \tag{8.52}$$

If T is the sampling period, then the time delay τ_d is expressed as

$$\tau_d = dT + \tau'_d, 0 < \tau'_d < T \tag{8.53}$$

where τ'_d is the fractional time delay and d the integer number of sampling periods included in the delay and corresponding to a sampled delay of d-periods. From Table 8.2, we can derive the transfer function of the corresponding sampled model (plant+ zero order hold)

$$H(z^{-1}) = z^{-d}(b_1 z^{-1} + b_2 z^{-2})/(1 + a_1 z^{-1}) \tag{8.54}$$

where

$$a_1 = -\exp(-T/\tau);\ b_1 = k_p[1 - \exp\{(\tau'_d - T)/\tau\}] \tag{8.55}$$

$$b_2 = k_p \exp(-T/\tau)\ [\exp(\tau'_d/\tau) - 1] \tag{8.56}$$

The effect of fractional time delay is evident in the presence of the coefficient in b_2 in the transfer function model. For example when $\tau'_d = 0$, we get $b_2 = 0$. On the other hand, if $\tau'_d = T$ and then we get $b_1 = 0$ resulting in an additional delay of one sampling period. For $\tau'_d = 0.5T_s$, we get $b_1 = b_2$. Hence, a zero is introduced in the pulse transfer function. For $\tau'_d > 0.5T$, we get $b_2 > b_1$ and the zero is unstable.

Our interest is not to reconstruct the signal perfectly but to provide sufficient information to the controller to achieve good dynamic performance. It is often possible to sample much less frequently than specified by Shannon's theorem and still achieve good control performance. If the signal has substantial high-frequency components with significant amplitudes, the continuous signal may have to be filtered.

There are many options for using the sampled values to reconstruct the signal approximately. The most common methods, zero-order hold and first-order hold, will be considered here. As discussed earlier, the zero order hold assumes that the variable is constant between the samples. Whereas the first order hold assumes that the variable changes in a linear fashion as predicted from the most recent two samples. These two methods are compared in Figs. 8.4(a) and (b), where the main difference is the amplification in the magnitude caused by the first-order hold. The first-order hold has a larger phase lag, which is undesirable for closed loop control. For both of these reasons, the simplest zero-order hold is used almost exclusively for process control.

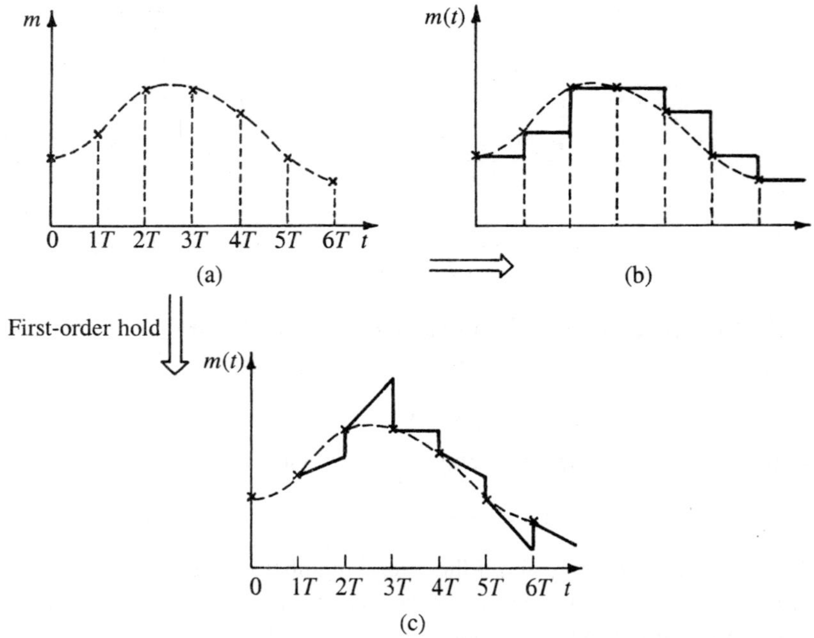

Fig. 8.4 (a) Sequence of impulses; (b) reconstruction of continuous signal with zero-order hold and (c) reconstruction with first order hold

The effect of zero order hold on the dynamics can be seen clearly if we reconstruct the original signal as shown in Fig. 8.5, the reconstructed signal is a smooth curve through the midpoint of the zero order hold. It is clear that the reconstructed signal after the zero-order hold is identical to the original signal after being passed through the dead-time of $T/2$, where T is the sampling time. The rule of thumb is that the major effect on the stability and control performance of sampling can be estimated by adding $T/2$ to the dead time of the system. Since any additional delay due to sampling is undesirable for feedback control and process monitoring, feedback control performance is degraded as the sampling time increases.

Since the sampling introduces a delay in the feedback loop, the controller tuning should be detuned to account for the sampling. The tuning rules used for continuous (analog) controller can be used for sampled process also by adding $0.5T$ to the time delay of the process. The effective delay should be used in the controller tuning formulae.

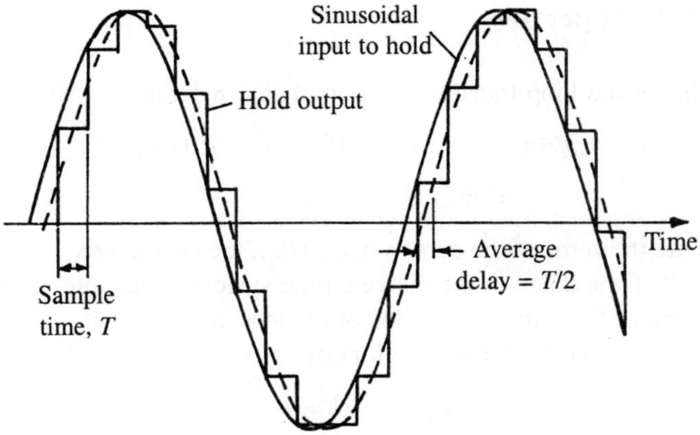

Fig. 8.5 Effective delay introduced by a zero-order hold

Since $z = \exp(Ts)$, it can be checked that the entire left half of S-plane is converted into a unit circle in z-domain. The discrete system is stable if and only if all the poles of the discrete closed loop transfer function lie within the unit circle. The corresponding relation between S- and Z-plane given in Table 8.3 is not one-to-one mapping. Many points in the S-plane correspond to a given point in the Z-domain, since $\exp(Ts)$ is periodic function with respect to ω. The effect of real poles in the z-plane on the system response is more complicated than that in the s-plane. This can be found from the information given in Table 8.3. Since the entire left half of S-plane is mapped into a unit circle in Z-Domain, (a small region) a great accuracy in the calculations of the roots of the system characteristic equations is required. The stability of the closed loop system can be found out by checking that all the roots of the characteristic equation $1 + G_c(z)HG_p(z) = 0$ should be ≤ 1.

Table 8.3 Relationship between S- and Z-plane

S-plane	Z-plane	Comment
Right half s-plane	Outside the unit circle	Unstable region
Left half s-plane	Inside the unit circle	Stable region
Origin	$z = 1$	Integral action
Negative real axis	$0 < z < 1$	Non-oscillatory convergent response
Positive real axis	$1 < z < \infty$	Monotonic divergent response
Points $\sigma + j(\omega + 2\pi k/T)$	One point $\exp(\sigma + j\omega)T$	Many points in the s-plane are mapped to one point in z-plane
$\sigma + j(\omega + 2\pi k/T)$ $\sigma < 0, k = 0, \pm 1, \ldots$	$-1 < z < 0$	Oscillatory Convergent response
$\sigma + j(\omega + 2\pi k/T)$ $\sigma > 0, k = 0, \pm 1, \ldots$	$-\infty < z < -1$	Oscillatory Divergent response

8.7 Some General Properties

Poles and zeros: The closed loop transfer function of Fig. 8.3 can be written as

$$y(z)/r(z) = G_c(z)HG_p(z)/[1 + G_c(z)HG_p(z)] \tag{8.57}$$

$$= B(z)/H(z) \tag{8.58}$$

The poles of a system are the zeros of the denominator $H(z)$, the characteristic polynomial. The zeros are the roots of $B(z) = 0$. Time delay in the discrete time system gives pole at the origin. The order of the system is the same as the number of poles of the system.

The forward operator denoted by q has the property

$$qy(k) = y(k + 1) \tag{8.59}$$

Here $(k + 1)$ denotes the next sampling instance. Similarly, the backward shift operator (inverse forward shift operator) is defined as

$$q^{-1}y(k) = y(k - 1) \tag{8.60}$$

Here $(k - 1)$ denotes the previous sampling instance. The shift operator is used to simplify the manipulation of higher-order difference equations. It is important for the range of the operator to be doubly infinite sequences, otherwise the inverse of the forward shift operation may not exist.

The variable z is a complex variable and should be distinguished from the operator q. When manipulating difference equation, we can use either z-transform or q operator, where q acts on sequences and z is a complex variable. (This is similar to what we use d/dt for s in continuous time system.) In z-transform calculus, we can divide the terms with arbitrary expression, but this is not allowed in shift-operator calculus. That is, cancellation of poles takes place in z-transform calculus (if the canceled pole is unstable one, then the system cannot be stabilized).

A continuous time system with a rational transfer function is nonminimum phase if it has right half plane zeros or time delays. A discrete time system is often defined to be nonminimum phase if it has zeros outside the unit disc (the time delay does not make the system nonminimum phase). Time delays do not pose the same severe problems as they do for continuous-time systems.

A discrete time system has an unstable inverse if it has zero outside the unit disc. It is possible, a continuous time system with a stable inverse may become a discretized system with an unstable inverse when it is sampled particularly when the pole excess of the continuous time system is larger than two and if the sampling time is significantly short. A continuous time, nonminium phase system will not always become a discrete time system with an unstable inverse.

The continuous time output of the process may have oscillations that are not seen at the sampling points. These are called hidden oscillations or intersample ripple. By simulating the continuous time system with the controller gives the nature of the response in between the samples also. The use of a modified z-transform (Stephanopoulos, 1984) can also be used to calculate the continuous time response between the sampling instants.

The intersample ripple is essentially determined by the open loop dynamics because the system operates in open loop between the sampling points:

(i) Open loop or closed loop systems where there is an oscillation in the continuous time output which cannot be seen in the control signal. This type of intersampling oscillations may occur if cancellation of poles and zeros in the system occurs due to sampling. These may be hidden oscillations if the continuous time open loop system has oscillatory modes and if the sampling period matches the frequency of these nodes. This type of hidden oscillation occurs only for certain values of the sampling period.

(ii) Oscillations between the sampling points caused by oscillations in the control signal. This occurs if there are poorly damped zeros in the open loop system that are canceled by the controller. The oscillations can be seen in the control signal (u).

Design of Controllers

We shall discuss two methods for designing digital controllers:

(1) Design of the controllers based on transfer function models in s-domain and converting it to discrete time domain and (2) direct design of controllers in the z-domain.

8.8 Converting Continuous Time Controllers to Discrete Time Domain

Conversion of a continuous time controller works well when the sample rate is high. The z-transform techniques are preferable at lower sampling rates. The z-domain technique has some disadvantages. The design must take place within the limited confines of the unit circle. The roots should be calculated more accurately. Bode diagram techniques that correlate gain cross over frequency and phase margin with closed loop behaviour do not apply to the z-transform. These techniques apply to transfer functions whose stability is the left-half S-plane.

The w-transform is a technique that allow us to apply all s-domain design methods to digital systems. In the techniques, we transform $H(z)$ into $H(w)$ using the relation

$$w = (2/T)(z - 1)/(z + 1) \tag{8.61}$$

The interior of the unit circle in the Z-plane maps into the left side of the w-plane. All the controller design method for the plant $G(s)$ can be applied to $G(w)$. Given $G(s)$, we use ZOH approximation to convert $G(s)$ into $G(z)$. We then use

$$z = (1 + 0.5wT)/(1 - 0.5wT) \tag{8.62}$$

to convert $G(z)$ to $G(w)$. Once the controller is designed for $G(w)$, then we finally convert the controller design back to the z-plane since we need to implement controller as difference equations.

Example: Given $H(s) = 1/(s + 1)$, find $H(w)$.

The ZOH approximation using $T = 0.5s$, we get

$$H(z) = 0.393/(z - 0.607) \tag{8.63}$$

Using the w-transform on the original process we get

$$H(w) = 1/\{[(1 + 0.25w)/(1 - 0.5w)] + 1\} \tag{8.64}$$

$$= (1 - 0.25w)/(1 + 1.023w) \tag{8.65}$$

We note that the poles of $H(w)$ and the original $H(s)$ are very close. This will be function of sampling period. We also note that $H(w)$ has a RHP zero that does not exist in the original $H(s)$.

8.8.1 Modification of the Discrete Controller

Since approximately a delay of $T/2$ is introduced by sampling, we can modify the controller in order to produce closed loop behaviour similar to the continuous system. The transfer function of the process is multiplied by exp $(-Ts/2)$ to account for the effect of sampling. Using Pade's approximation we get

$$\exp(-Ts/2) = [1 - (Ts/4)]/[1 + (Ts/4)] \quad (8.66)$$

Let us substitute

$$s = (2/T)(z - 1)/(z + 1) \quad (8.67)$$

Eq. (8.66) becomes

$$\exp(-Ts/2) = 2(z + 3)/(7z + 1) \quad (8.68)$$

We can derive another approximation as follows:

Using eq. (8.67) in $\exp(-Ts/2)$ we get

$$\exp(-Ts/2) = \exp[-(z - 1)/(z + 1)] \quad (8.69)$$

Using the Pade's approximation for the right side of the above equation, we get

$$\exp(-Ts/2)] = (z + 3)/(3z + 1) \quad (8.70)$$

Let $D'(z)$ be the discrete controller designed for the system $G_vG_p(s)\exp(-Ts/2)$. The controller $D(z)$ to be implemented after taking into the sampling effect compensation, is given by

$$D(z) = D'(z)C(z) \quad (8.71)$$

where

$$C(z) = 0.5(7z + 1)/(z + 1) \quad \text{or} \quad = (3z + 1)/(z + 3) \quad (8.72)$$

Now let us review the method of designing PID controllers for continuous time transfer function model with time delay. Most of the processes are adequately described by first order plus time delay model. An excellent review of the methods for designing PID controllers is given by Astrom and Hagglund (1995). Since most of the systems are adequately represented by first order plus time delay models (FOPTDs), Ziegler and Nichols (1942) have proposed simple equations based on practical experiences for designing P/PI/PID controllers. Table 8.4 gives such tuning formulae. The response for set point changes usually gives a larger overshoot. Hence, these settings should be further tuned to get an improved performance. An improved tuning formulae were proposed by Chidambaram (1995).

Table 8.4 Controller settings by Ziegler-Nichols for delay plus first order model $k_p \exp(-Ls)/(\tau s + 1)$

Controller	$k_{c,\text{des}}$	τ_I	τ_D
P	$\tau/(LK_p)$	∞	0
PI	$0.9\tau/(LK_p)$	$3.33L$	0
PID	$1.2\tau/(LK_p)$	$2.0L$	$0.5L$
Improved Ziegler-Nichols method (Chidambaram, 1995)			
P	$0.5[1 + \{1.6(\tau/L\}^2]^{0.5}$	∞	0
PI	$0.4 + 0.665(\tau/L)$	$3.4L$	0
PID	$0.45 + 1.8(\tau/L)$	$2.4L$	$0.38L$
PD	$0.5 + 1.74(\tau/L)$	∞	$0.36L$

8.8.2 Pole Placement Controllers

One of the methods of designing PI/PID controllers is to assume the poles locations for the closed loop system and then calculate the values for the PI or PID settings. Clement and Chidambaram (1997) have proposed simple tuning formulae for the PI/PID settings based on Pole placement method.

PI controller:

$$k_c k_p = 3.113 - 5.73\varepsilon + 3.24\varepsilon^2 \quad (8.73)$$

$$\tau_I/\tau = 0.413 + 0.8\varepsilon \quad (8.74)$$

where $\varepsilon = L/\tau$.

PID controller:

$$k_c k_p = 4.105 \exp(-1.763\varepsilon) \quad (8.75)$$

$$\tau_I/\tau = 0.311 + 1.372\varepsilon - 0.545\varepsilon^2 \quad (8.76)$$

$$\tau_D/\tau = 1/[16.016 - 11.702\varepsilon] \quad (8.77)$$

8.8.3 Controller Tuning by Haalman's method

Haalman (1965) has assumed (i) that controller and process can be matched in such a way that the open loop transfer function is the same for all combinations, and (ii) a purely integral controller yields a good control of a pure dead time process. It can be easily shown that a pure delay process with an integral controller gives a minimum squared error when τ_I = 1.5 L. Hence the open loop transfer function is (2/3) exp $(-Ls)/(Ls)$. Therefore, the controller settings of any delay plus dynamic process can be found out. Table 8.5 gives the controller settings for different process transfer functions.

Table 8.5 Controller model and settings by Haalman's method

Transfer function model	k_{ckp}	τ_I	τ_D
$k_p \exp(-Ls)$	–	τ_I^*	0
$k_p \exp(-Ls)/s$	$1/\tau_I^*$	∞	0
$k_p \exp(-Ls)/(\tau s + 1)$	$k_c^* \tau$	τ	0
$k_p \exp(-Ls)/[(\tau s + 1)s]$	k_c^*	∞	τ
$k_p \exp(-Ls)/[(\tau_1 s + 1)(\tau_2 s + 1)]$	$k_c^*(\tau_1 + \tau_2)$	$(\tau_1 + \tau_2)$	$\tau_1\tau_2/(\tau_1 + \tau_2)$
$k_p \exp(-Ls)/[(s^2/\omega_n^2) + 2(\zeta s/\omega_n) + 1)]$	$k_c^*(2\zeta/\omega_n^2)$	$2\zeta/\omega_n$	$1/(2\zeta\omega_n)$

where $\tau_I^* = 1.5L$.

Tuning formulae for PID controllers for FOPTD systems are given by Sung et al. (1995) and Zhuang and Atherton (1993) separately for servo problem and regulatory problems in order to get improved performances. These formulae are rather complicated equations.

8.8.4 Control of Unstable Systems

Open loop instability means that the system will move away from the steady-state for even a small perturbation of the system. Linearization of mathematical model of such nonlinear systems around the unstable operating point will give a transfer function which has at least one pole in the right half of s-plane (RHP). The response of such transfer function models for a small perturbation will be ever increasing but in the case of nonlinear systems, the perturbation will cause the system to move to other but a stable steady-state. Sometimes, time delay is introduced into the transfer function description of such systems due to measurement delay or by the approximation of higher order dynamics of such systems into a lower order plus time delay system. One of the methods of designing the controller is to assume the poles locations for the closed loop system and then calculate the values for the PI or PID settings.

Pole placement method: The following equations are proposed by Clement and Chidambaram (1997a, b)

PI

$$k_c k_p = 2.695 \exp(-1.277\varepsilon) \tag{8.78}$$

$$\tau_I/\tau = 0.4015 \exp(5.8\varepsilon) \quad \text{for } 0 < \varepsilon \le 0.6 \tag{8.79}$$

PID

$$k_c k_p = 2.121 - 1.906\varepsilon + 0.945\varepsilon^2 \tag{8.80}$$

$$\tau_D/\tau = 0.176 + 0.360\varepsilon \tag{8.81}$$

$$\tau_I = \tau_D/\alpha \tag{8.82}$$

where

$$\alpha = 0.179 - 0.324\varepsilon + 0.161\varepsilon^2 \quad \text{for} \quad \varepsilon < 0.6 \tag{8.83a}$$

$$= 0.04 \quad \text{for} \quad 0.6 < \varepsilon < 0.8 \tag{8.83b}$$

$$= -0.1\varepsilon + 0.12 \quad \text{for} \quad 0.8 < \varepsilon < 1.0 \tag{8.83c}$$

Controller settings can be proposed for minimizing performance criteria like ISE, IAE and ITAE. Jhunjhunwala and Chidambaram (1998) have evaluated PI controller performance by minimizing ISE, IAE and ITAE and found that the controller settings based on minimizing ITAE give least oscillation in the response both for servo and regulatory problems. Jhunjhunwala and Chidambaram (1998) have proposed the following equation based on ITAE criteria.

$$k_c k_p = 10.7572\text{-}35.1\varepsilon \quad \text{for} \quad \varepsilon \le 0.2 \tag{8.84}$$

$$= 0.876\varepsilon^{-0.898} \quad \text{for} \quad 0.2 \le \varepsilon \le 0.6 \tag{8.85}$$

$$\tau_I/\tau = 0.324 \exp(6.46\varepsilon) \quad \text{for} \quad 0 < \varepsilon \le 0.6 \tag{8.86}$$

Jhunjhunwala and Chidambaram (1999) have also proposed the following correlation for PID settings based on minimizing ISE criterion which gives least overshoot when compared to the minimization of IAE and ITAE criteria:

PID setting based on ISE criteria:

$$k_c k_p = 13.0\text{-}39.712\varepsilon \tag{8.87}$$

$$= 1.397(1/\varepsilon) \quad \text{for} \quad 0.2 < \varepsilon \le 1.4 \tag{8.88}$$

$$\tau_I/\tau = 0.856 \exp(2.044\varepsilon) \quad \text{for} \quad 0 < \varepsilon < 1 \tag{8.89}$$

$$= 0.344 \exp(2.9595\varepsilon) \quad \text{for} \quad 1 < \varepsilon \le 1.4 \tag{8.90}$$

$$\tau_D/\tau = 0.5643\varepsilon + 0.0075 \quad \text{for} \quad 0 < \varepsilon \le 1.4 \tag{8.91}$$

The maximum error for the fitted equations is less than 2%. Tuning formulae for PID controllers for unstable systems in order to get improved performance are given by Huang et al. (1999). The equations have a single adjustable parameter. However, these equations are rather complicated.

Chidambaram et al. (1999) have extended Haalman's method for selecting the suitable controller mode and the associated settings for systems with positive and negative zero (Table 8.6).

Table 8.6 Controller mode and settings for systems with a zero

Transfer function model	$k_c k_p$	τ_I	τ_D
$k_p(1 \pm \tau_n s)\exp(-Ls)$		τ_I^*	0
$k_p(1 \pm \tau_n s)\exp(-Ls)/s$	$1/\tau_I^*$	∞	0
$k_p(1 \pm \tau_n s)\exp(-Ls)/(\tau s + 1)$	$k_c^* \tau$	τ	0
$k_p(1 \pm \tau_n s)\exp(-Ls)/[(\tau s + 1)s]$	k_c^*	∞	τ
$k_p(1 \pm \tau_n s)\exp(-Ls)/ [(\tau_1 s + 1)(\tau_2 s + 1)]$	$k_c^*(\tau_1 + \tau_2)$	$(\tau_1 + \tau_2)$	$\tau_1\tau_2/(\tau_1 + \tau_2)$
$k_p(1 \pm \tau_n s)\exp(-Ls)/ [(s^2/\omega_n^2) + 2(\zeta s/\omega_n) + 1]$	$k_c^*(2\zeta/\omega_n^2)$	$2\zeta/\omega_n$	$1/(2\zeta\omega_n)$

where -ve sign in $(1 \pm \tau_n s)$ is for nonminimum phase systems

$$k_c^* = 1/\tau_I^*$$

For nonminimum phase systems: $\tau_I^* = [1 + a\pi\tau_n/L^2]^{0.5}/(b\pi/L)$, where $b = 0.5\,a$.
For smaller L/τ_n, $a = 1$ and for larger L/τ_n, $a = 1.5$.

8.8.5 Set Point Weighted PI/PID Controllers

Since most of the systems are adequately represented by first order plus time delay transfer function models, the PI/PID controllers are designed by using the open loop Ziegler-Nichols tuning method. The response of the closed loop system using the Ziebler-Nichols tuning method gives a good regulatory response whereas a significant overshoot occurs for servo problems. Astrom and Hagglund (1995) have shown that use of set point weighting reduces the overshoot.

$$u(t) = k_c\left[e_p + (1/\tau_I)\int e\,dt - \tau_D(dy/dt)\right] \tag{8.92}$$

where

$$e_p = (\beta y_r - y) \tag{8.93}$$

$$e = y_r - y \tag{8.94}$$

The time derivative is taken on the output variable y and not on the set point value. Here β is the set point weighting parameter ($0 \le \beta \le 1$). Chidambaram (1999) has proposed a method to select the value of β for first order plus time delay systems. In order to minimize the overshoot, Chidambaram (1999) has shown that β is to be obtained from the equation

$$\beta = -\zeta^2(\tau/\tau_I)/\sigma \tag{8.95}$$

Here ζ is the damping coefficient and σ the real part of the dominant pole of the closed loop system. Given the process transfer function model parameters (k_p, τ and L) and the PID controller settings, we can claculate the value of β. The relation is fitted for a stable FOPTD system with the Ziegler-Nichols settings by a simple relation.

$$\beta = 0.3238 + 0.4332(L/\tau);\quad (L/\tau) \le 0.8 \tag{8.96}$$

For pole placement controller settings given by Clement and Chidambaram (1997), the set point weighting parameter is given by the fitted equation:

$$\beta = 0.793 - 3.149\varepsilon + 8.405\varepsilon^2 - 8.431\varepsilon^3 \quad \text{for} \quad \varepsilon \le 0.5 \tag{8.97}$$

$$\beta = 0.0584 + 0.44\varepsilon \quad \text{for} \quad 0.5 \le \varepsilon \le 0.9 \tag{8.98}$$

Similarly for PID controller design for an unstable FOPTD model by the ISE method, the expression for set point weighting parameter β is proposed by Jhunjhunwala and Chidambaram (1999) as

$$\beta = 0.4772 \exp(-1.58\varepsilon) \tag{8.99}$$

The expression for β for PI controller using Eqs (8.95) for unstable FOPTD systems can be derived as

$$\beta = [\tau_I + k_c k_p(\tau_I - L)]/(2k_c k_p) \tag{8.100}$$

8.8.6 Digital Computer Implementation of PI/PID Controllers

The PID control law is given by

$$u(t) = k_c\left[e(t) + (1/\tau_I)\int e d\tau + \tau_D de/dt\right] + u_s \tag{8.101}$$

By using backward finite difference scheme for the derivative term and using sum term for the integral term we get the control algorithm as

$$u(n) = k_c\left[e(n) + (T/\tau_I)\sum_{k=0}^{n} e(k) + (\tau_D/T)\{e(n) - e(n-1)\}\right] + u_s \tag{8.102}$$

where u_s is the controller output when the error is zero (the initial valve position). The above equation is called position form of the algorithm.

Velocity form of algorithm: Let us write the position algorithm for n and $(n - 1)$th sampling instants:

$$u_n = k_c[e(n) + (T/\tau_I)\sum_{k=0}^{n} e(k) + (\tau_D/T)\{e(n) - e(n-1)]\tag{8.103}$$

$$u_{n+1} = k_c[e(n-1) + (T/\tau_I)\sum_{k=0}^{n-1} e(k) + (\tau_D/T)\{e(n-1) - e(n-2)\}] \tag{8.104}$$

Subtracting second term from the first

$$\Delta u(n) = u(n) - u(n-1) \tag{8.105}$$

$$= k_c[1 + (T/\tau_I) + (\tau_D/T)]e(n) - k_c[1 + (2\tau_D/T)]e(n-1) + (k_c\tau_D/T)e(n-2) \tag{8.106}$$

The above equation is called velocity form of the PID algorithm.

The velocity form of the algorithm has certain advantages over the position form:

(i) It does not need initialization. The value of control output when time t = 0 is not required. When the switchover from manual to automatic is to be made, then the knowledge of u (at $t = 0$) is to be fed to the controller for the implementation of position form of algorithm. Usually the value of u at $t = 0$ may not be available. For velocity form of algorithm, the value of u ($t = 0$) is not required.

(ii) The velocity form protects against integral windup. Presence of integral mode makes the controller output changing as long as there is a nonzero error. Usually, the errors cannot be eliminated quickly and they produce larger and larger values for the integral term, which increases the control action (u) until its saturated (i.e., the valve is fully closed or fully opened). This condition is known as an integral windup. Then, even if the error returns to zero, the control action will remain saturated. The velocity form is protected from the integral windup since as soon as the error changes sign, the control action returns within the control range in one sampling period.

The velocity form protects the control system against computer failure. When the computer fails, the velocity form algorithm keeps the last calculated value, whereas the position algorithm does not give any value for $u(k)$.

Position algorithms

The equation for PID controllers given by Eq. (8.102) is called position algorithm because the output of the algorithm is the control variable. In some cases, the control algorithm generate the rate of change of the control signal. Such a control law is called a velocity algorithm. Windup protection and bumpless transfer are easily implemented by using the velocity algorithm. However, a velocity form of algorithm cannot be used directly for a controller without integral action because such a controller cannot keep the stationary value. The system will have an unstable mode, an integrator, that is canceled. Set point weighting cannot be used in the velocity form of algorithm.

8.8.7 Selection of Sampling Period

The selection of sampling period is an important but complicated issue. For a FOPTD process, the sampling period is to be taken between $0.2L$ and L (i.e, $0.2 < (T/L) < 1$). If L is too small then $T < 0.1\tau$. Here L is the process time delay and τ the time constant. The value of the sampling period depends on whether an integral action or a derivative action is present in the controller. Tuning of PID controller may require adjusting the sampling period also. Several rules of thumb for choosing the sampling period for a digital PID regulator are given in the literature. There is a significant difference between PI and PID regulators. For PI regulators, the sampling period is related to the integral time. A typical rule of thumb is

$$T/\tau_I = 0.1 \text{ to } 0.3 \tag{8.107}$$

When Ziegler-Nichols tuning is used this equation implies

$$T/L = 0.3 \text{ to } 1 \tag{8.108}$$

$$T/P_u = 0.1 \text{ to } 0.3 \tag{8.109}$$

With PID control the critical issue is that the sampling period must be so short that the phase lead is not adversely affected by the sampling. This implies that the sampling period should be chosen so that the number TN/τ_D is in the range of 0.2 to 0.6. With $N = 10$ this means that for Ziegler-Nichols tuning the ratio of T/L is between 0.01 and 0.06. This gives

$$TN/\tau_D = 0.2 \text{ to } 0.6 \tag{8.110}$$

Significantly shorter sampling periods are thus required for controllers with derivative action.

However, it should be noted that in many of the concentration measurement, particle size analyzers etc., the sampling time is dictated by the measurement techniques or instruments used. If the sampling time is large, then the effect of sampling time on the design of digital controller should be taken into account as discussed earlier. Table 8.7 gives typical sampling times for few process variables.

Table 8.7 Choice of sampling period for digital control systems

Type of variable	Sampling period (sec)
Flow rate	1–3
Level	5–10
Pressure	1–5
Temperature	10–45

8.9 Design of Controllers Based on Discrete Time Model

So far we have discussed the design of digital controller based on continuous time controller. If the sampling time is small, then the performance of digital controller can approach to that of the continuous time controller. By designing the digital controller based on the discrete model will however, give improved performance than that of the method designed by converting continues time PID controllers. In this section let us discuss the method of designing such digital controllers. The design of RST controllers, pole placement controllers, deadbeat controllers, Dahlin controllers, internal model controllers, simplified model predictive controllers will be considered in the section.

8.9.1 Design of RST Controllers

In the conventional block diagram representation for continuous systems, the process is described by a ratio of numerator polynomial in s divided by a denominator polynomial in s. The controller is also to be defined by similar form. In discrete system, since the process is described by $A(q^{-1})y_k = B(q^{-1})u_k$, the controller will be described by

$$S(q^{-1})u_k = R(q^{-1})y_{r,k} - R(q^{-1})y_k \tag{8.111}$$

To make it more flexible we usually write $T(q^{-1})$ instead of $R(q^{-1})$ associated with $y_{r,k}$. The block diagram representation is shown in Fig. 8.6. Hence, the closed loop transfer function is given by

$$H_{cl}(q^{-1}) = B(q^{-1})R(q^{-1})/[A(q^{-1})S(q^{-1}) + B(q^{-1})R(q^{-1})] \tag{8.112}$$

$$= B(q^{-1})R(q^{-1})/P(q^{-1}) \tag{8.113}$$

in which

$$P(q^{-1}) = [A(q^{-1})S(q^{-1}) + B(q^{-1})R(q^{-1})] \tag{8.114}$$

$$= 1 + P_1 q^{-1} + P_2 q^{-2} + \ldots \tag{8.115}$$

defines the poles of the closed loop system. The objective of digital controller design first becomes specifying the desired closed loop locations and the polynomial A and B from which the polynomial R, S and T are to be calculated. The closed loop poles are generally chosen as a second order polynomial corresponding to the discretization of a second order continuous time system having a given damping coefficient ζ and the settling time t_s.

In Eq. (8.113), the product $B(q^{-1})R(q^{-1})$ defines the closed loop zeros. The digital PID controller, in general does not simplify the plant zeros [unless $B(q^{-1})$ is chosen as a factor of $P(q^{-1})$] and thus can be used for the regulation of plants having a discrete time model with unstable zeros. The digital PID controller introduces additional zeros defined by $R(q^{-1})$, which will depend on $A(q^{-1})$, $B(q^{-1})$, and $P(q^{-1})$, and therefore which cannot be specified *a priori*. In certain situations, these zeros may give rise to undesirable overshoots during the transient.

If $T(q^{-1})$ is considered as $R(1)$ instead of $R(q^{-1})$, the closed loop system will not introduce any additional zero. Then the closed loop pulse transfer function is given by

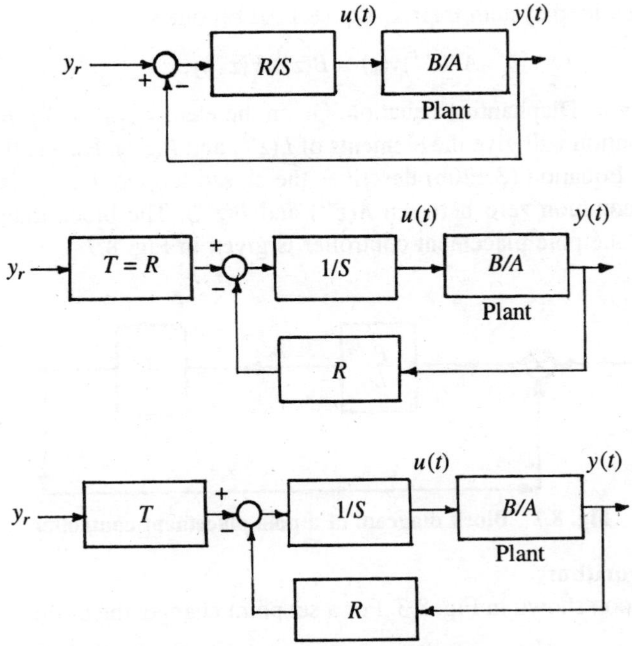

Fig. 8.6 Canonical structures of digital controller

$$H_{cL}(q^{-1}) = [P(1)/B(1)]\ [B(q^{-1})/P(q^{-1})] \tag{8.116}$$

The term $P(1)/B(1)$ in the above equation ensures a unity gain between the reference and the output in steady-state. The controller will eliminate the overshoot.

8.9.2 Pole Placement Algorithm

The process model is given by

$$A(z^{-1})y(z) = B(z^{-1})u(z) \tag{8.117}$$

The feedback controller is designed in order to have the closed loop system with desired pole location, which are specified by the zeros of $A^*(z^{-1})$.

Let us choose the control law as:

$$L(z^{-1})u(z) = P(z^{-1})\{y_r(z) - y(z)\} \tag{8.118}$$

where $y_r(t)$ is the desired reference input. Our interest is now to find $L(z^{-1})$ and $P(z^{-1})$.

Substitute the control law Eq. (8.118) in (8.117), after multiplying on Eq. (8.117) by $L(z^{-1})$ both sides, we get

$$A(z^{-1})L(z^{-1})y(z) = B(z^{-1})P(z^{-1})\{y_r(z) - y(z)\} \tag{8.119}$$

i.e.,

$$[A(z^{-1})L(z^{-1}) + B(z^{-1})P(z^{-1})]y(z) = B(z^{-1})P(z^{-1})y_r(z) \tag{8.120}$$

Let us define

$$A^*(z^{-1}) = A(z^{-1})L(z^{-1}) + B(z^{-1})P(z^{-1}) \tag{8.121a}$$

where A^* is the closed loop system matrix. Eq. (8.120) becomes

$$A^*(z^{-1})y(z) = B(z^{-1})P(z^{-1})y_r(z) \tag{8.121b}$$

Eq. (8.121b) is known as Diaphantine equation. Given the element of $A(z^{-1})$, $B(z^{-1})$ and $A^*(z^{-1})$, the solution of above equation will give the elements of $L(z^{-1})$ and $P(z^{-1})$. Hence, the control law (8.118) can be implemented. Equation (8.120b) describes the closed loop system. It is assumed that in Eq. (8.121), there is no common zero between $A(z^{-1})$ and $B(z^{-1})$. The block diagram of the feedback system incorporating the pole placement controller is given in Fig. 8.7.

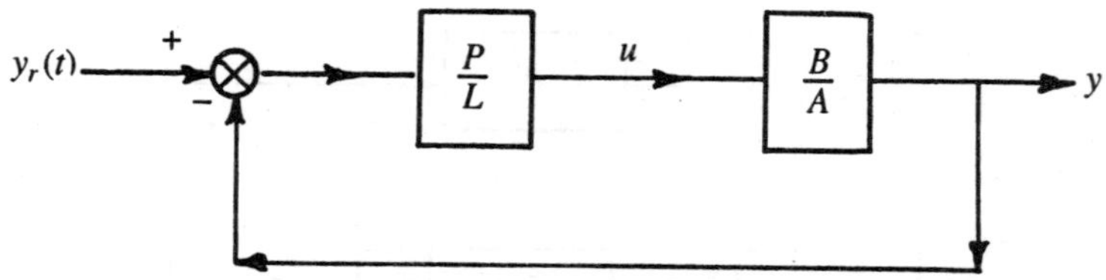

Fig. 8.7 Block diagram of a pole placement controller

8.9.3 Deadbeat Algorithm

Consider the control loop shown in Fig. 8.3. For a set point change, the closed loop transfer function is given by

$$y(z) = HG_p(z)D(z)/[1 + HG_p(z)D(z)] \tag{8.122}$$

Therefore
$$D(z) = [1/HG_p(z)][y(z)/y_r(z)]/[1 - \{y(z)/y_r(z)\}] \tag{8.123}$$

The deadbeat control algorithm requires there is no error in the output after $(n + 1)$ sampling instant, where n is the time delay of the process. For a unit step input in y_r, we have $y_r(z) = 1/(1 - z^{-1})$.

Therefore, we have

$$y(z) = z^{-N-1}/(1 - z^{-1}) \tag{8.124}$$

that is
$$y(z)/y_r(z) = z^{-N-1} \tag{8.125}$$

Hence the expression for the controller is given by

$$D(z) = [1/HG_p(z)]z^{-N-1}/(1 - z^{-N-1}) \tag{8.126}$$

Since the specification of the closed loop system is given at the sampling instance, the response of the output in between the sampling period is usually very large (i.e., larger overshoot in the response). Since the process output reaches in $N + 1$ sampling instant, the control value (u) also exhibit overshoot. However, the output reaches the desired value at the steady-state condition. The deadbeat method is not applicable for unstable systems. Such systems have to be first stabilized by a simple proportional controller and then deadbeat controller can be designed for the resulting stabilized system to get an improved performance.

Example: Design a deadbeat controller for the system

$$HG_p(z) = 10z^{-3}/(1-0.1z^{-1}) \tag{8.127}$$

Since the numerator of the system should be of the form $z^{-1}z^{-k}$, we observe for the present system $k = 2$. Hence, the deadbeat controller Eq. (8.126)

$$G_c(z) = X(z)/E(z) = [1/HG_p(z)]z^{-k-1}/(1 - z^{-k-1}) \tag{8.128}$$

becomes

$$X(z)/e(z) = (1 - 0.1z^{-1})/(1 - z^{-3}) \tag{8.129}$$

Hence

$$x(k) = 0.1[e(k) - 0.1e(k - 1) + 10x(k - 3)] \tag{8.130}$$

8.9.4 Dahlin's Algorithm

In this method the closed loop response is specified first in the continuous time domain and then the corresponding behaviour in discrete time domain is given. The closed loop response desired is given by that of a first order plus time delay system. We have

$$y(s) = \exp(-\theta s)/[(\mu s + 1)s] \tag{8.131}$$

where μ is the time constant of the desired response and θ is the dead time of the response assumed to be $\theta = kT$. Thus, we get the specification in discrete time domain as

$$x(z)/e(z) = \{1/HG_p(z)]az^{-N-1}/[1 - bz^{-1} - az^{-N-1}] \tag{8.132}$$

where

$$a = 1 - b; \quad b = \exp(-T/\mu) \tag{8.133}$$

It should be noted that when $\mu = 0$, the Dahlin's algorithm reduces to the deadbeat algorithm. Dahlin's algorithm is not applicable for unstable systems. The unstable system should first be stabilized by a simple proportional controller. Then, Dahlin's algorithm can be designed for the stabilized system to get an improved performance.

In discrete systems, the controller output sometimes oscillates with a period twice the sampling period, even through the process output response is good. This oscillation in u is called ringing. If the deadtime θ is an integral multiple of sampling time T, then ringing will not occur even when there exists uncertainty in the plant model. If the dead time θ is not an integer multiple of sampling time T, then $G(z)$ must have one z-plane zero lying on the negative real axis. This is the pole of the controller. This will introduce oscillation in the controller output.

To eliminate ringing, Dahlin has suggested that the problem pole of the controller is replaced by a pole at the origin with an equivalent steady-state gain to assure transfer function settling behaviour. This can be achieved by taking $z = 1$ in the problematic term.

8.9.5 Internal Model Controller

Let us first explain the method for continuous time systems. The internal model control (IMC) structure is shown in Fig. 8.8. The IMC scheme makes use of a process model to infer the effect of unmeasurable disturbances on the process output and then counteracts that effect. The controller consists of an inverse of the process model. It can be easily shown that (Fig. 8.8)

$$y(s) = y_r(s) \quad \text{for all } t > 0 \tag{8.134}$$

when

$$G_m(s) = G_p(s) \tag{8.135}$$

and

$$G_c(s) = [G_m(s)]^{-1} \tag{8.136}$$

i.e., perfect control could be obtained in the presence of any changes in the set point and load disturbances. When the model is exact, the stability of both controller and plant is sufficient for over all system stability. However, the realization of perfect control is not likely to occur in practice for the following reasons:

(i) a perfect model is seldom available, especially when the process is nonlinear and or time-varying. Modeling error is more likely to occur in practice.

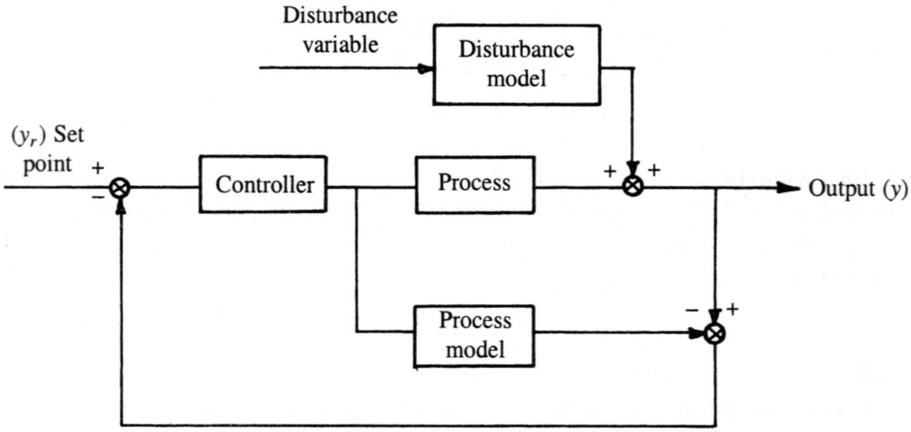

Fig. 8.8 Internal model control of system with disturbance

(ii) the exact inverse of a process model can never be physically realizable for systems with dead time and for nonminimum phase systems.

Usually a suitable filter is added to the controller to compensate modeling error. The design procedure for the IMC is as follows:

(i) factor the process model into two parts

$$G_p(s) = G_p^+(s)G_p^-(s) \tag{8.137}$$

such that the inverse of $G_p^-(s)$ is stable and causal and

$$G_p^+(0) = 1 \tag{8.138}$$

(ii) combine $G_p^-(s)$ with a filter to form the controller

$$G_c(s) = F(s)[G_p^-(s)]^{-1} \tag{8.139}$$

Usually $F(s)$ is a low pass filter with the filter time constant as an adjustable parameter. The filter is required in IMC system to make the controller realizable and to make the closed loop system robust to modeling errors. In particular, in the absence of modeling error, the filter determines the speed of response, and in the presence of modeling errors, the problem of robustness has to be addressed by adjusting the filter time constant. Several case studies involving simulation and experimental studies of linear IMC have shown that good performance is obtained over the conventional PI controller (Garcia and Morari, 1982; Arkun et al., 1986). Simple tuning formulae for analog PID controllers with a single adjustable parameter based on IMC principles are given by Rivera et al. (1986) and Lee et al. (1998).

Let us discuss the method of designing IMC for an example of a second-order plus dead time discrete time process

$$G_p(z^{-1}) = (\beta_1 z^{-1} + \beta_2 z^{-2})z^{-N}/(1 + \alpha_1 z^{-1} + \alpha_2 z^{-2}) \tag{8.140}$$

The predictive model is employed in the structure

$$G_{pr}(z^{-1})G_D(z^{-1}) = G_m(z^{-1}) = \sum_{i=1}^{n} \beta_i z^{-1}/[1 + \alpha_1 z^{-1} + a_2 z^{-2}] \tag{8.141}$$

The process model $G_m(z^{-1})$ is equivalent to an over-parameterized discrete transfer function for a process with dead time variations. The IMC structure is shown in Fig. 8.9. Figure 8.9 is modified into Fig. 8.10. The controller G_I is given by

$$G_I(z^{-1}) = \frac{G_c(z^{-1})}{1 + [G_c(z^{-1})(\Sigma \beta_i)z^{-1}/(1 + \alpha_1 z^{-1} + \alpha_2 z^{-2})]} \tag{8.142}$$

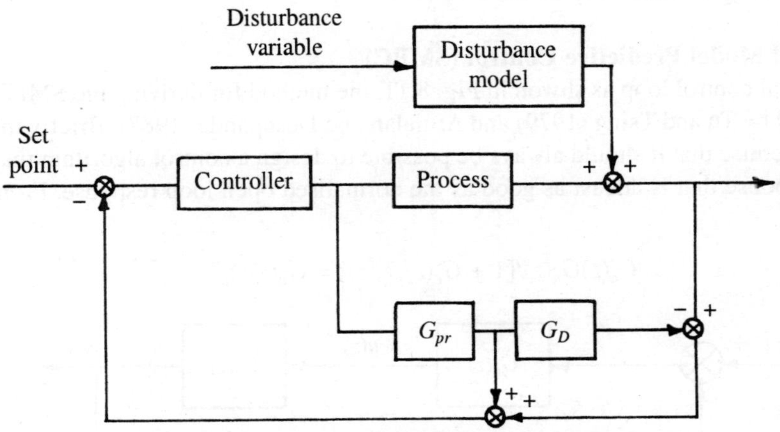

Fig. 8.9 Vogal-Edgar dead time compensator

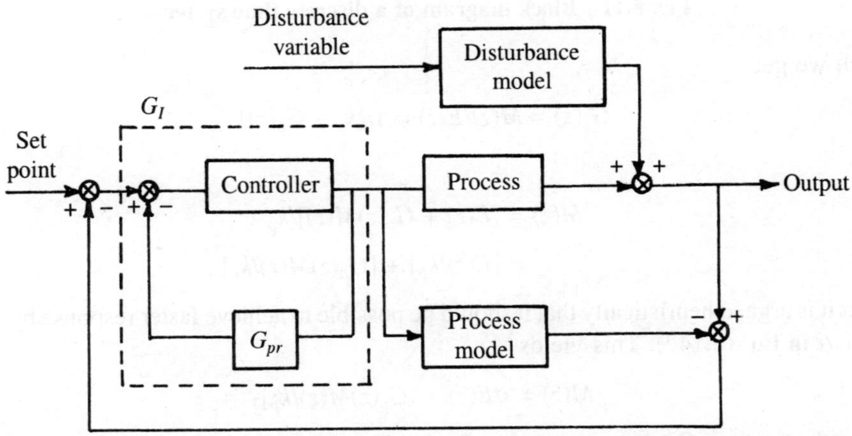

Fig. 8.10 Reconstructed Vogel-Edgar dead time compensator with a view point of IMC

Dahlin's controller can be assumed for $G_c(z^{-1})$ as:

$$G_c(z^{-1}) = \{1 - \exp(-T/\lambda)/(1 - z^{-1})\}(1 + \alpha_1 z^{-1} + \alpha_2 z^{-2})/(\sum_{i=1}^{r} \beta_i) \tag{8.143}$$

Substituting the Dahlin's controller for G_c in Eq. (8.142) we get the expression for IMC as

$$G_I(z^{-1}) = \frac{[1 + \alpha_1 z^{-1} + \alpha_2 z^{-2}][1 - \exp(-T/\lambda)]z^{-1}}{(\sum_{i=1}^{r} \beta_i)z^{-1}[1 - \exp(-T/\lambda)z^{-1}]} \tag{8.144}$$

where T is the sampling period and λ is the tuning constant in Dahlin's algorithm. The controller in the IMC structure corresponds to the inverse of the predictive model of Eq. (8.141) plus a first order filter with an adjustable constant. The above method is valid for stable processes. Methods of Designing PI/PID analog controllers based on IMC approach for unstable FOPTD and SOPTD systems without any zero are given by Rotstein and Lewin (1991). Simple tuning formulae with a single tuning parameter are given by Rotstein and Lewin (1991).

8.9.6 Simplified Model Predictive Control (SMPC)

For a typical digital control loop as shwon in Fig. 8.11, the method for deriving the SMPC algorithm has been provided by Tu and Tsing (1979) and Arulalan and Deshpande (1987). Briefly the approach is based on the premise that it should always be possible to design a control algorithm that will yield a closed loop response that is atleast as good as the normalised open loop response. From Fig. 8.11 we can write:

$$G_c(z)G_p(z)/[1 + G_c(z)G_p(z)] = G_p(z)/k_p \tag{8.145}$$

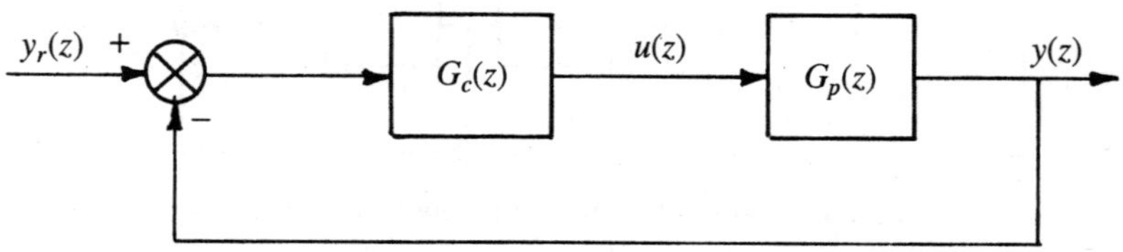

Fig. 8.11 Block diagram of a discrete time system

From which we get

$$G_c(z) = M(z)/E(z) = 1/[k_p - G_p(z)] \tag{8.146}$$

and hence

$$M(z) = [E(z) + G_p(z)M(z)]/k_p \tag{8.147a}$$

$$= [E(z)/k_p] + [G_p(z)M(z)/k_p] \tag{8.147b}$$

At this point it is argued heuristically that it should be possible to achieve faster response by introducing a parameter α in Eq. (8.147). This yields

$$M(z) = \alpha E(z) + [G_p(z)M(z)/k_p] \tag{8.148}$$

representing the process $G_p(z)$ in terms of its impulse response coefficients h_i as

$$G_p(z) = \sum_{i=1}^{N} h_i z^{-i} \tag{8.149}$$

Then the time domain equivalent can be written as

$$M_n = \alpha E_n + (1/k_p)[h_1 M_{n-1} + h_2 M_{n-2} + \ldots + h_N M_{n-N}] \tag{8.150}$$

Here αk_p varies from 1 to ∞. The above equation is the final form of the SMPC algorithm with α as the only tuning parameter.

As stated earlier that the SMPC is based on the premise that a control algorithm should yield a closed loop set point response that is at least as good as the normalized open loop response. There are in general two spectrum of control quality. At the one end there is the notion of perfect control; it is not possible for any controller to do any better. At the other end, there is the notion of open loop control. In IMC one begins with the notion of perfect control and the absence of modeling errors. Then, robustness is introduced in the presence of modeling errors by employing a filter with an adjustable parameter. In SMPC, in the absence of modeling errors we get a closed loop response that is same as the normalized open loop response. Since the controller does not involve an inverse of a process model, it would be more robust. The method is not applicable as such for unstable systems.

Summary

Sampling introduces delay in the controlled system. The relation between s variable and z variable is given by $z = \exp(Ts)$. Adding a delay of $0.5T$ to the process delay we can design PI/PID controllers based on continuous time domain method. Some of the methods for design of PI/PID controllers for stable and unstable continuous time systems are presented. Controller design methods specifically meant for discrete time systems are given by Deadbeat algorithms, Dahlin's algorithms, Pole placement algorithms, internal model controllers and simplified model predictive controllers. Each method has a separate tuning parameter to get an improved closed loop performance.

References

Albertos, P., R. Strietzel and N. Mort, Control Engg. Solutions, A Practical approach, IEEE Control Engg. Series No. 54, 1997.

Anand, D.K. and R.B. Zmood, Introduction to control systems, III Ed., Asian Books Pvt Ltd, New Delhi, 1995.

Arkun, Y., Hollett, J., Canney, W.M. and Morari, M., Experimental study of internal model control, Ind. Eng. Chem. Proc. Des. Dev. 25 (1986) 102–108.

Arulalan, G.R. and P.B. Deshpande, Simplified model predictive control, Ind. Eng. Chem. Res., 26 (1987) 51–59.

Astrom, K.J. and B. Wittenmark, Computer Controlled Systems, Theory and Design, Prentice-Hall, New Jersey (1991).

Astrom, K.J. and B. Wittenmark, Adaptive Control, Addison Wesley, 1989.

Astrom, K.J. and T. Hagglund, PID Controllers: Theory, design and Tunining, 2nd Ed., ISA Pub., North Carolina, (1995).

Chidambaram, M., S.N. Jyothi and S. Arvind, Design of PI/PID controller for systems with zero, Trends in Industrial Measurement and Automation, (Ed.) S. Renganathan, Allied Pub, New Delhi, 1999, pp. 298–301.

Chidambaram, M., Applied Process Control, Allied Pub, New Delhi, 1998.

Chidambaram, M., Design formula for a PID controller, Indian Chemical Engineer, 37 (1995) 90–93.

Chidambaram, M., Set point weighted PI/PID controllers, Chem. Engg. Commun., 179 (2000)1–13.

Clement, C. and M. Chidambaram, PID control of Unstable first order plus time delay systems, Chem. Engg. Commun., 162 (1997) 63–74.

Clement, C. and M. Chidambaram, PI control of Unstable first order plus time delay systems, Indian Chem. Engr., 39 (1997) 9–14.

Dahlin, E.B., Designing and tuning digital controllers, Instru. & Control Sys., 41 No. 6 (1968) 77–83.

Garcia, C.E. and M. Morari, Internal Model Control, 1. A unifying review and some new results. Ind. Eng. Chem. Proc. Des. Dev., 21, (1982) 308–323.

Gajic Z. and M. Lelic, Modern Control Engineering, Prentice-Hall, London,1996.

Haalman, A., Adjusting controllers for dead time processes, Control Engg., 12 (1965) 71–73.

Huang, H.P. and C.-C. Chen, Auto-tuning of PID controllers for second order unstable process having dead time, J1. Chem. Eng. Japan, 32 (1999) 486–497.

Isermann, R., Digital Control Systems, 2nd Ed., Springer-Verlag, 1991.

Jhunjhunwala, M.K. and M. Chidambaram, Tuning of PI and PID controllers for Unstable first order plus time delay systems, Chem. Eng. Commun., 185 (2001) 91–113.

Landau, I.D., System identification and control design, Prentice-Hall, New Jersey, 1990.
Lee, Y., S. Park, M. Lee and C. Brosilow, PID controller tuning for desired closed loop responses for SI/SO systems, AICHE J., 44 (1998) 106–115.
Marlin, T.E., Process control, designing processes and control systems for dynamic performances, McGraw Hill, New York 1995.
Neuman, C.P. and C.S. Baradello, Digital transfer functions for microcomputer control, IEEE Trans.-Sys. Man. Cyber., SMC-9 (12) 856-(1979).
Rao, M. and H. Qiu, Process control engineering, Gordon and Breach Science Pub., Germany, 1993.
Rivera, D.E., M. Morari and S. Skogestad, Internal model controller, 4. PID controller design, Ind. Eng. Chem. Res. Dev., 25 (1986) 252–260.
Rotstein, G.E. and D.R. Lewin, Simple PI and PID tuning for open loop unstable system, Ind. Eng. Chem. Res. 30 (1991) 1864–1869.
Shahian, B. and M. Hassul, Control system design using MATLAB, Prentice-Hall, Englewood Cliffs, New Jersey, 1993.
Smith, O.J.M., Close control of loops with dead-time, Chem. Eng. Prog., 53 (1957) 217–219.
Tu, C.Y.F. and Y.H.J. Tsing, Synthesizing a digital algorithm for optimizing control, Intech, 26 (5) (1979) 52–59.
Vogel, E.F. and T.F. Edgar, An adaptive dead-time compensation for process control, Proc. JACC, San Fransisco, CA, Paper TP5-E (1980).
Zhang, W.-D., Y.-X. Sun and X.-M. Xu, Robust digital controller design for processes with dead times: New results, IEEE Proc., CTA-145, 1998, 159–164.
Ziegler, J.G. and N.B. Nichols, Optimum settings for automatic controllers, Trans. ASME, 64 (1942) 759–768.

Review Questions

1. Explain with a block diagram a computer control loop.
2. For the system
   $y(k + 3) + 0.5y(k + 2) - 0.5y(k + 1) - 0.5y(k) = 2x(k - 1) + 0.5x(k)$ with zero initial conditions, derive the expression for $y(z^{-1})/x(z^{-1})$.
3. Find the inverse of z-transform of $G(z) = (3 - z^{-1})/[(2 - z^{-1})(1 - 0.6z^{-1})]$.
4. Given $z = \exp(Ts)$, derive the different approximation relations between z and s.
5. What is zero order hold equivalence?
6. For a first order system with delay of 11 secs and time constant of 9 secs and sampling time of $T = 2$ secs, derive the transfer function of the discrete time system (plant + zero order hold).
7. Explain the difference between zero order hold and first order hold elements. Which one will be recommended?
8. Compare the S- and Z-planes.
9. Given $H(s) = 1/[(2s + 1)(4s + 1)]$ with a sample time of 0.2 secs, find the expression for $H(w)$.
10. How do you modify a discrete time controller to take into account the effect of delay due to sampling?
11. Design a suitable discrete time controller via design of continuous time PID controller by Ziegler-Nichols method for the system $\exp(-2s)/(2s + 1)$. Use a sampling time of 0.5 secs.
12. Repeat Question (11) by using Haalman's method for the system $\exp(-2s)/[(2s + 1)(4s + 1)]$.
13. Discuss the two algorithms for implementing PID controller in a PC.
14. Design a suitable RST controller for the system
    $A(q^{-1})y(k) = q^{-1}Bu(k)$; $y(0) = 1$.
    where $A(q^{-1}) = 1 + 2q^{-1} + q^{-2}$; $B(q^{-1}) = (1 + 0.5q^{-1} + 0.5q^{-2})$
15. Design a suitable deadbeat algorithm for the system
    $$2.0 \exp(-2s)/(4s + 1); \text{ sampling time} = 1.$$
16. Design a suitable Dahlin's algorithm for the system in Q 15.

17. Design a IMC for the discrete system of the system in Q 15.
18. Design an SMPC for the system in Q 15.
19. Compare the IMC and SMPC strategies.
20. For the system $dy/dt + ay = bu$, use a suitable finite difference scheme method to convert it into a discrete time model.
21. For the system $dy/dt + ay^2 = bu$, first linearize the system around an operating point $(\bar{y}, \bar{u})$. Then apply the finite difference scheme to get the discrete time model.
22. For the nonlinear system in Q 21, apply the difference scheme to get a nonlinear discrete time model.

9

Design of Feed Forward Controllers

If an external distrurbance v of a process can be measured before it acts on the output variable Y, then the control performance with respect to this disturbance can often be improved by feed forward control (Fig. 9.1). Here, immediately after a change in the disturbance v, the process input u is manipulated by a feed forward controller G_{FF} which does not wait as with feedback control until the disturbance has affected the control variable Y. Significant improvement in control performance, however, can only be obtained for a restricted manipulated range if the process behaviour G_P is not slow compared with the disturbance behaviour G_V. When designing a control system one should always try to control the effects of measurable disturbances using feed forward, leaving the incompletely feed forward controlled parts and the effect of unmeasurable disturbances on the controlled variable to the care of the feedback controller.

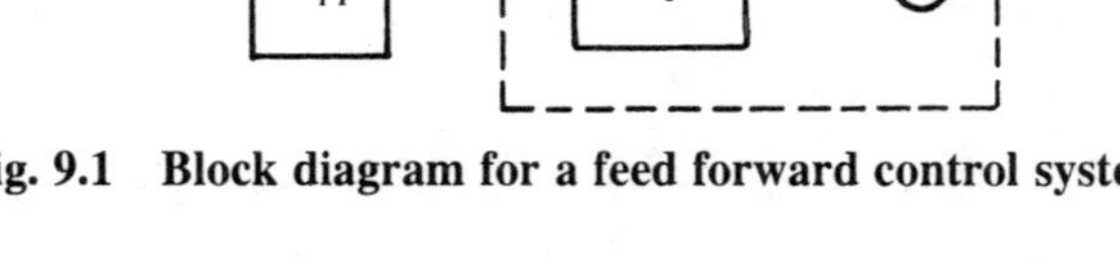

Fig. 9.1 Block diagram for a feed forward control system

9.1 Block Gigram

It is assumed that the process can be described by

$$y(z) = G_P(z)u(z) + G_V(z)v(z) \tag{9.1}$$

where $v(z)$ is a measurable disturbance signal and

$$G_P(z) = y(z)/u(z) = B(z^{-1})\, z^{-d_P}/A(z^{-1}) \tag{9.2}$$

$$= (b_1 z^{-1} + b_2 z^{-2} \ldots + b_n z^{-n})\, z^{-d_p} / (1 + a_1 z^{-1} + a_2 z^{-2} + \ldots + a_n z^{-n}) \tag{9.3}$$

$$G_V(z) = y(z)/v(z) = D(z^{-1})\, z^{-d_v} / C(z^{-1}) \tag{9.4}$$

$$= (d_1 z^{-1} + d_2 z^{-2} + \ldots + d_m z^{-m})\; z^{-d_v} / (1 + C_1 z^{-1} + C_2 z^{-2} + \ldots + C_m z^{-m}) \tag{9.5}$$

The feed forward controller has the transfer function

$$G_{FF}(z) = u(z)/v(z) = S(z^{-1})/R(z^{-1}) \tag{9.6}$$

$$= (s_0 + s_1 z^{-1} + \ldots s_p z^{-p})/(1 + r_1 z^{-1} + r_2 z^{-2} + \ldots + r_q z^{-q}) \tag{9.7}$$

Equation (9.1) is also given by the structure

$$A^*(z^{-1}) Y(z) = B^*(z^{-1})\, z^{-d} u(z) + D^*(z^{-1})\, z^{-dv} v(z) \tag{9.8}$$

where

$$A^*(z^{-1}) = 1 + \alpha_1 z^{-1} + \alpha_2 z^{-1} + \ldots + \alpha_n z^{-n} \tag{9.9}$$

$$B^*(z^{-1}) = \beta_1 z^{-1} + \beta_2 z^{-2} + \ldots + B_n z^{-n} \tag{9.10}$$

$$D^*(z^{-1}) = \delta_1 z^{-1} + \delta_2 z^{-2} + \ldots + \delta_n z^{-n} \tag{9.11}$$

$A^*(z^{-1})$ is the common denominator of $G_P(z)$ and $G_V(z)$ · $B^*(z^{-1})$ and $D^*(z^{-1})$ are the corresponding extended numerators. The expressions for *FFC* will be given in the next section both in terms of the polynomial *A*, *B*, *C* and D^*, and the corresponding extended polynomials A^*, B^* and D^* (where $A^* = AC$; $B^* = BC$; $D^* = DA$).

9.2 Feed Forward Control (FFC) Algorithm

9.2.1 Completely Dynamic FFC

The transfer function of the controlled system is given by

$$G(z) = Y(z)/v(z) \tag{9.12}$$

$$= [D(z^{-1})\, z^{-d_v} / C(z^{-1})] + [G_C(z)\, B(z^{-1})\, z^{-d_p} / A(z^{-1})] \tag{9.13}$$

The condition $G(z) = 0$ yields the transfer function of the complete dynamic feed forward controller given by

$$G_{FF,CD}(z) = -\, D(z^{-1})\, A(z^{-1})\, z^{-(d_v - d_p)} / [C(z^{-1})\, B(z^{-1})] \tag{9.14}$$

$$= -\, D^*(z^{-1})\, z^{-(d_v - d_p)} / B^*(z^{-1}) \tag{9.15}$$

The *CD* feed forward controller is realizable and applicable if

(i) $d = d_p - d_v \le 0$ (9.16)

(ii) the transfer function $G_{CD,\,FF}$ is stable (this means that the disturbance filter should be stable and the process transfer function G_P should be minimum phase).

In the following sections, some feed forward controller algorithm will be derived for the systems which do not fulfill conditions (i) and/or (ii) without loss of generality only the case, $d_v = 0$, $d_p = d \geq 0$, is considered. For $d_v > d_p$ all *FFC* algorithms, as derived for the case $d = d_p - d_v = 0$, can be applied by delaying the control signal u for $(d_v - d_p)$ steps.

9.2.2 Static FFC

The easiest way to overcome the difficulties pointed out in the last section is the application of the static feed forward control. Here, for the design of *FFC*, only the steady-state process description is considered. The static controller is given by

$$G_{ST,FF} = -D(1)\,A(1)/[C(1)\,B(1)] \tag{9.17}$$

$$= -D^*(1)/B^*(1) \tag{9.18}$$

This compensation causes only approximate cancellation of disturbance effect and it causes offset-free only at steady state condition. This controller should be applied for step change in the disturbances, Since it gives a simple proportional control action and it is also applicable for nonminimum phase processes.

9.2.3 Partial Static FFC

In this case we substitute $z = 1$ only for the problematic part of the transfer function. For nonminimum phase processes and $d = 0$, the partial static feed forward control may be applied to the system according to:

$$G_{PS,FF} = -D(z^{-1})\,A(z^{-1})/[C(z^{-1})\,B(1)] \tag{9.19}$$

$$= -D^*(z^{-1})/C^*(z^{-1}) \tag{9.20}$$

This control avoids the undesired instability of *FFC* due to nonminimum phase processes. The controller performance will be better than given in Case 9.2.2.

9.2.4 Minimum Time FFC for Step Change in *v*

The minimum time feed forward controller is characterized by the minimum time settling of the controlled system Y after a step change of the disturbance v, i.e.,

$$v(z) = 1/(1 - z^{-1}) \tag{9.21}$$

For a minimal settling time of d steps, the transfer function of the controlled system has to fulfill the condition

$$G(z) = y(z)/v(z) = F^*(z) \tag{9.22}$$

$$= D(z^{-1})/C(z^{-1}) + [G_{FF}\,B(z^{-1})z^{-d}/A(z^{-1})] \tag{9.23}$$

$$= F^*(z) \tag{9.24}$$

where

$$F^*(z^{-1}) = f_1^* z^{-1} + f_2^* z^{-2} + \ldots + f_{d+1}^* z^{-(d+1)} \tag{9.25}$$

and it contains a zero.

This gives the controller transfer function as

$$G_{MT,FF} = -A(z^{-1})\,L(z^{-1})/[B(z^{-1})\,C(z^{-1})] \tag{9.26}$$

$$= -L^*(z^{-1})/B^*(z^{-1}) \tag{9.27}$$

$L(z^{-1})$ and $L^*(z^{-1})$ result from the polynomial identities as

$$D(z^{-1}) = F^*(z^{-1})\, C(z^{-1}) + z^{-d}\, L(z^{-1}) \tag{9.28}$$

$$D^*(z^{-1}) = F^*(z^{-1})\, A^*(z^{-1}) + z^{-d}\, L^*(z^{-1}) \tag{9.29}$$

with the additional condition

$$F^*(1) = 0 \tag{9.30}$$

and

$$L(z^{-1}) = 1_1 z^{-1} + 1_2 z^{-2} + \ldots + 1_{n+1} z^{-(n+1)} \tag{9.31}$$

i.e.,

$$L^*(z^{-1}) = L(z^{-1})\, A(z^{-1}) \tag{9.32}$$

The minimum time feed forward control is applicable and stable if condition (ii) in section (9.2.1) is fulfilled (i.e., the system should be minimum phase and the disturbance filter should be stable). The system output Y is settled after d steps after a step change in the disturbance variable v, while the process input u continues to oscillate with decreasing amplitude.

9.2.5 Deadbeat FFC

The deadbeat feed forward control is characterized by the finite time settling of the process output Y, and process input u after a step change of the disturbance variable v. Then, the responses of the system input u and the process output Y have to fulfill the equations

$$y(z) = Q(z^{-1}) = q_1 z^{-1} + q_2 z^{-2} + \ldots + q_N z^{-N} \tag{9.33}$$

$$u(z)\,(1 - z^{-1}) = H(z^{-1}) = h_0 + h_1 z^{-1} + \ldots + h_M z^{-M} \tag{9.34}$$

which together with $v(z) = 1/(1 - z^{-1})$ gives for the controlled system

$$G(z) = y(z)/v(z) \tag{9.35}$$

$$= Q\,(z^{-1})(1 - z^{-1}) = Q^*(z^{-1}) \tag{9.36}$$

$$G_{FF} = u(z)/v(z) \tag{9.37}$$

$$= H(z^{-1})(1 - z^{-1})/(1 - z^{-1}) \tag{9.38}$$

$$= H(z^{-1}) \tag{9.39}$$

The order of $H(z)$ is represented by the notation N_h.

Replacing $G(z)$ and $G_{FF}(z)$ by the above two equations in the transfer funcion of the controlled system, we get

$$G(z) = Y(z)/v(z) \tag{9.40}$$

$$= [D^*(z^{-1})/A^*(z^{-1})] + [G_{FF}(z)\, B^*(z^{-1})\, z^{-d}/A^*(z^{-1})] \tag{9.41}$$

The above equation gives the polynomial identity

$$A^*(z^{-1})\, Q^*(z^{-1}) - B^*(z^{-1})\, H(z^{-1})\, z^{-d} = D^*(z^{-1}) \tag{9.42}$$

The parameters of $H(z^{-1})$ and $Q^*(z^{-1})$ can be determined from the above equation for the polynomial orders:

$$N = n + d;\ M = n + d - 1 \quad \text{and} \quad N_h = n \tag{9.43}$$

and the additional condition

$$Q^*(1) = 0 \tag{9.44}$$

The condition for the *FFC* to be realizable is that the $A^*(z^{-1})$ and $B^*(z^{-1})$ do not contain common zeros. Then the deadbeat *FFC* is given by

$$G_{DB,FF} = H(z^{-1}) \tag{9.45}$$

After a step change in the disturbance υ, the process input u is settled after n steps, and the process output Y after $(n + d - 1)$ steps, if deadbeat *FFC* can be applied.

In deadbeat and also in minimum time feedforward control algorithm we come across the following type of polynomial equation

$$AX + BY = C \tag{9.46}$$

where A, B and C are known polynomials in z^{-1}. X and Y are unknown polynomial in z^{-1} which are to be evaluated. Both the order of the polynomials and the coefficient of X and Y polynomials are to be evaluated. This type of equation is called Diophantine equation (also known as Aryabhatta equation). The following condition is to be satisfied for the existence of the solution:

greatest common factor of A and B divides C

The following condition is to be satisfied for unique solution

$$\deg(X) < \deg(B)$$

or

$$\deg(Y) < \deg(A) \tag{9.47}$$

The values of the coefficients are calculated by solving algebraic equations obtained by equating coefficients of the corresponding powers of the polynomial equation.

9.2.6 Minimum Variance FFC

For stochastic disturbances υ with $E[\upsilon(k)] = 0$, a feed forward control strategy can be derived which minimizes the criterion

$$I(k + d + 1) = E[\, y^2(k + d + 1)] \tag{9.48}$$

Here, operator E is an expected operator. $E[\upsilon(k)]$ means the expected value of $\upsilon(k)$. The minimization of $I(k + d + 1)$ yields the minimum variance *FFC* as

$$G_{M\upsilon 1,FF} = -\,[A(z^{-1})\, L(z^{-1})]/[B(z^{-1}) C(z^{-1})] \tag{9.49}$$

$$= -\,L^*(z^{-1})/B^*(z^{-1}) \tag{9.50}$$

$L(z^{-1})$ and $L^*(z^{-1})$ are calculated respectively from the following polynomial identities

$$D(z^{-1}) = F(z^{-1})\, C(z^{-1}) + z^{-d} L(z^{-1}) \tag{9.51}$$

$$D^*(z^{-1}) = F(z^{-1})\, A^*(z^{-1}) + z^{-d}\, L^*(z^{-1}) \tag{9.52}$$

where $F(z^{-1})$, $L(z^{-1})$ or $L^*(z^{-1})$ are defined as

$$F(z^{-1}) = f_1 z^{-1} + f_2 z^{-2} + \ldots + f_d z^{-d} \tag{9.53}$$

$$L(z^{-1}) = l_1 z^{-1} + l_2 z^{-2} + \ldots + l_d z^{-d} \tag{9.54}$$

$$L^*(z^{-1}) = \lambda_1 z^{-1} + \lambda_2 z^{-2} + \ldots + \lambda_d z^{-d} \tag{9.55}$$

The minimum variance controller yields the following transfer function of the controlled system:

$$G_{MV1,FF} = y\ (z)/v\ (z) \tag{9.56}$$

$$= [D\ (z^{-1})/C\ (z^{-1})] - [L\ (z^{-1})\ A\ (z^{-1})\ B\ (z^{-1})C\ (z^{-1})]\ [B\ (z^{-1})/A\ (z^{-1})]\ z^{-d} \tag{9.57}$$

$$= F\ (z^{-1}) \tag{9.58}$$

i.e., the impulse response is zero after d steps and the step change response is settled with static error after the same time (minimum time impulse control). For a stable *MV*1 *FFC* still condition (ii) (Section (9.2.1)) (i.e., system minimum phase and disturbance filter stable) has to be fulfilled.

9.2.7 Extended Minimum Variance FFC (MV2)

The minimization of a second criterion

$$I(k + d + 1) = E[\,y^2(k + d + 1) + ru^2\ (k)] \tag{9.59}$$

yields the feed forward controller transfer function

$$G_{MV2,FF} = A\ (z^{-1})L\ (z^{-1})/[B(z^{-1})\ C(z^{-1}) + (r/b_1)A(z^{-1})C(z^{-1})\ z^{-1}] \tag{9.60}$$

$$= L^*(z^{-1})/[B^*(z^{-1}) + (r/b_1)\ A^*(z^{-1})\ z^{-1}] \tag{9.61}$$

The polynomials $L(z^{-1})$ or $L^*(z^{-1})$ is calculated from eq (9.52).

The introduction of weighting factor r not only gives the possibility of reducing the control effort but also changes the poles of the controller transfer function so that a stable control for sufficiently larger r may be achieved although condition (ii) (Section (9.2.1)) is not fulfilled. Table 9.1 gives the advantages and limitations of FFC over FBC.

Table 9.1 Advantages and limitations of FFC over FBC

Advantages
- (i) rapid corrective action
- (ii) delay in feedback action is avoided
- (iii) always stable for open loop stable process
- (iv) simple design procedure

Limitations
- (i) disturbance must be measurable
- (ii) system dynamics (G_p and G_v) should be known
- (iii) for each disturbance, one controller and one measuring element required
- (iv) the required transfer function of the controller must be realizable (it should not require any prediction)
- (v) FFC can not stabilize the open loop unstable system

9.3 Feed Forward Controllers for Servo Response

The problem in tracking control system design is to make the controlled variable y follow the command input y_r as closely as possible. If the model G_P of a stable process is known exactly and if there is no other disturbance, this problem could be solved by a feed forward control system (Fig. 9.2). Ideally, one could require that the output y follows the input y_r exactly. This would be the case if

$$G_{FF}(z) = G_{FF}^{0}(z) = 1/G_p(z) \tag{9.62}$$

If $G_{FF}^{0}(z)$ is realizable, this feed forward element would "compensate" the process completely since it has reciprocal transfer function behaviour.

Fig. 9.2 Feed forward controller for a servo problem

For processes with time lags, however, the feed forward element is not realizable and one has to add a "realizability term"

$$G_{FF}(z) = G_{FF}^{0}(z)\, G_{FF}^{R}(z) = [1/G_p(z)]\, G_{FF}^{R}(z) \tag{9.63}$$

which leads to deviations between y_r and y. In considering the cancellation of poles and zeros of the feed forward element and the process, the effects discussed earlier have to be taken into account.

Summary

The feed forward controller reduces the effects of measurable disturbances which have significant effects on the output. Various discrete-time feed forward controllers are completely dynamic FFC, static FFC, partial static FFC, minimum variance FFC, minimum time FFC and deadbeat FFC. For systems exhibiting nonminimum phase behaviour, partial static FFC and static FFC are alone applicable.

References

Isermann, R., Digital control systems, Springer-Verlag, Berlin (1980).

Schumann, R. and H. Christ, Adaptive feedforward controllers for measurable disturbances. Joint Auto. Control Conf., Denver (1979) 500–506.

Sternad, M. and T. Soderstrom, LQG-optimal feedforward regulators, Automatica 24 (1988) 557–561.

Review Questions

1. Design: (i) dynamic FFC, (ii) static FFC, (iii) minimum time FFC and (iv) deadbeat FFC for the system

$$G_P(z^{-1}) = (4 - z^{-1})/[(2 - z^{-1})(3 - z^{-1})$$
$$G_V(z^{-1}) = z^{-1}/[(1.5 - z^{-1})(4 - z^{-1})]$$

2. Design a suitable partial static FFC for the system:

$$G_P(z^{-1}) = (0.8 - z^{-1})/[(2 - z^{-1})(3 - z^{-1})]$$

G_V is the same as in $Q(1)$

3. Herbst et al. (Powder Technology, 1992, V 69, 21–32) have presented a feed forward supervisory control of a flotation circuit. Give a write up about this work.
4. Rix et al. (Chem. Engg. Commun., 1997, V 159, 105–118) have presented an experimental implementation of feedforard control for a distillaion column. Give a write up about the work.
5. Bhat et al. (Int. J. Contr., 1991, *V*51, 237–243) have presented a simulation study on an adaptive deadbeat FFC for a nonlinear system. Give a sumary of their study.

10

Model Identification

The basic concepts in model identification using the input output data of the system are brought out in this chapter. The least square method and its modifications are also discussed.

10.1 Modeling and Identification

Before going into the identification techniques, some distinctions will be established among the terms *identification, modeling and estimation*. The term *modeling* usually refers to the description of the system by set of mathematical equations—a mathematical model-of the system, based on physical principles or empirical relation or other *a priori information*. Estimation refers to the process of obtaining the parameters of a given model. Models, usually take the form of linear difference equations. The term *identification* refers to the process of selecting a parameterized model. from a set of them, and then obtaining its parameters. That is, identification refers to modeling followed by estimation. Modeling from physical and chemical laws besides needing a good knowledge of the process, will eventually require knowledge of physical and chemical process constants like mass transfer coefficient, heat transfer coefficient, reaction rate constants, etc., which may be difficult to obtain with reasonable accuracy. The model may also need the measurement of all the state variables.-

The identification of discrete time parametric model (such as Eq. 8.13) offers following advantages:

- parametric identification methods allow the modeling of process disturbances
- relatively easy to construct and use
- nonparametric models such as step or frequency response type may be obtained accurately from the parametric models.
- recursive on-line identification algorithms may be used if required (i.e., in adaptive control)

However, the model obtained by system identification has limited validity unless the model is updated frequently. The identified model gives little physical insight, since in most cases the parameters of the model have no direct physical insight. The parameters are used only as tools to give a good description of the system's overall behaviour around the operating point.

10.2 System Identification

System identification is the development of mathematical models of a dynamic system based on measured data. The outputs are also affected by noise signal through measured signal. Such 'unmeasured' signal will be called disturbance signal or noise. The input-output or transfer function model is familiar to control engineers in the frequency domain with Laplace transforms. Similarly, in the discrete time case (Chapter 8), it is called a pulse transfer function, and it is given by

$$H(z^{-1}) = B(z^{-1})/A(z^{-1}) \tag{10.1}$$

where A and B are polynomials in z^{-1}

$$A(z^{-1}) = 1 + az^{-1} + a_2 z^{-2} + \ldots + a_n z^{-n} \tag{10.2}$$

$$B(z^{-1}) = b_0 + b_1 z^{-1} + b_2 z^{-2} + \ldots + b_m z^{-m} \tag{10.3}$$

If we consider the equivalent time description, then we have a difference equation relating the output sequence $\{y(t)\}$ to the input sequence $\{u(t)\}$. Here we use the backward-shift operator q with the property that $q^{-n}\, y(k) = y(k-n)$

$$A(q^{-1})y(k) = B(q^{-1})u(k) \tag{10.4}$$

$$y(k) + a_1 y(k-1) + \ldots + a_n y(k-n) = b_0 u(k) + b_1 u(k-1) + \ldots + b_m u(k-m) \tag{10.5}$$

Since there is difference between the model and actual data, the above model is usually written in the general form called auto regressive Moving-Average model with exogenous inputs (ARMAX)

$$A(q^{-1})y(k) = B(q^{-1})u(k) + C(q^{-1})e(k) \tag{10.6}$$

where A, B, C are polynomials in q^{-1} and $\{e(k)\}$ is noise (known as white noise). Equation (10.6) is of the form

$$y(k) = -\sum_{j=1}^{n_n} a_j y(k-j) + \sum_{j=0}^{n_b} b_j u(k-j) + \sum_{j=0}^{n_c} c_j e(k-j) \tag{10.7}$$

In a typical process application, this model would relate the process response sequence $\{y(k)\}$ to the input sequence $\{u(k)\}$, called an exogenous input in the statistical time series literature. In the process control applications, this input is usually deterministic and represents the time varying value of a manipulated process variable. The 'disturbance sequence' or 'noise sequence' $\{e(k)\}$ represents the combined effects of measurement noise, modeling errors (neglected nonlinearities), unmeasured disturbances etc.

With $n_a = 0$ and $n_b = \infty$, the model corresponds to the convolution model with $\{b_j\}$ representing the impulse response of the system. When $n_a = 0$ the model is called moving average model. When $n_a = \infty$ and $n_b = 0$ it represents an infinite order autoregressive model. Some linear systems are best approximated by 'low order autoregressive models', while other systems are best approximated by 'low order moving average models'.

The ARMAX model is a general form for the following subclass models:

1. Pulse transfer function model (infinite impulse response (IIR) model): $C(.) = 0$
2. Finite impulse response (FIR) model: $A(.) = 1$, $C(.) = 0$
3. Autoregressive model with exogenous input (ARX): $C(.) = 1$
4. Autoregressive (AR) model: $B(.) = 0$, $C(.) = 1$
5. Moving average (MA) model: $A(.) = 1$, $B(.) = 0$
6. Autoregressive moving average (ARMA) model:. $B(.) = 0$

All these models are known as time-series models ('time-series' here simply means a sequence of data ordered in time). The sequences $\{y_k\}$ and $\{u_k\}$ are called discrete time series. In many cases, the model parameters (A and B polynomials) can be identified by using a linear regression approach. We also have some more sub models known as Box-Jenkins model and output error model, respectively, given by

$$y(k) = [B(q^{-1})/F(q^{-1})]u(k-m) + [C(q^{-1})/D(q^{-1})]e(k) \quad (10.8)$$

$$y(k) = [B(q^{-1})/F(q^{-1})]u(k-m) + w(k) \quad (10.9)$$

The AR model is effective for model harmonics confounded with noise. The moving average (MA) model is popular in signal processing as a basis for filter design (FIR) and the identification of truncated impulse responses. The ARMA model is effective in modeling disturbance spectra with spectral peaks as well as zeros and therefore used in model based spectral analysis.

10.3 Interpretation of Noise Source

There are many sources and causes for the disturbances term given in Eq. (10.6), e.g. (i) measurement noise: the sensors that measure the signals are subject to noise and drift and (ii) uncontrollable noise: the system is subject to signals that have the character of inputs, but are not controllable by the user. The character of the disturbances could also vary within wide ranges. Usually in deterministic control study, the step and/or pulse changes are considered disturbance variables. In stochastic control the disturbances are modeled as realization of stochastic processes. The disturbances in some cases can be separately measurable, but in typical situation they are noticable only via their effect on the output. Sometimes the measurement of the input to the system may also be noise corrupted. In such cases we regard the measurable input values as the actual inputs $u(t)$ to the process, and their deviations from the time stimuli will be propagated through the system and lumped into the disturbance.

The disturbance can be written as:

$$v(k) = R(q^{-1})e(k) \quad (10.10)$$

where $\{e(k)\}$ is a white noise with variance λ. Here $R(q^{-1})$ is a rational function and represented by

$$R(q^{-1}) = C(q^{-1})/A(q^{-1}) \quad (10.11)$$

where

$$C(q^{-1}) = 1 + c_1 q^{-1} + \ldots + c_n q^{-n} \quad (10.12)$$

$$A(q^{-1}) = 1 + a_1 q^{-1} + \ldots + a_n q^{-n} \quad (10.13)$$

The equation can be written as

$$v(k) + a_1 v(k-1) + \ldots + a_{na} v(k-n_a) = e(k) + c_1 e(k-1) + \ldots + c_{nc} e(k-n_{ac}) \quad (10.14)$$

Such a representation of a stochastic process is known as an ARMA (autoregressive moving average) model. If $n_c = 0$, it is called AR (autoregressive) model

$$v(k) + a_1 v(k-1) + \ldots + a_{na} v(k-n_a) = e(k) \quad (10.15)$$

and if $n_a = 0$, we have a MA (moving average) model

$$v(k) = e(k) + c_1 e(k-1) + \ldots + c_{nc} e(k-n_c) \quad (10.16)$$

In many cases of system identification, the effect of noise on the output is insignificant compared to those of the input. With good signal to noise ratio [(SNR) > 10], it is less important to get a accurate noise model. The noise model contains all the influences on the measured y, known and unknown, that are not contributed in the input u. Even if an experiment is repeated with the same input, the output signal will typically be somewhat different. The noise source $e(t)$ need not have physical significance. Describing the effects of unknown variable arising from a white noise source via a transfer function H is just a convenient way of capturing the characteristic.

If the noise model is used for simulation, i.e., the response to various inputs are to be studied, then the noise model plays no immediate role. Since the noise source $e(t)$ for new data will be unknown, it is taken as zero in the simulations so as to study the effects of input alone (noise free simulation). Making another simulation with $e(t)$ being arbitrary with white noise will reveal how reliable the results of the simulation are, but it will not give a more accurate simulation result for the actual system's response.

To calculate the disturbance influence, a random number generation (in the computer) is used to produce a sequence of numbers $e(k)$, $k = 1, 2, \ldots, n$ that can be considered as a realization of a white noise stochastic process with variance λ. Then the disturbance is calculated as

$$v(k) = H(q^{-1})e(k)$$

The parameters of $H(q^{-1})$ are to be obtained by least square methods as will be discussed later. If the noise spectrum is found out experimentally, then the transfer function model $H(q^{-1})$ can be obtained, provided the noise spectrum is rational.

The Gaussian discrete time white noise is a sequence of independent equally distributed random variables of zero mean value and variance σ^2. This sequence will be denoted by $\{e(t)\}$ and will be characterized by $(0, \sigma)$, where the first number indicates the mean value and the second number indicates the standard deviation.

10.4 Equation Error Model

The most single input-single output relationship is represented by a linear equation

$$y(k) + a_1 y(k-1) + \ldots + a_{na}\, y(k-n_a) = b_1 u(k-1) + \ldots + b_{nb} u(k-n_b) + e(k) \quad (10.17)$$

Since white noise term $e(t)$ enters as a direct error in the difference equation, the model Eq. (10.17) is called equation error model structure. The adjustable parameters are

$$\theta = [a_1, a_2, \ldots, a_{na}, b_1, \ldots b_{nb}]^T \quad (10.18)$$

Equation (10.17) is rewritten in the form of Eq. (10.4) by defining

$$A = 1 + a_1 q^{-1} + \ldots + a_{na} q^{-n_a} \quad (10.19)$$

$$B = b_1 q^{-1} + \ldots + b_{nb} q^{-n_b} \quad (10.20)$$

The system model is given by

$$y(k) = G(q^{-1})u(k) + H(q^{-1})e(k) \quad (10.21)$$

where

$$G(q^{-1}) = B(q^{-1})/A(q^{-1}) \quad (10.22)$$

$$H(q^{-1}) = 1/A(q^{-1}) \quad (10.23)$$

This model is called ARX model, where AR refers to the autoregressive part $A(q^{-1})y(k)$ and X to the extra input $B(q^{-1})u(k)$ (called exogeneous variable in econometrics).

10.5 ARMAX Model Structure

The basic disadvantage with a simple model like ARX is the lack of adequate freedom in describing the properties of the disturbance term. By describing the equation error as a moving average of white noise, we get the noise model as

$$y(k) + a_1 y(k-1) + \ldots + a_{na}\, y(k-n_a) = b_1 u(k) + \ldots + b_{nb}\, u(k-n_b) + e(k) + c_1 e(k-1) + \ldots + c_{nc} e(k-n_c) \tag{10.24}$$

This model can be rewritten as Eq. (10.6)

$$A(q^{-1})y(k) = B(q^{-1})u(k) + C(q^{-1})e(k) \tag{10.25}$$

The adjustable parameters are given by

$$\theta = [a_1, \ldots, a_{na}, b_1, \ldots, b_{nb}, c_1, \ldots, c_{nc}]^T \tag{10.26}$$

In view of the moving average (MA) part $C(q^{-1})e(k)$, the model is called ARMAX. A version with an enforced integration in the system representation is the ARIMA model (I for integration). ARIMAX is useful to describe systems with slow disturbances. The model is obtained by replacing $y(t)$ by $\Delta y(k)$, where $\Delta y(k) = y(k) - y(k-1)$.

10.6 Parameter Estimation

There are two approaches for parameter estimation: (i) off-line and (ii) on-line. Off-line methods use all the data availabe prior to analysis. They usually give parameters with large precision and set no time limit for the process analysis. The record of input and output data set first collected and the parameter estimates are based on entire data record. The on-line method gives estimates recursively (sequentially) as the measurements are obtained within the time limiting imposed by the sampling period. The parameter estimates are recursively calculated for every data set so that the new data is used to update existing estimates. The on-line methods are used in adaptive control, especially in the case of time varying processes. Clearly, if the updating process can be made very fast it becomes possible to get parameter estimates of time varying system with reasonable accuracy (on-line real time identification).

Basic steps of system identification are:

1. Collect the input-output data
2. Use a set of candidate models (the model structure: ARX, ARMA etc.)
3. A criteria to select the particular model in the set, based on the data
4. Examine the obtained model parameters
5. If the model is good enough, then stop; otherwise we have to assume different model structure and repeat the identification procedure.

When recording experimental data occasionally large measurement errors may occur. Such errors can be caused by disturbances, conversion failures etc. The corresponding abnormal data points are called 'outliers'. If no specific action is taken, the outliers will influence the estimates model considerably. The outliers tend to appear as spikes in the sequence of prediction errors, and will give large contributions to the loss function.

We have to examine the data, polish it so as to remove the trends and outliers, select the useful portion of the original data and apply filtering to enhance important frequency range. (Real data are also subjected to possible large disturbances such as temperature sensor failure or transmitter failure. It is important that such outliers are not allowed to affect the models too strongly.)

It is generally recommended that atleast the mean value for the data be removed before the estimation phase. By filtering the input and output signals through a linear filter (same filter for all signals), we can focus on whether the model is fit to the system in the specific frequency ranges. Pre-filtering is a good alternative to de-trending (by cutting out low frequencies from the pass band). We can also make sure that the model concentrates on the important frequency ranges.

Parameter estimation procedure involves the pre-selection of certain algorithm parameters. For discrete time models, the first issue is the sampling period selection. This will depend either on the time constants of the process (if the aim is just to model it) or on the desired time constants of the controlled closed loop system if the main goal is to model it in order to design an appropriate controller.

Data conditioning: The real data obtained from the process must be properly conditioned, prior to its application in the estimation algorithms. Three steps are involved, namely high frequency signal filtering, offset and drift removal, and scaling of inputs and outputs. The first step involves filtering out high frequency signal components so as to avoid sampling aliases. Additional filters may be used if enhancement of a specific data frequency range is desired.

The parametric structures to be identified correspond to dynamic models, expressing the variation of process variables around the operating point. Therefore the offset levels and non-stationary DC components (drift) must be removed from the data. If off-line estimation is required, data are available as a set, and computation can be done on the whole set. In case of on-line estimation, input-output data are available for each sampling period, and recursive calculations are required. The following steps are carried out for offset and drift removal.

Off-line estimation

(i) Offset levels can be eliminated by subtracting the mean signal level and (ii) drifts can be eliminated using filtered input-output variations

$$y_f(k) = [\,y(k) - y(k-1)]/[1 + f_1 q^{-1}] \tag{10.27}$$

$$u_f(k) = [u(k) - u(k-1)]/[1 + f_1 q^{-1}] \tag{10.28}$$

with $-0.5 \le f_1 \le 0.0$.

The drift removal method can also be used for on-line identification.

On-line estimation

In addition to the previous method for drift removal, the following, which can remove both stationary and non-stationary DC components, are also available.

(i) Use of data differences between sampling

$$y_f(k) = y(k) - y(k-1) \tag{10.29}$$

$$u_f(k) = u(k) - u(k-1) \tag{10.30}$$

(ii) Calculation of the mean level approximately through a recursive first-order filter

$$\bar{y}(k) = (1 - \beta)\bar{y}(k-1) + \beta y(k) \tag{10.31}$$

$$y_f(k) = y(k) - \bar{y}(k) \tag{10.32}$$

(iii) Use of identification algorithms modified so as to estimate the DC components.

Scaling of inputs and outputs
If the levels of inputs and outputs are very different from each other, then the convergence speed of the estimated parameters associated with the model polynomials will be different. One solution to this problem consists of scaling data by multiplying inputs or outputs by the appropriate constant. Obviously the stationary gain identified will not be correct, and the estimated parameters of $B(z^{-1})$ will have to be rescaled accordingly.

10.7 Model Structure Selection

A correct model structure has to be determined for each of the process loops to be identified. All the parametric identification methods can be grouped into two groups:

(i) Identification methods based on the whitening of the prediction error. The model structure is given by

$$A(z^{-1})y(k) = B(z^{-1})z^{-d}u(k) + e(k)$$

The order of the polynomials A and B are respectively n_a and n_b.

(ii) Identification methods based on the de-correlation of the prediction error and the regression or observation vector

$$A(z^{-1})y(k) = B(z^{-1})z^{-d}u(k) + C(z^{-1})e(k)$$

The determining model structure involves: (i) determining the disturbance structure and (ii) determining the process structure, i.e., the correct orders for polynomials A and B and the correct delay d.

Let us consider the model structure given by Eq. (10.6). A set of least square algorithms corresponding to different values of n_a, n_b and d is selected, and the sum of the squared prediction errors, i.e. the loss function, is computed over a data set for each of the runs. The structure giving the least value of the loss function, while maintaining simplicity is chosen.

It should be taken into account that if a model structure is evaluated on the same data set on which it was estimated, the loss function (i.e., sum of squared errors) will always decrease as the model structure orders (n_a and n_b) are increased (whereas this is not true for the delay d). However for the given value of the number of parameters $n_a + n_b$ upwards, the decrease in the loss function becomes small, indicating that an increase in the number of parameters does not improve performance.

If there is no *a priori* knowledge about the process, then the typical values for n_a can be assumed as 2 or 3. The typical value for n_b is 2.

If the adaptive controller was applied from the very beginning, we would face the problem of a lack of estimates of the process parameters, which in turn would produce undetermined and possibly unacceptable control actions. Therefore, a prior estimation of the process model order, delay and parameters is made by open-loop testing for stable processes.

10.8 Estimation Method

Estimation methods minimize certain appropriately defined error criterion as a means to optimally fit the system data. The error criteria can be defined by three ways.

(i) Error can be defined as a deviation of the parameter estimates from the true values (parameter error).

$$\theta_k = \theta_k - \hat{\theta}_k \tag{10.33}$$

This method is known as *parameter error method.* Since true parameters values are unknown this method may be difficult to formulate.

(ii) The error can be defined as difference between the output of the system and that of the model

$$e_k = y_k - \hat{y}_k \tag{10.34}$$

where (for example)

$$y_k = ay_{k-1} + bu_{k-1} \tag{10.35}$$

and

$$\hat{y}_k = \hat{a}\hat{y}_{k-1} + \hat{b}u_{k-1} \tag{10.36}$$

This method is known as output error method (Fig. 10.1a). The same value of input (u_k) is used. In this method the solution of the model equation for the given sequence of u should be known. The model parameters usually appears nonlinearily in the resulting solution of the model equation. Hence the estimation of the parameters will require solution of nonlinear algebraic equations which require initial guess for the parameter values. The possibility of getting local minima rather than global minima may also arise.

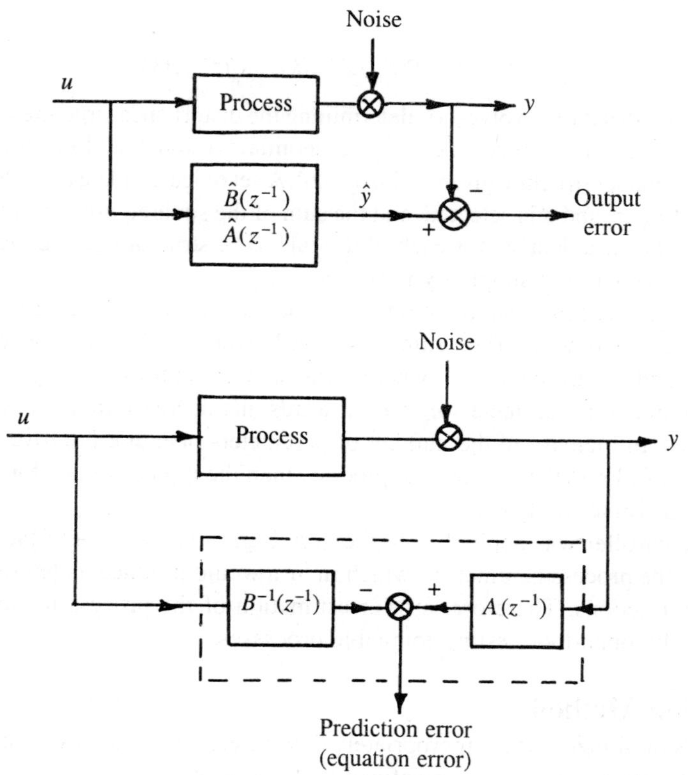

Fig. 10.1 Error models for identification of discrete time model: (a) output error (b) equation error (prediction error)

For example consider the model equation

$$y_k = ay_{k-1} + u_k \tag{10.37}$$

with given initial condition on $_{-1}$. For a unit step change in u_k, the solution of the model equation can be shown by taking z-transform and then followed by inverse z-transform as

$$y_k = a^{k+1}y_{-1} + [1/(1-a)] - [a^{k+1}/(1-a)] \tag{10.38}$$

If error is formulated based on output error then we find that the model parameter 'a' present as nonlinear. Hence, the resulting optimizaton problem is not easy to solve.

(iii) The error can be defined as a discrepancy between the model equation and measured input and output data ($e = y_k - \hat{y}_k$, for the example case we have $y_k = ay_{k-1} + bu_{k-1}$ and $\hat{y}_k = ay_{k-1} + bu_{k-1}$). This method is known as equation error method or prediction error method. Here there is no solution of model equations required and the model parameters appear linearily in the error. We substitute the experimental data sequences $\{y_k, u_k\}$ in the model equations and find the expression for the resulting error of the equation. Hence, the method is called identification based on equation error.

The prediction error method compares the behaviour of the model and the model based prediction with experimental data and tries to adjust the model parameters to obtain a better fit (Fig. 10.1b).

Output error method

$$y_k = [B(z^{-1})/A(z^{-1})]u_k + w_k \tag{10.39}$$

The difference in model equation between equation error (prediction error) analysis and output error analysis can be demonstrated by the two predictors

$$\hat{y}_k = \hat{a}\hat{y}_{k-1} + \hat{b}u_{k-1}; \quad \text{output error method} \tag{10.40}$$

$$\hat{y}_k = \hat{a}y_{k-1} + \hat{b}u_{k-1}; \quad \text{prediction error} \tag{10.41}$$

The output error method thus relies more on the accuracy of future output modeling. As stated earlier, the output error identification is a nonlinear estimation problem, whereas the equation error identification is a linear estimation problem. Because of nonlinear estimation problem of output error method, the parameter may exhibit local minima.

An iterative solution to the problem of output error identification starts by applying standard least-squares identification to find an initial estimates of F and B from the model

$$F(z^{-1})y_k = B(z^{-1})u_k + v_k \tag{10.42}$$

A second step is to filter the data according to

$$y_k^F = 1/\hat{F}(z^{-1}) \tag{10.43}$$

$$u_k^F = 1/\hat{F}(z^{-1}) \tag{10.44}$$

with a subsequent least squares estimation F and B from the model

$$F(z^{-1})y_k^F = B(z^{-1})u_k^F + v_k \tag{10.45}$$

The filtering and estimation steps are continued until the residual noise sequence ($\hat{y}_k - y_k$) is minimized in the least squares sense, and until the parameter estimation has converged. The procedure of regressor filtering is often called pre-whitening with a filter $\hat{F}(z^{-1})$. The term is justified in cases where the filtering results in white noise residuals. A standard use of pre-whitening allows effective reduction of bias caused by correlated noise, and thus the principle of pre-whitening is incorporated into several estimation algorithms.

10.9 Least Squares Method of Estimating the Model Parameters

Let the variables Y be related to set of n variables

$\mathbf{x} = [x_1, x_2, x_3 \ldots x_n]^T$ through a set of linear parameters

$$y(t) = a_1 x_1 + a_2 x_2 + a_3 x_3 + \ldots + a_n x_n \tag{10.46}$$

where $\theta = [a_1, a_2, a_3 \ldots a_n]^T$ is a set of constant parameters which are to be identified. Given the experimental data on y and the vector $\mathbf{x}$ for various time t, we have to estimate the model parameter vector θ in some optimal way.

Writing the model equation for time (t_1), we get

$$y(t_1) = a_1 x_1(t_1) + a_2 x_2(t_1) + a_3 x_3(t_1) + \ldots + a_n x_n(t_1) \tag{10.47}$$

we can write the equation for convenience as

$$y(1) = a_1 x_1(1) + a_2 x_2(1) + a_3 x_3(1) + \ldots + a_n x_n(1) \tag{10.48}$$

For any time t, we can write

$$y(i) = a_1 x_1(i) + a_2 x_2(i) + a_3 x_3(i) + \ldots + a_n x_n(i) \tag{10.49}$$

The sample number, i, is an integer quantity and symbolizes discrete time. An increment of one in discrete time corresponds to a change of T in real time.

We can write all the equations for time $t = T$ to nT, to get

$$\begin{aligned} y(1) &= a_1 x_1(1) + a_2 x_2(1) + a_3 x_3(1) + \ldots + a_n x_n(1) \\ y(2) &= a_1 x_1(2) + a_2 x_2(2) + a_3 x_3(2) + \ldots + a_n x_n(2) \\ y(3) &= a_1 x_1(3) + a_2 x_2(3) + a_3 x_3(3) + \ldots + a_n x_n(3) \\ y(m_m) &= a_1 x_1(m) + a_2 x_2(m) + a_3 x_3(m) + \ldots a_n x_n(m) \end{aligned} \tag{10.50}$$

The above equations can be written in a matrix form as

$$\mathbf{y} = \mathbf{x}\theta \tag{10.51}$$

where

$$\mathbf{x} = \begin{bmatrix} x_1(1) & x_2(1) & \ldots & \mathbf{x}_n(1) \\ x_1(2) & x_2(2) & \ldots & \mathbf{x}_n(2) \\ \ldots & \ldots & \ldots & \ldots \\ \ldots & \ldots & \ldots & \ldots \\ x_1(m) & x_2(m) & \ldots & \mathbf{x}_n(m) \end{bmatrix} \tag{10.52}$$

$$\mathbf{y} = [\, y(1)\ y(2) \ldots y(m)]^T; \quad \theta = [a_1, a_2 \ldots a_m]^T \tag{10.53}$$

Since the model is only an approximation of the actual system and the measurement may be corrupted by noise, it is more appropriate to collect a large number of experimental data than the minimum data point required (for estimating n parameters we need only n data points, provided the model is exactly of the actual system and the measurement is noise free and accurate).

We define error (or residual) ε as

$$\varepsilon = [\varepsilon_1, \varepsilon_2 \ldots \varepsilon_m]^T \quad (10.54)$$

$$\varepsilon = \mathbf{y} - \mathbf{x}\theta \quad (10.55)$$

The parameter vector θ is to be calculated in order to minimize the scalar objective function

$$J = \sum_{i=1}^{N} \varepsilon_i^2 \quad (10.56)$$

$$= \varepsilon^T \varepsilon \quad (10.57)$$

$$= (\mathbf{y} - \mathbf{x}\theta)^T(\mathbf{y} - \mathbf{x}\theta) \quad (10.58)$$

$$= y^T y - \theta^T \mathbf{x}^T \mathbf{y} - \mathbf{y}^T \mathbf{x}\theta + \theta^T \mathbf{x}^T \mathbf{x}\theta \quad (10.59)$$

For J to be minimized, we require the necessary and sufficient conditions

$$\partial J / \partial \theta = 0 \quad (10.60)$$

$$\partial^2 J / \partial \theta^2 > 0 \text{ (i.e., positive value)} \quad (10.61)$$

Differentiating Eq. (10.59) with respect to vector θ, we get

$$\partial J / \partial \theta = -2\mathbf{x}^T \mathbf{y} + 2\mathbf{x}^T \mathbf{x}\hat{\theta} = 0 \quad (10.62)$$

Here $\hat{\theta}$ is the estimated value of θ and is called the least square estimate of θ. This equation gives

$$\mathbf{x}^T \mathbf{x}\hat{\theta} = \mathbf{x}^T \mathbf{y} \quad (10.63)$$

Hence, we get

$$\hat{\theta} = (\mathbf{x}^T \mathbf{x})^{-1} \mathbf{x}^T \mathbf{y} \quad (10.64)$$

Equation (10.62) becomes

$$\partial^2 J / \partial \theta^2 = 2\mathbf{x}^T \mathbf{x} \quad (10.65)$$

which is a positive value. Thus $\hat{\theta}$ is the least square error value of the parameter θ. An easy way to remember Equation (10.64) is to premultiply both sides of Eq. (10.51) by $\mathbf{x}^T$ and get directly (10.64).

10.10 Least Square Estimation of ARX Model

The ARX model is given by

$$y(i) = -\sum_{i=1}^{n} a_1 y(k-i) + \sum_{i=1}^{n} b_i u(k-i) + e(k) \quad (10.66)$$

Let us define

$$\mathbf{x}(k) = [-y(k-1), -y(k-2) \ldots -y\,(k-n), u(k), u(k-1) \ldots u(k-n)]^T$$

The size of x is $1x(2n+1)$

$$\theta = [a_1, a_2 \ldots a_n, b_1, b_2 \ldots b_n]^T \tag{10.68}$$

We can rewrite Eq. (10.66) as

$$\mathbf{y}(k) = \mathbf{x}(k)\theta + \varepsilon(k) \tag{10.69}$$

where

$$\mathbf{y} = [y(n+1), y(n+2), \ldots y(n+N)]^T \tag{10.70}$$

$$\mathbf{e} = [e(n+1), e(n+2), \ldots e(n+N)]^T \tag{10.71}$$

$$\mathbf{x} = \begin{bmatrix} \mathbf{x}^T(n+1) \\ \mathbf{x}^T(n+2) \\ \ldots \\ \mathbf{x}^T(n+N) \end{bmatrix} = \begin{bmatrix} -y(n), \ldots -y(1), & u(n+1), \ldots u(1) \\ -y(n+1), \ldots -y(2), & u(n+2), \ldots u(2) \\ \ldots & \\ -y(n+N-1), \ldots y(N), & u(n+N), \ldots u(N) \end{bmatrix} \tag{10.72}$$

Using the data set $\{y(k), u(k)\}$ for $k = 1, 2, \ldots (N+n)$, we can evaluate matrix $\mathbf{x}$ and vector $\mathbf{y}$. Hence by least square method we can estimate the parameters vector $\hat{\theta}$ as

$$\hat{\theta} = (\mathbf{x}^T\mathbf{x})^{-1}\mathbf{x}^T\mathbf{y} \tag{10.73}$$

The solution exists if matrix $[\mathbf{x}^T\mathbf{x}]$ is invertible. That is, the matrix is non-singular. This method of estimating parameters was proposed by Kalman (1958). The least squares solutions to the linear regression problems have excellent properties in cases where the disturbances at different times are uncorrelated. However, the presence of outliers or noise with non-zero mean or otherwise correlated disturbances may give rise to characteristic systematic errors and bias in the parameter estimates.

The above derivation can also be explained without matrix theory when the total number of parameters (total of a's and b's) in the model equations are less than or equal to 3. Let us write the model equation as

$$y(k) = -a_1 y(k-1) - a_2 y(k-2) + b_1 u(k-1)$$

as stated earlier parameters a_1, a_2 and b_2 are to be estimated by minimizing the sum of the error squared over all the experimental data set.

$$p = \sum_{k=1}^{N} [y(k) + a_1 y(k-1) + a_2 y(k-2) - b_1 u(k-1)]^2 \tag{10.74}$$

The necessary conditions for minimization of the performance index are

$$\partial P/\partial a_1 = \sum_{k=1}^{N} 2[y(k) + a_1 y(k-1) + a_2 y(k-2) - b_1 u(k-1)][y(k-1)] = 0 \tag{10.75}$$

$$\partial P/\partial a_2 = \sum_{k=1}^{N} 2[y(k) + a_1 y(k-1) + a_2 y(k-2) - b_1 u(k-1)][y(k-2)] = 0 \tag{10.76}$$

$$\partial P/\partial b_1 = \sum_{k=1}^{N} 2[y(k) + a_1 y(k-1) + a_2 y(k-2) - b_1 u(k-1)][-u(k-1)] = 0 \tag{10.77}$$

Solving the above linear equations we get the linear parameters a_1, a_2 and b_1. The index used for evaluating the sum [Σ (.)] term can be from 1 or from 3. If the index is zero, then the values of y (0), y(−1) and u(0) are all zero. If the index is 3, then the corresponding values of y(0), y(1), u(1) are to be used. Here N is the total number of experimental data points.

The next step is to deal with the disturbance model selection. The least square algorithm aims at whitening the residuals (prediction error) sequence. Therefore, if the correlation function of residuals is calculated, noise being estimated as the prediction error in the identification procedure, the correlation function of the residuals should vanish except for the auto correlation. In practice, values of the correlation function will not vanish for time shifts greater than one, but should be negligible. Statistical criteria may be used to evaluate the correlation functions. If the disturbance present in the process can be modeled as white noise, then

$$\text{corr}[e(k), e(k)] = \sigma_e^2$$

$$\text{corr}[e(k), e(k + \tau)] = 0, \ \tau > 0$$

If noise is colored, there will be v terms in the second expression which are non-zero, v being the order of the colored noise which models the process disturbance.

Signal excitability

For a dynamic model to be correctly estimated, input signals to the process must be sufficiently excited. More precisely, all possible dynamic modes in the frequency range of interest must be persistently excited. Hence, process input signals should have a frequency content sufficiently rich to span a range wide enough to correctly identify both the fast dynamic modes and the stationary gain. Pseudo-random binary sequence signals (PRBS) are widely used for this purpose.

If the process input is not sufficiently exciting, numerical problems will arise in the inverse of variance matrix, leading to poor estimation.

10.11 Least Square Estimation of Weighting Functions

There are number of advantages of the weighting function model from the viewpoint of system identification. First, the model requires less *a priori* knowledge than the parameter models. For example, the system order, time delay needed for parametric models do not arise. Second, the model can be identified more satisfactorily in the presence of noise. However the method is applicable only for stable system and not applicable for unstable systems.

We assume linear systems. All noise effects are accounted by a single additive noise source at the system output (Fig. 10.2). Here $u(t)$ is observed, $w(t)$ is inaccessible and $y(t)$ is measured one.

$$w(t) = \int_0^t g(t - \lambda)u(\lambda)d\lambda \tag{10.78}$$

where

$$y(t) = w(t) + v(t) \tag{10.79}$$

Let $y(t)$ and $u(t)$ sampled periodically with a sample period T secs. If T is sufficiently small, we can take piecewise constant approximation as

$$u(t) = u(kT),\ g(t) = g(kT) \quad \text{for} \quad kT \le t < (k + 1)T \tag{10.80}$$

Then $w(t)$ can be approximated at $t = kT$ as

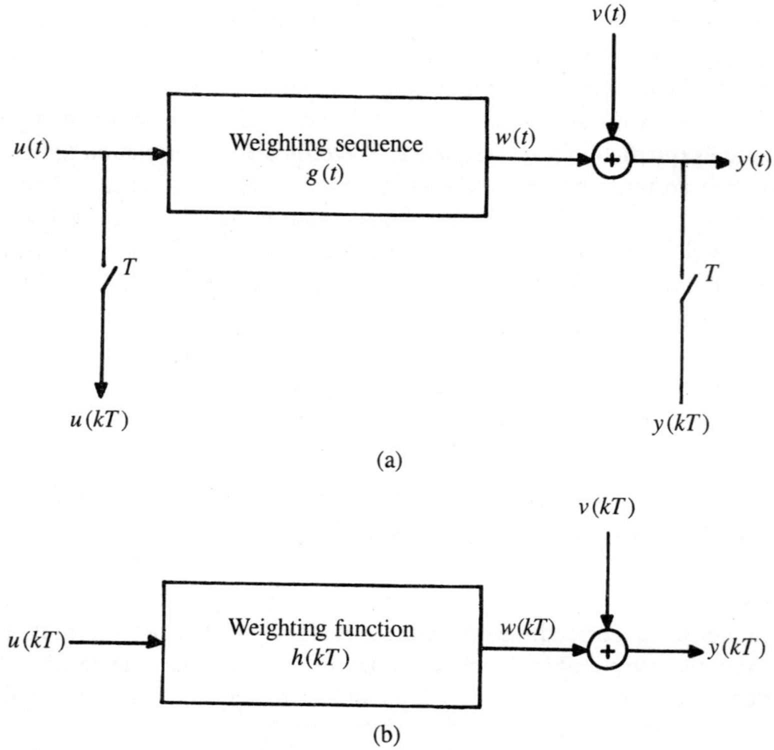

Fig. 10.2 (a) Basic configuration for system identification and (b) equivalent discrete time system

$$w(kT) = \sum_{i=-\infty}^{k} Tg(kT - iT)u(iT) \tag{10.81}$$

Let

$$h(kT) = Tg(kT) \tag{10.82}$$

$$w(kT) = \sum_{i=-\infty}^{k} h(kT - iT)u(iT) \tag{10.83}$$

Let us assume that the system is stable and the settling time is finite so that $g(t) = 0$ for $t > pT$, where p is an integer. Therefore, we get

$$y(kT) = \sum_{i=k-p}^{k} h(kT - iT)u(kT) + v(kT) \tag{10.84}$$

Identification problem: Given $\{u(kT), y(kT)\}$ sequences and the integer p, we have to estimate the weighting sequence $\{h(kT)\}$, $k = 0, 1, 2 \ldots p$. Then the desired weighting function samples $g(kT)$ are obtained from $g(kT) = h(kT)/T$. Using the observed input-output data for $k = 1, 2 \ldots (m + p)$, in Eq. (10.84), we can write the set of $m + 1$ equations written in vector form as

$$\mathbf{y} = \mathbf{Uh} + \mathbf{v} \tag{10.85}$$

where

$$\mathbf{y} = [\, y(p), y\,(p + 1), \ldots, y\,(p + m)]^T \tag{10.86}$$

$$\mathbf{v} = [v(p), v(p+1), \ldots, v(p+m)]^T \tag{10.87}$$

$$\mathbf{h} = [h(0), h(1), \ldots, h(p)]^T \tag{10.88}$$

$$\mathbf{U} = \begin{bmatrix} u(p) & u(p-1) & \ldots & u(0) \\ u(p+1) & u(p) & \ldots & u(1) \\ \ldots & \ldots & \ldots & \ldots \\ \ldots & \ldots & \ldots & \ldots \\ u(p+m) & u(p+m-1) & \ldots & u(m) \end{bmatrix} \tag{10.89}$$

Since $m > p$, the unknown parameter $\mathbf{h}$ can be estimated by the least square method discussed earlier by

$$\hat{\mathbf{h}} = (\mathbf{U}^T\mathbf{U})^{-1}\mathbf{U}^T\mathbf{y} \tag{10.90}$$

Then the samples of the weighting function $g(t)$ can be calculated from

$$\hat{g}(kT) = \hat{h}(kT)/T, \qquad k = 0, 1, 2, \ldots, p \tag{10.91}$$

For the above method we have to select the settling time parameter p. For fixed T, the larger the value of p, the better the sequence $\{h(kT)/T\}$ approximates $g(kT)$. However, the larger p, the computational efforts is more, since we have to invert a large size matrix. First we have to start with a small value for p such that $g\ (t > pT) = 0$. Then the procedure is repeated with progressively increasing p values until a satisfactory fit is obtained.

10.12 Extended Least Squares Method

The least squares estimation of model parameters are biased when applied to data generated by Eq. (10.66) where the errors are correlated. One possible method is to model the correlation disturbances and to estimate the parameters describing the correlation. That is, the extended least squares (ELS) is a straight forward extension of LS with an extended prediction model in place of Eq. (10.66)

$$A(q^{-1})y(t) = B(q^{-1})u(t-k) + \varepsilon(t) \tag{10.92}$$

so that the estimated prediction errors are included as another input signal. In the ELS method the estimates are found as in the LS method. However, it has been observed that ELS algorithms are not as robust as RLS algorithms.

Let us consider the model

$$A(q^{-1})y(t) = B(q^{-1})u(t) + C(q^{-1})e(t) \tag{10.93}$$

where $A(q^{-1})$, $B(q^{-1})$, and $C(q^{-1})$ are polynomials in the shift operator and $\{e(t)\}$ is white noise. The parameters of polynomial $C(q^{-1})$ describe the correlation of the disturbance. Since the variables $\{e(t)\}$ are not known, the model Eq. (10.93) cannot be converted into a regression model. However, if the predicted value of $e(t)$ is used for $e(t)$, then the regression model can be obtained. Let the predicted value of $e(t)$ denoted by $\varepsilon(t)$ be calculated from

$$\varepsilon(t) = y(t) - \phi^T(t-1)\,\hat{\theta}(t-1) \tag{10.94}$$

where

$$\theta = [a_1, a_2, a_3 \ldots, a_n, b_1, b_2, \ldots, b_n, c_1, c_2, \ldots, c_n]^T \quad (10.95)$$

$$\phi^T = [-y(t-1), \ldots, -y(t-n+1), u(t), \ldots, u(t-n+1), \varepsilon(t), \varepsilon(t-n+1)] \quad (10.96)$$

Hence the regression model is obtained as

$$y(t) = \phi^T(t-1)\theta \quad (10.97)$$

for this model, standard least squares method can be applied to estimate the vector θ. This technique is called extended least square methods.

10.13 Real-Time Identification

Most identification methods based on a set of measurements are not suitable for real time operation and must be reformulated suitably for real-time identification. The recursive identification algorithm are estimators of the type

$$\hat{\theta}_k = \hat{\theta}_{k-1} + P_k \phi_k \varepsilon_k \quad (10.98)$$

$$\varepsilon_k = y_k - \phi_k^T \hat{\theta}_{k-1} \quad (10.99)$$

$$P_k = P_{k-1} - [P_{k-1}\phi_k \phi_k^T P_{k-1})/(1 + \phi_k^T P_{k-1}\phi_k)] \quad (10.100)$$

with the parameter estimate $\hat{\theta}_k$, the regressor ϕ_k, the prediction error ε_k, and the matrix P_k, which are also evaluated at time $k = 1, 2, 3. \ldots$

10.13.1 Derivation of Recursive Least Squares Identification

Equation (10.98) can be derived from the ordinary least squares estimator as follows.

We have regressor ϕ_i and observations y_i which are collectively described by the following matrices

$$\Phi = [\phi_1^T, \phi_2^T, \phi_3^T, \ldots \phi_k^T]^T \quad \text{and} \quad y_k = [y_1, \ldots, y_k]^T \quad (10.101)$$

The ordinary least square method gives

$$\hat{\theta}_k = (\Phi_k^T \Phi_k)^{-1} \Phi_k^T y_k \quad (10.102)$$

$$= \left[\sum_{i=1}^{k} (\phi_i \phi_i^T)\right]^{-1} \left[\sum_{i=1}^{k} \phi_i y_i\right] \quad (10.103)$$

Introduce the matrix

$$P_k = \left(\sum_{i=1}^{k} \phi_i \phi_i^T\right)^{-1} = (\Phi_k^T \Phi_k)^{-1} \quad (10.104)$$

A recursive updating is given by

$$p_k^{-1} = P_{k-1}^{-1} + \phi_k \phi_k^T \quad (10.105)$$

$$\hat{\theta}_k = P_k\left(\sum_{i=1}^{k-1} \phi_i y_i + \phi_k y_k\right) \quad (10.106)$$

$$= P_k (P_{k-1}^{-1}\hat{\theta}_k + \phi_k y_k) \quad (10.107)$$

$$= \hat{\theta}_{k-1} + P_k\phi_k (y_k - \phi_k^T \hat{\theta}_{k-1}) \quad (10.108)$$

It is also feasible to calculate martrix P_k instead of its inverse. Notice that

$$P_k = (\Phi_k^T \Phi_k)^{-1} \quad (10.109)$$

$$= (\Phi_{k-1}^T \Phi_{k-1} + \phi_k\phi_k^T)^{-1} \quad (10.110)$$

$$= (P_{k-1}^{-1} + \phi_k\phi_k^T)^{-1} \quad (10.111)$$

In real time operation it is usually difficult and computationally expensive to do matrix inversion. Application of the matrix inversion relation (refer to Appendix for proof)

$$(A + BC)^{-1} = A^{-1} - A^{-1}B\,(I + CA^{-1}B)^{-1}CA^{-1} \quad (10.112)$$

to the above expression for P_k gives

$$P_k = (P_{k-1}^{-1} + \phi_k\phi_k^T)^{-1} \quad (10.113)$$

$$= P_{k-1} - P_{k-1}\phi_k\,(I + \phi_k^T P_{k-1}\phi_k)^{-1}\,\phi_k^T P_{k-1} \quad (10.114)$$

Hence, the expression for $\hat{\theta}_k$ can be written as Eq. (10.108) which requires the initial matrix P_0. A systematic solution is to make an initial estimate by means of ordinary least square identification, using an initial block of few data points which then provide the initial value for $\hat{\theta}_0$ and P_0. Recursive identification then provides a procedure for updating the parameter estimates.

10.14 Modification for Time-Varying Parameters

The recursive least squares estimation in Eq. (10.108) gives equal weighting to old data and new data. In is desirable to pay less attention to old data in many applications when time-varying parameters should be tracked. For this purpose, a forgetting factor λ and hence modified performance criterion of least squares estimation are introduced as

$$J(\hat{\theta}_k) = \sum_{i=1}^{k}\lambda^{k-1}(y_i - \phi_i^T\hat{\theta}_k)^2, 0 < \lambda < 1 \quad (10.115)$$

This results in the recursive algorithm

$$\hat{\theta}_k = \hat{\theta}_{k-1} + P_k\phi_k\varepsilon_k \quad (10.116)$$

$$\varepsilon_k = y_k - \phi_k^T\hat{\theta}_{k-1} \quad (10.117)$$

$$P_k = (1/\lambda)\left[P_{k-1} - \frac{P_{k-1}\phi_k\phi_k^T P_{k-1}}{(\lambda + \phi_k^T P_{k-1}\phi_k)}\right] \quad (10.118)$$

This algorithm emphasis the filling of recent data and reduces the influence of old data. However, there are some undesirable effects of λ: (i) as λ decreases, noise sensitivity increases and (ii) as k grows P_k matrix may increase if the input is such that the magnitude of $P_{k-1}\phi_k$ is small (P-matrix explosion or covariance matrix explosion).

The value for λ is chosen as a trade-off between the required ability to track a time-varying parameter and the noise sensitivity involved. A low value for λ results in a system with a good ability to track time-varying parameters but may also give variations in the estimated parameters caused by disturbances. A value of λ close to 1 is less sensitive to disturbance but compromises the ability to track rapid variations in the parameters. The recommended value of λ is $0.97 \leq \lambda \leq 0.995$. The number of data points kept in 'memory' can roughly be calculated as $1/(\lambda - 1)$, which corresponds to the time constant associated with λ.

10.15 Parameter Tracking and Forgetting Factor

Process parameter changes are not unusual and three main causes can be considered: (i) drifts in the process dynamic behaviour caused by temperature etc.; (ii) changes in the process physical configuration, leading to changes in its dynamic behaviour (e.g., opening or closing of valves in liquid tanks). The changes in the dynamic behaviour may be rather fast in this case; and (iii) large reference excursions in nonlinear systems. In this case, the linearized transfer function around the new operating point may differ significantly from the previous ones. An important issue when the estimated model has to be used on-line, and needs to be updated when possible, is that of parameter tracking.

Algorithm fading: Use of recursive identification algorithm is insufficient to ensure tracking of process parameter changes. The algorithm becomes insensitive to new data and also to possible process dynamic changes. This effect is known as algorithm fading. To monitor algorithm fading, the trace of the inverse covariance matrix can be used, since it gives an indirect measure of the magnitude of its components. When the value of the trace of P falls below a certain threshold, two basic actions can be taken: (i) matrix P re-initialization, this step has the disadvantage of modifying the algorithm search direction with corresponding transient behaviour and (ii) the new data should be weighted more than old data. This can be achieved by using forgetting factor.

However, the use of forgetting factor may introduce problems. If the process inputs are not sufficiently exciting, then the quadratic term in Eq. (10.118) becomes small, and thus

$$P(k+1) \geq P(k) \tag{10.119}$$

In fact, it may become negligible, so that

$$P(k+1) = P(k)/\lambda \tag{10.120}$$

If λ is small, the gain term in Eq. (10.118) will eventually have large fluctuations, leading to similar effects in the estimated parameters, which is known as 'blow-up'.

If the forgetting factor is too low, noise tracking by the estimated parameters will appear as a side effect. Therefore, relatively high values are used (0.96 to 0.98 for exponential windows), although they may be temporarily lowered so as to improve process parameter tracking.

10.16 Recursive Algorithm for Extended Least Squares Method

The regression model is

$$y_k = \phi_k^T \theta + v_k \tag{10.121}$$

$$\theta_k = [a_1, \ldots, a_{nA}, b_1, \ldots, b_{nB}, c_1, \ldots, c_{nC}]^T \tag{10.122}$$

The recursive equation is given by

$$\hat{\theta}_k = \hat{\theta}_{k-1} + K_k \varepsilon_k \tag{10.123}$$

$$K_k = P_{k-1}\phi_k / (1 + \phi_k^T P_{k-1} \phi_k) \tag{10.124}$$

$$\varepsilon_k = y_k - \phi_k^T \hat{\theta}_{k-1} \tag{10.125}$$

$$P_k = P_{k-1} - \frac{P_{k-1}\phi_k \phi_k^T P_{k-1}}{(1 + \phi_k^T P_{k-1}\phi_k)} \tag{10.126}$$

The regression vector that would be natural to use for linear regression is

$$[-y_{k-1}, \ldots, y_{k-na}, u_{k-1}, \ldots, u_{k-nb}, v_{k-1}, \ldots, v_{k-nc}]^T$$

Since the sequence $\{v_k\}$ is not available for measurement, their predicted values

$$\varepsilon_k = y_k - \phi_k^T \theta \tag{10.127}$$

are used in the regression vector as

$$\phi_k = [-y_{k-1}, \ldots, -y_{k-na}, u_{k-1}, \ldots, u_{k-n}, \varepsilon_{k-1}, \ldots, \varepsilon_{k-nc}]^T \tag{10.128}$$

This method is also called pseudo linear regression since the disturbance components of the desired regression vector have been replaced by estimated variables. Pseudo linear regression is an approximate linear regression method as the unknown regression elements v_k are replaced by their residuals ε_k. Such a method is known to work well for ARMAX models with moderate noise correlations (i.e, ARMAX models whose C polynomial are close to 1). The method is also called recursive maximum likelihood method as it tries to use the disturbance components as regressor elements and approximates these unknown elements by the calculated predicted error (ε_k). The method may also be modified to include a forgetting factor for identification of time-varying systems. The algorithm may also be modified to iterate for the best possible ε within each recursive step.

10.17 Closed Loop Identification

There are two methods available for closed loop identification: (i) Indirect process identification: a model of the closed loop is identified using the reference value and process output data set and after wards, assuming that the controller is known, the process model is calculated by de-convolution. (ii) Direct process identification: the process model is directly identified based on the input and output data set of the process (u and y). The controller need not be known.

Within direct process identification, two methods are available if the process input and output are measured on-line: (a) no external disturbance is applied. In this case a set of identifiability condition must be met to ensure convergence. (b) an external disturbance (measurable or not) is applied. If the additional external disturbance added to the process input is sufficiently exciting and uncorrelated with the process noise e, only the order of the process must be known, and no additional identification condition is required.

10.17.1 Iterative Identification in Closed Loop

The plant model estimates are carried out at every sampling interval and usually the controller parameters are also updated at every sampling instant. However, updating the controller parameters only at every N sampling instants is desirable for some situations for the following reasons: (i) possibility of getting better parameter estimates for controller design; (ii) the eventual re-initialization of the plant parameters estimation algorithm after controller updating; and (iii) the possibility of using a more sophisticated control design procedure (in particular robust controller design) requiring a large amount of computation.

The technique can be used for: (i) re-tuning and re-designing of an existing controller without opening the loop and (ii) re-tuning from time to time in order to take into account possible change in the model parameters. The controller designed based on the closed loop identification method often improves the controller performance than the controller designed based on open loop identification. The reason for this is that identification in closed loop is made with an effective plant input which corresponds to the external excitation filtered by a sensitive function. This sensitive function will enhance the signal energy in the frequency range around the band pass of the closed loop and therefore will allow a more accurate model to be obtained in this region, which is critical for the design.

Summary

Least square estimation method is a popular method used for estimating model parameters of linear discrete time systems. The experimental data should be filtered and conditioned before applying the least square method. For on-line parameter estimation method, recursive version of the least square method is to be used. For identifying time-varying parameters, a forgetting factor should be used in the recursive least square method. For identifying linear systems with significant noise level, an extended least square method should be used.

References

Anand, D.K. and R.B. Zmood, Introduction to control systems, IIIrd Ed. Asian Books Pvt Ltd, New Delhi, 1995.
Astrom K.J. and Wittenmark, Adaptive Control System, Addison Wesley, 1989.
Pico, J., P. Albertos and M. Martinez, Process model identification, in: Control Engg. Solution: a practical approach, (Ed.) P. Athertos et. al., IEE publication, (1997) pp 1–12.
Hsia, H.C., System identification: Least square methods, Lexington Books, Lexington, 1977.
Iserman, R., Digital control systems, Springer-Verlag, 1995.
Johanson, R., System modeling and identification, Prentice Hall, Englewood Cliffs, New Jersey, 1993.
Landau, I.D., R. Lozano nad M.M'saad, Adaptive Control, Springer Verlag, Berlin, 1998.
Ljung, L., System Identification Tool Box, The Matlab Inc., Mass., 1988.
Ljung, L., System Identification: Theory for the Users, Prentice-Hall, Englewood Cliffs, New Jersey, 1987.
Paraskevopoulos, Digital Control Systems, Prentice-Hall of India Pvt. Ltd., New Delhi, 1999.
Stephanopoulos, G., Chemical Process Control: An introduction to theory and practice, Prentice-Hall of India, Pvt. Ltd., New Delhi, 1984.

Review Questions

1. Discuss briefly the different linear discrete time models.
2. From ARMAX models what are the sub-class model one can get?
3. Write the steps involved in system identification.

4. What are the two error models used for discrete time system identification?
5. Give an interpretation of noise term in the discrete time model.
6. Explain the least square method of parameter estimation for

$$y(k) = ax(k) + bx(k - 1)$$

by using the given data set in $y(k)$ and $x(k)$.
7. What is meant by system excitability for identification of discrete time systems?
8. What is an extended least square method?
9. Explain the recursive least square method.
10. Discuss a method of estimating time varying parameters.
11. What is a forgetting factor?
12. What is meant by 'algorithm fading' in recursive LS method?
13. Write down the recursive algorithm for the extended least square method.
14. The input-output data set is given as follows:

k	–1	0	1	2	3	4	5	6
$u(k)$	1	1	1	1	1	1	1	1
$y(k)$		1	1/2	3/4	5/8	11/16	21/32	43/64

Using the LS method, get an estimate of the parameters of the model

$$y(k) = a_1 y(k - 1) + b_1 u(k - 2)$$

15. Using the data set in Q14, estimate the parameters of the model:

$$y(k) = -0.5y(k - 1) + bu(k - 1)$$

16. In Q14 assume a new data is available $y(7) = 1$ and $u(7) = 1$, find the estimate of a and b using recursive LS method.
17. Re do Q14 for the model $y(k + 1) + ay^2(k) = u(k)$.

11

Adaptive and Predictive Control

For linear systems with time-invariant dynamics, a PID controller with fixed settings gives good performances. However for a system whose dynamics changes significantly with time, retuning of the PID controller settings is required. Adaptive control systems automatically adjust the controller settings in real time to compensate for unanticipated changes in the process conditions or the environment. Prediction is to determine the future value of an output variable based on available information. This prediction can be used in the design of control laws for better performance of the control systems. Such controllers are called predictive controllers. In this chapter, methods of designing adaptive and related predictive controllers will be reviewed.

11.1 Adaptive Control

An adaptive control is a regulator that can modify its behaviour in response to changes in the dynamics of the process and disturbances. In an adaptive controller there is some kind of feedback from the performance of the closed loop system. A control system coupled to an identification procedure is often called an adaptive control system. In a wider sense, adaptive control may mean modification of the parameters of some control mechanism or control actions: Excellent reviews are given by Seborg et. al. (1986), Chalam (1990), Astrom and Wittenmark (1989) and Landau et. al. (1998).

Gain scheduling: In some systems, there are auxiliary variables that relate well to the characteristics of the process dynamics. If these variables can be measured, they can be used to change the regulator parameters. This approach is called gain scheduling (Fig. 11.1) because the scheme was originally used to accommodate changes in process gain.

Gain scheduling is an open-loop compensation and can be viewed as a system with feedback control in which feedback gains are adjusted by feed forward compensation. There is no feedback from the performance for an incorrect schedule. In process control, the production rate can often be chosen as a scheduling variable, since time constants and time delays are often invariably proportional to production rate. Gain scheduling is a very useful technique for reducing the effects of parameter variations.

The regulator parameter can be quickly changed in response to changes in plant dynamics (it is convenient to implement this method if the plant dynamics depend in a well-known manner on a relatively few, easily measurable variables). However, being an open-loop compensation technique, its design is time consuming. Thus it is not advisable to use this method for those processes where the process dynamics or the process distubances are not accurately known.

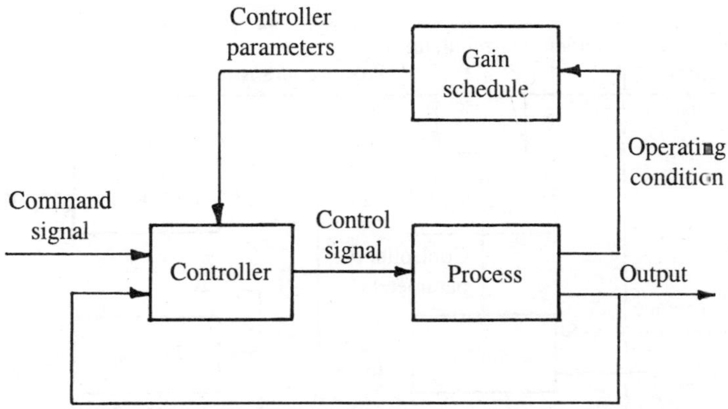

Fig. 11.1 Block diagram of a system with gain scheduling

The use of gain scheduling can have limitations such as: (i) it may require additional expensive transducers; (ii) it may take a long time and numerous experiments in order to establish the desired relationship between environment measurements and controller parameters; and (iii) the controller parameter change may be abrupt across the region boundaries, which may result in unsatisfactory or even unstable performance across the transition region.

11.2 Self-Tuning Regulators (STRs)

In non-adaptive control design (e.g., pole placement method), one computes the parameters of the controllers from those of the plant. If the plant parameters are not known, it is intuitively reasonable to replace them by their estimated values, as provided by a parameter estimator. A procedure is said to have self-tuning if its performance coincides with the performance that would be obtained if the system parameters were known exactly. A controller thus obtained by coupling a controller with an on-line parameter estimator is called a self-tuning controller. Thus, parameter estimation and the design of control law emerges as two separate activities. This is known as a separation principle. A block diagram of STR is shown in Fig. 11.2. The adaptive regulator consists of two loops. The inner loop consists of the process and an ordinary feedback regulator. The parameters of the regulator are adjusted by the outer loop, which is composed of a recursive parameter estimator and design calculations. To obtain good estimates, it may also be necessary to introduce perturbation signals alongwith the input variable.

It may be noted that it is not a part of the self-tuning property that the parameter estimates should converge to their true values; this may not be necessary. Astrom and Wittenmark (1973) have shown that in some circumstances, a combination of simple least squares estimation and minimum variance control (or some classes of pole placement methods) has the self-tuning property. This is true even if the system is represented by Eq. (10.7) with C not equal to 1, when least squares estimation would be expected to give biased estimates. The effect of the control algorithm is somehow to 'unbias' the estimates. This is a very attractive result since, least squares being by the far the simplest form of recursive identification, it opens up the possibility of designing self-tuning controllers of very modest computational complexity.

The block labelled 'controller design' represents an on-line solution to a controller design problem for a system with known parameters. This is the underlying design problem. To evaluate adaptive control schemes, it is often useful to find the underlying design problem, because, it will give the characteristics of the system under ideal conditions, where the parameters are known exactly.

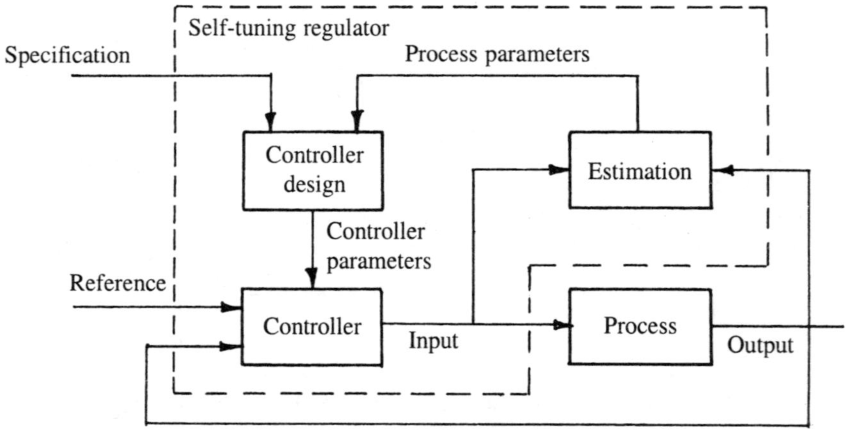

Fig. 11.2 Block diagram of a self-tuning regulator (STR)

In the basic approach to self-tuning control, one estimates the plant parameters and then computes the controller parameters. Such a scheme is called indirect adaptive control (because of the need to translate estimated parameter to controller parameters). The regulator parameters are updated indirectly via the design calculations. It is sometimes possible to re-parameterize the process model so that the model can be expressed in terms of the regulator parameters. The estimation method gives the regulator parameters. The disadvantage of this approach is that there is no guarantee of stability for the system. During the learning phase of the control, the input signal can infinitely change and may not be realizable. Another problem is that the method requires an assumption of the structure of the system being controlled. If the assumed structure does not match with the actual system, then the least square estimates may not completely describe the system and the closed loop system performance may not match design specifications.

In adaptive control systems, the controller parameters or the process parameters are estimated in real time. The estimates are then used as if they are equal to the true ones (i.e., the uncertainty of the estimates are not considered). This is called the *certainty equivalence principle.* In many estimation schemes, it is also possible to get a measure of the quality of the estimates. This uncertainty may then be used to modify the controller.

Parameter estimation can be understood simply as the process of finding a set of parameters that fits the available input-output data from a plant. This is different from parameter adaptation in Model Reference Adaptive Control (MRAC) systems (Sec. 11.3), where the parameters are adjusted so that the tracking errors converges to zero. For linear plants, many techniques are available to estimate the unknown parameters of the plant.

In the presence of model parameter variations, the conventional feedback control is to be designed by a robust control design method in order to get a satisfactory level of performance. The information needed for the design of robust feedback controller are: (i) a nominal model of the process and (ii) a size of the uncertainty, specified in the parameter domain or in the frequency domain. The robust controller design method gives fixed controller settings that will satisfactorily work for all the specified variations of plant model parametes. The range of uncertainty domain for which satisfactory performance can be achieved depends on the problem. Sometimes a larger domain of uncertainty can be tolerated, while in other cases, the uncertainty to tolerance range may be very small. If the desired performance cannot be achieved for the full range of possible parameter variations, adaptive control has to be considered in addition to a robust controller design. Robust controller design will benefit from the use of adaptive control in terms of performance improvements and extension of the range of operation. Using an underlying robust design for building an adaptive control system may drastically improve the performance of the adaptive controller.

Use of direct adaptive control schemes is limited by the hypothesis related to the underlying linear design in the case of known parameters. While the performance can, in many cases, be specified in terms of a reference model, the conditions for the existence of a feasible controller allowing for the closed loop to match the reference model are restrictive. Usually it is assumed that the plant model has in all the situations stable zeros, which in the discrete time case is quite restrictive. [Fractional delay, larger than half sampling period, leads to unstable zeros; high frequency sampling of continuous time systems with difference of degree between denominator and numerator larger or equal to two leads to unstable zeros.] The problem becomes even more difficult in the multi-input multi-output case. Hence, the direct adaptive control cannot always be used.

Direct adaptive control of systems with significant delay will require the use of a controller with larger number of parameters than the ones of the plant model (if d is the integer number of sampling periods representing the delays, the controller will have additional d parameters with respect to the number of parameters of the plant model). In such case, it is better to use an indirect adaptive control scheme, since the estimation of the plant model parameters followed by the computation of the controller at each sample may require a smaller number of operations and may lead to shorter adaptation transients.

11.3 Model Reference Adaptive Control System (MRAC)

The model reference adaptive (Fig. 11.3) was originally proposed to solve a problem in which the specifications are given in terms of a reference model that tells how the process output ideally should respond to the command signal. The parameters of the regulation are adjusted by the outer loop in such a way that error (e) between process output y and model output y_m becomes small. The key problem is to describe the adjustment mechanisms so that a stable system, which brings the error to zero is obtained.

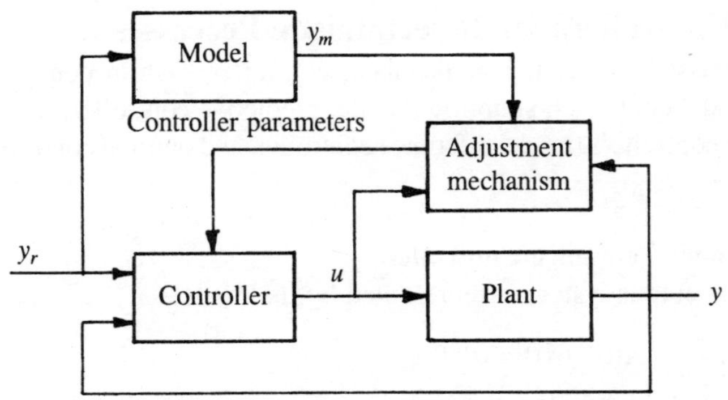

Fig. 11.3 Block diagram of a model reference adaptive control system (MRAC)

For this system, being based on Lyapunov's method, an implicit proof of closed loop stability exists. The controller is designed in such a way that the overall closed loop system is asymptotically stable i.e., the system being controlled eventually tracks the reference model with zero error. The transients during the adaptive or learning phase of the control are guaranteed within the bounds. The disadvantage of MRAC is that the stability is not guaranteed if the model of the system does not match the actual ones.

The reference model is used to specify the ideal response of the adaptive control system to the external command. The choice of the reference model is part of the adaptive control system design, and it has to satisfy two requirements:

(i) It should reflect the performance specification in the control tasks, such as rise time, settling time, overshoot or equivalent frequency domain characteristics.

(ii) The ideal behaviour specified by the reference model should be achievable for the adaptive control system.

The controller is usually parameterized by a number of adjustable parameters. Usually the control law is linear in terms of the adjustable parameters. Existing adaptive control designs normally require linear parameterization of the controller in order to obtain adaptive mechanisms with guaranteed stability and tracking convergence.

11.4 Relationship Between MRAC and STR

The MRAC originated from a deterministic servo problem and the STR from a stochastic regulation problem. In spite of these different origins it is clear from Figs. 11.2 and 11.3, that they are clearly related. Both have two feedback systems. The methods for design of the inner loop and the techniques used to adjust the parameters in the outer loop are different. The two methods (MRAC and self-tuning controller) can be quite different in terms of analysis and implementation. Compared with MRAC controllers, Self tuning controllers are more flexible because of the possibility of coupling various estimators with various controllers (i.e., the separation of estimation and control). However, the stability and convergence of self tuning controllers are generally quite difficult to guarantee, often requiring the signals in the system to be sufficiently rich so that the estimated parameters converge to the true parameters. If the signals are not very rich (e.g., if the reference signal is constant or zero), the estimated parameters may not be close to the true values. In this situation, one must either introduce perturbation signals in the input, or somehow modify the control law. In MRAC systems, however, the stability and tracking error converges are usually guaranteed regardless of the richness of the signals.

11.5 Design of Controllers for Deterministic Processes

In Chapter 8 we discussed the design of deadbeat controller, Dahlin controller, pole placement controller, internal model controller, simplified model predictive controller, In this section, methods of designing two more controllers (model reference controller, and optimal controller) for deterministic processes are presented.

11.5.1 Design of Model Reference Controller

The single input single output system is represented by the model

$$A(q^{-1})y(t) = q^{-d}B(q^{-1})u(t) \tag{11.1}$$

where

$$A(q^{-1}) = 1 + a_1q^{-1} + \ldots + a_nq^{-n} \tag{11.2}$$

$$B(q^{-1}) = b_0 + b_1q^{-1} + \ldots + b_nq^{-n} \tag{11.3}$$

It is assumed that $A(q^{-1})$ and $B(q^{-1})$ do not have common zeros (that is, the polynomials are monic).

A feedback controller is designed such that the output $\{y(t)\}$ of the process, when driven by a common signal $\{r(t)\}$, tracks a desired reference model output $\{y^*(t)\}$ driven by the same command input $\{r(t)\}$. The reference model is assumed to be of the form

$$J(q^{-1})y^*(t) = q^{-d} H(q^{-1})r(t) \tag{11.4}$$

where

$$H(q^{-1}) = h_0 + h_1 q^{-1} + \ldots + h_l q^{-l} \tag{11.5}$$

$$J(q^{-1}) = 1 + j_1 q^{-1} + \ldots + j_l q^{-l} \tag{11.6}$$

The transfer function model for the reference model is therefore given by

$$G(z^{-1}) = z^{-d}H(z^{-1})/J(z^{-1}) \tag{11.7}$$

The process model, Eq. (11.1) can be written as a prediction form with $J(q)$, $y(t + d)$ as its output, and then a suitable control law is selected so that $J(q^{-1})\, y(t + d)$ is equal to $H(q^{-1})\, r(t)$. Consequently, the process output will track the reference model output.

Equation (11.4) can be written in the predictive form as

$$J(q^{-1})y(t + d) = \alpha(q^{-1})y((t) + \beta(q^{-1})u(t) \tag{11.8}$$

where

$$\alpha(q^{-1}) = G(q^{-1}) \tag{11.9}$$

$$\beta(q^{-1}) = F(q^{-1})B(q^{-1}) \tag{11.10}$$

It can be shown that the polynomials $F(q^{-1})$ and $G(q^{-1})$ are unique and satisfy the condition

$$J(q^{-1}) = F(q^{-1})A(q^{-1}) + q^{-d}G(q^{-1}) \tag{11.11}$$

where $F(q^{-1})$ and $G(q^{-1})$ are of order $(l + d - 1)$ and $(l + n - 1)$ respectively. Given $A(q^{-1})$ and $J(q^{-1})$, we can calculate the elements of F and G.

To derive Eq. (11.8), Eq. (11.1) is multiplied by $F(q^{-1})$ and substituting Eq. (11.11) in the resulting equation, we get

$$J(q^{-1})y(t) = q^{-d}G(q^{-1})y(t) + q^{-d}F(q^{-1})B(q^{-1})u(t) \tag{11.12}$$

that is,

$$J(q^{-1})y(t + d) = G(q^{-1})y(\mathrm{t}) + F(q^{-1})B(q^{-1})u(t) \tag{11.13}$$

$$= \alpha(q^{-1})y(t) + \beta(q^{-1})u(t) \tag{11.14}$$

Let us choose the control law as

$$\alpha(q^{-1})y(t) + \beta(q^{-1})u(t) = H(q^{-1})r(t) \tag{11.15}$$

Hence

$$J(q^{-1})y(t + d) = H\,(q^{-1})r(t) \tag{11.16}$$

$$= J(q^{-1})y^*(t + d) \tag{11.17}$$

Hence $y(t)$ follows $y^*(t)$. Figure 11.4 shows the block diagram representation of the control system.

11.5.2 Adaptive MRC Algorithm

The control system is an indirect MRC formed by an adaptive model reference with an algorithm of parameters adoption. The adaptive model, for a SISO system, is described by the equation

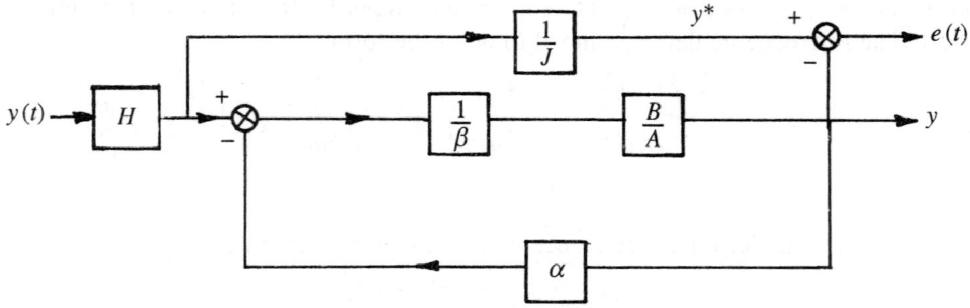

Fig. 11.4 Block diagram of a model reference controller

$$y_m(k) = \sum_{i=1}^{H} A_i(k-1)y(k-i-R) + \sum_{i=1}^{F} B_i(k-1)u(k-i-R) \qquad (11.18)$$

The model calculates, at each time step k, a value of the output variable, $y_m(k)$. This is the reference model value of the output that the actual plant output, $y(k)$, should asymptotically approach. Coefficients A_i and B_i are the adaptive parameters; $y(k-i-R)$ is past value of the output, and $u(k-i-R)$ is a past value of the input. The delay of the system is represented by parameter R. Parameters A_i and B_i can be calculated by diverse algorithms. The gradient algorithm is of great simplicity and has good convergence properties. This algorithm gives the following law of adaptation of A_i and B_i at each time step k

$$A_i(k) = A_i(k-1) + \beta_A\alpha(k)e(k)y(k-i-R) \quad i = 1, H \qquad (11.19)$$

$$B_i(k) = B_i(k-1) + \beta_B\alpha(k)e(k)u(k-i-R) \quad i = 1, F \qquad (11.20)$$

where the prediction error $e(k)$, is given by

$$e(k) = y(k) - y_m(k) \qquad (11.21)$$

The asymptotic factor, $\alpha(k)$, is given by

$$\alpha(k) = 1 \Bigg/ \left[1 + \sum_{i=1}^{H} \beta_A y(k-i-R)^2 + \sum_{i=1}^{F} \beta_B u(k-i-R)^2 \right] \qquad (11.22)$$

The adaptive control action, $u(k)$, is calculated by solving Eq. (11.8) for a future value of the output variable equal to the set point, $y_m(k + R + 1)$

$$u(k) = B_1^{-1}(k)y_m(k+R+1) - B_1^{-1}(k)\sum_{i=1}^{H} A_i(k)y(k-i+1)$$

$$- B_1^{-1}(k)\sum_{i=2}^{F} B_i(k)u(k-i+1) \qquad (11.23)$$

The controller requires simulation and/or experimentation for tuning its five parameters H, F, R, β_A/β_B and β_A. A sensitivity analysis has to be made by numerical simulation to find which parameters are most relevant. The parameters H and F can be tuned from 1 to 5 (integer values).

The MRAC has several advantages for one-line implementation, i.e., the algorithms require simple codes and computer memory requirements (this means shorter computing time), and the number of variables to measure is low (occasionally, during the system functioning, it is only required to measure the controlled variable).

The MRAC is a good alternative to PID, but is limited and requires tuning for each particular process. The tuning can be more or less easy, depending on the presence of lag, delay, and other factors. For processes not well known, the simulation could not be possible as it has been made usually for assumed model equations, and it would only be possible to tune the controller experimentally. This could be, evidently, a disadvantage from a commercial or business point of view, but actually this type of controller always requires particular adjustments for its implementation. In general, extensive simulation work is done on tuning the MRAC controllers based on assumed model equations (with appropriate values for the model parameters) of the process. Further on implementing the controller on a real plant requires additional tuning of the controllers (Palence et al. 1996).

11.5.3 Design of Optimal Controllers

Consider the following discrete time system

$$y(k) = \sum_{i=1}^{N} a_i y(k-i) + \sum_{i=1}^{M} b_i u(k-i) \tag{11.24}$$

The following performance index is considered

$$I(k) = [y(k+1) - y_r]^2 + Q'[dy(k+1)/dt]^2 + r_1[u(k) - u(k-1)]^2 \tag{11.25}$$

where $[Q', r_1]$ are prescribed positive weighting constants. The above performance index takes into consideration deviation of the system output and its derivative as well as that of the control effort $[u(k) - u(k-1)]$. The following approximation is used for the derivative term

$$dy(k+1)/dt = [\,y(k+1) - y(k)]/T \tag{11.26}$$

The performance index becomes

$$I(k) = [y(k+1) - y_r]^2 + Q[y(k+1) - y(k)]^2 + r_1[u(k) - u(k)]^2 \tag{11.27}$$

where $Q = Q'/T^2$. The optimal controller signal which minimizes $I(k)$ of Eq. (11.25) can be readily derived by first replacing $y(k+1)$ in Eq. (11.25) by that obtained from Eq. (11.24) and then setting $\partial I(k)/\partial u(k)$ to zero to yield

$$u(k) = [b_1 Q y(k) - S(k) - b_1 y_r + r_1 u(k-1)]/[r_1 + b_1^2(1+q)] \tag{11.28}$$

where

$$S(k) = b_1(1+Q)\sum_{i=1}^{N} a_i y(k-i+1) + \sum_{i=1}^{M} b_i u(k-i+1) \tag{11.29}$$

If $Q' = 0$ and $R = 0$, then it is called minimum variance controller.

If $Q' = 0$, then it is a conventional strategy that is followed. If r_1 is not equal to zero, then integral control action is imported. Thus, the general form with the weighting in the control action is more robust. The performance of general optimal control is better than that obtained without the derivative action. Experimental comparison on liquid level is discussed by Lim (1990, 1989).

11.6 Model Predictive Control

The term model predictive control designates a range of control methods which make an explicit use of a model of the process to obtain the manipulated variable by minimizing an objective function. Various MPC algorithms differ amongst themselves only in the model used to represent the process and the noises and the objective function to be minimized. Excellent reviews on the reported work on model predictive control are given by Garcia et al. (1989), Kanjilal (1995) and Camacho and Bordons (1999). The basic steps involved in designing a MPC are:

(i) Explicit use of a model to predict the process ouput at future time instants (called horizon): The future outputs for a determined horizon N, called the prediction horizon, are predicted at each instant t using the process model. These predicted outputs $y(t + k/t)$ for $k = 1, 2 \ldots, N$ calculated using the known values of y upto instant t (past inputs and outputs) and on the future control signal $u(t + k/t)$, $k = 0, 1, 2, \ldots, N - 1$, which are those to be sent to the system, are to be calculated.

(ii) Calculation of a control sequence minimizing an objective function: The set of future control signals is calculated by optimizing a determined criterion in order to keep the process as close as possible to the reference trajectory $w(t + k)$ (which can be the set point itself or a close approximation of it). The criterion may be a quadratic function of the errors between the predicted output signal and the predicted reference trajectory. The control effort is included in the objective function in most cases. If the model used is linear and the objective function is quadratic then an explicit solution for the control law can be obtained, otherwise an iterative optimization method is used.

(iii) Receding strategy: The manipulated variable (control signal), $u(t)$ is sent to the process whilst the next control signals calculated are rejected, because at the next sampling instant $y(t + 1)$ is already known and step 1 is repeated with this new value and all the sequences are brought up to date. Thus $u(t + 1/t + 1)$ is calculated (which in principle will be different to $u(t + 1/t)$ because of the new information available) using the receding horizon concept.

MPC has proved itself to be economically profitable by reducing operating costs or increasing production and is one of the most successful advanced control techniques used in industry. The capability of MPC to optimize and treat constraints is the reason for its success.

Optimization of operating conditions: MPC optimizes a cost function which can be formulated in order to minimize operating costs or any objective with economic implications.

Optimization of transitions: The MPC objective function can be formulated in order to optimize a function which measures the cost of moving the process from one operating point to another, with faster process start-ups or commissioning times.

Minimizing of error variance: An MPC can be formulated in order to minimize the variance of the output error. A smaller variance will produce economic benefits for the following reasons:

- a smaller variance may increase the quality of the product as well as its uniformity.
- it will allow the process to operate closer to the optimal operating conditions: As most processes are constrained, the optimal operating points usually lie at the intersection of some constraints. Processes operating with smaller error variance can operate closer to the operating points.

Although the resulting control law is easy to implement and requires little computation, its derivation is more complex than that of the classical PID controllers. If the process dynamics do not change, the derivation of the controller can be done beforehand, but in adaptive control case for varying process dynamics, all the computation has to be carried out at every sampling time. When constraints are considered, the extensive computation is required (because of the optimization step).

The implementation of MRC in industry is not a trivial matter. An MPC application is more costly, time consuming and requires personnel with better training in control than when implementing classical PID controllers. To start, a model of the process has to be found, and this requires a significant number of plant tests, which in most cases implies taking the plant away from its nominal operating conditions. The control equipment needs, in some cases, a more powerful computer and better instrumentation, and commercial MPC packages are expensive. Furthermore, control personnel need appropriate training for commissioning and using MPC.

The different strategies of MPC can use various models to represent the relationship between the outputs and the measurable inputs, some of which are manipulated variables and others can be considered to be measurable disturbances which can be compensated for by a feed forward action. A disturbance model can also be taken into account in order to describe the behaviour which is not reflected by the process model error, including the effect of non-measurable inputs, noise and model errors. The model can be separated into two parts: the actual process model and the disturbances model. Both parts are needed for prediction.

Although there are several companies that make use of technology developed in-house, that is not offered externally, the ones listed below can be considered representative of the current state of the art of MPC technology. The product names and acronyms are:

- **DMC Corp.:** Dynamic Matrix Control
- **SCAP Europa:** Adaptive Predictive Control System (APCS)
- **Honeywell Profimatics:** Robust Model Predictive Control Technology (RMPCT) and Predictive Control Technology
- **Treiber Controls:** Optimum Predictive Control (OPC)

Each of the above product is not the algorithm alone, but is accompanied by additional packages, usually identification or plant test packages. The existing industrial MPC technology has the following limitations:

Over parameterized model: Most of the commercial products use the step or impulse response model of the plant, that are known to be over-parameterized. For instance, a first order process can be described by a transfer function model using only three parameters (gain, time constant and dead time) whereas the step response model will require more than 30 coefficients to describe the same dynamics. Besides, these models are not valid for unstable systems. These problems can be overcome by using an auto-regressive parametric model.

Tuning: The tuning procedure is not clearly defined since the trade-off between tuning parameters and closed loop behaviour is generally not very clear. Tuning in the presence of constraints may be even more difficult, and even for the nominal case, is not easy to guarantee closed loop stability; that is why so much effort must be made for prior simulations.

Sub-optimality of the dynamic optimization: Several packages provide sub-optimal solutions to the minimization of the cost function in order to speed up the solution time. It can be accepted in high speed applications where solving the problem at every sampling time may not be feasible, but it is difficult for justify for process control applications.

Model uncertainty: Controllers have to be detuned to improve robustness although the relation between performance and robustness is not clear. Only one product (RMPCT) uses model identification package to get the estimates of model uncertainty.

Constant disturbance assumption: It is assumed that disturbance will remain constant in the future. If the distribution of the disturbance is considered then that would give an improved performance.

Analysis: A systematic analysis of stability and robustness properties of MPC is not possible in its original finite horizon formulation. The control law is in general time-varying and cannot be represented in the standard closed loop form, especially in the constrained case.

Nonlinear model can also be used in the MPC strategy. However we have the following difficulties to use nonlinear MPC:

(i) The availability of nonlinear models due to lack of identification techniques for nonlinear processes.
(ii) Computational complexities for solving the model predictive control of nonlinear processes.
(iii) Lack of stability and robustness results for the case of nonlinear systems.

11.7 Details on Model Predictive Control

As stated earlier, the general strategy of model predictive control (MPC) is to utilize a model to predict the output in the future and minimize the difference between this predicted output and the desired one by computing appropriate control actions. Dynamic matrix control (DMC), Model algorithmic control (MAC) and Generalized predictive control (GPC) fall into the MPC category. These methods give good responses even for systems with unknown time delay and with unmodeled dynamics. However, these methods require a substantial amount of computation and require a microprocessor or on-line computer. An excellent review on MPC is given by Garcia et al. (1989). Let us discuss briefly the basic idea involved in each of the methods.

11.7.1 Dynamic Matrix Control

The major elements of DMC are: (i) the model, (ii) estimation of the disturbance and projection into the future and (iii) computation of the control inputs. The model used is a discrete truncated 'step response model' of the form

$$y(k) = \sum_{i=1}^{N} [a_i \Delta u(k-i)] + a_N u(k-N-1) + d(k) \tag{11.30}$$

where

k is the sampling time index
u the input variable
a_i step response coefficients
d the unmodeled or disturbance effects on ouput y
N number of step response coefficient needed to describe the process dynamics
$\Delta u(k)$ the change in the input from the previous value, defined by $u(k) - u(k-1)$

The present value of the disturbance effects can be estimated by the difference between the present measurement of the output $\{y_{meas}(k)\}$ and the effects of the past inputs on output, i.e.,

$$d(k) = y_{meas}(k) - \sum_{i=1}^{N} [a_i \Delta u(k-i)] - a_N u(k-N-1) \tag{11.31}$$

The linear estimate of the future output can be written in a matrix notation by writing Eq. (11.30) for $k = k+1$ to $k+p$

$$\mathbf{y}_{lin} = \mathbf{y}_{past} + \mathbf{A}\Delta u + \mathbf{d} \tag{11.32}$$

where

$$\mathbf{y}_{lin} = [\, y(k+1),\ y(k+2), \ldots\ y(k+p)\}^T \tag{11.33}$$

$$\mathbf{d} = [d(k+1), d(k+2), \ldots d(k+p)]^T \tag{11.34}$$

$$\mathbf{A} = \begin{bmatrix} a_1 & & & \\ a_2 & a_1 & & 0 \\ a_3 & a_2 & a_1 & \\ & a_M & a_{M+1} & \ldots a_1 \\ \ldots & \ldots & & \\ \ldots & \ldots & & \\ a_P & a_{p-1} \ldots & & a_{P-M+1} \end{bmatrix} \tag{11.35}$$

$$\Delta\mathbf{u} = [\Delta u(k), \Delta u(k+1), \ldots \Delta u(k+M-1)]^T \tag{11.36}$$

$$\mathbf{y}_{\text{past}} = a_N[u(k-N), u(k-N+1), u(k-N+2), \ldots u(k-N+M-1), \ldots, u(k-N+P-1)]^T$$
$$+ \mathbf{B}[\Delta u(k-1), \Delta u(k-2), \Delta u(k-3), \ldots \Delta u(k-N+1)]^T \tag{11.37}$$

$$\mathbf{B} = \begin{bmatrix} a_2 & a_3 & & & a_N \\ a_3 & a_4 & & a_N & 0 \\ a_4 & a_5 & a_{N-1} & 0 & 0 \\ \ldots & \ldots & \ldots & \ldots & \ldots \\ a_{P+1} & \ldots & \ldots & 0 & 0 \end{bmatrix} \tag{11.38}$$

The effects of the known past inputs on the future output is defined by the vector $\mathbf{y}_{\text{past}}$. $\mathbf{A}$ is the dynamic matrix composed of the step response coefficients. P denotes the length of the prediction horizon and M is the moving horizon of the number of future moves $\Delta u(k), \ldots, \Delta u(k+m-1)$ calculated by the DMC algorithm which will, be discussed later. The future behaviour of the disturbance is not known and is usually assumed to be equal to the currently estimated value, i.e., $d(k+i) = d(k)$ for $i = 1, 2, \ldots, P$ and $d(k)$ is calculated from Eq. (11.31). With these definitions, the future output can now be predicted from Eq. (11.32) for any given vector of future control moves $\Delta\mathbf{u}$.

The following optimization problem is used to calculate the control inputs (i.e., select Δu in order to minimize $\mathbf{E}$)

$$\min_{\Delta\mathbf{u}} \mathbf{E} \sum_{i=1}^{P} \gamma^2(i)[y_{\text{sp}}(k+i) - y_{\text{lin}}(k+i)]^2 + \sum_{J=1}^{M} \lambda^2(j)[\Delta u(k+M-j)]^2 \tag{11.39}$$

where $E(.)$ denotes the expectation operator; y_{sp} is the set point; γ and λ are the time varying weights on the output error and on the change in the input, respectively. The least square solution of the above optimization problem is given by

$$\Delta\mathbf{u} = [\mathbf{A}^T\Gamma^T\Gamma\mathbf{A} + \Lambda^T\Lambda]^{-1}\mathbf{A}^T\Gamma^T\Gamma(\mathbf{y}_{\text{sp}} - \mathbf{y}_{\text{past}} - \mathbf{d}) \tag{11.40}$$

where Γ and Λ are the diagonal matrices containing the weights $\gamma(i)$ and $\lambda(i)$, respectively. Usually only the first calculated move $\Delta u(k)$ is implemented and the calculations are repeated at the next sampling time to account for changing disturbances and to incorporate feedback. P, M, Γ and Λ are used as tuning parameters by the designer. The limitation of the method is that it cannot be applicable for open loop unstable systems.

Since most of the processes are described by simple FOPTD models, Shridhar and Cooper (1997) have proposed a tuning method for DMC for such models. The suggested tuning method is as follows:

1. Approximate the manipulated-to-measured output variable dynamics with FOPTD model

$$K_p \exp(-Ls)/(\tau s + 1).$$

2. Select the sampling interval as close as possible to

$$T = 0 \cdot 1\tau \quad \text{or} \quad T = 0.5L$$

whichever is larger.

3. Compute the prediction horizon, P, and the model horizon, N, as the process settling time in samples (rounded to the next integer)

$$P = N + (5\tau/T) + k, \quad \text{where} \quad k = (L/T) + 1$$

4. Select the control horizon, M, as an integer in the range

$$0.1(\tau/T) \leq M \leq 0.6(L/T) \quad \text{(usually from 1 to 6)}$$

5. Compute the move suppression coefficient

$$(\lambda^2/\gamma^2) = 0 \qquad \text{for } M < 1$$

$$= (M/500)[3.5(\tau/T) + 2 - 0.5(M - 1)]K_p^2 \qquad \text{for } M > 1$$

6. Implement DMC using the traditional step response matrix of the actual process and the above parameters (T, N, M, P, λ^2/γ^2).

11.7.2 Model Algorithmic Control (MAC)

MAC differ from DMC in the following three aspects:

(1) Instead of the step response model involving Δu, an impulse response model involving u is employed. The truncated impulse response model is given by

$$y(k) = \sum_{i=1}^{N} h_i u(k - i) + d(k) \tag{11.41}$$

where the coefficient, h_i, are related to the step response model coefficient a_i

$$h_i = a_i - a_{i-1} \tag{11.142}$$

If the input is penalized in the quadratic objective, then the controller does not remove offset. This can be corrected by a static offset compensator. If input u is not penalized, then the control algorithm cannot handle nonminimum phase system.

(2) The number of input moves m is not used for tuning ($M = P$).

(3) The disturbance vector estimate is filtered. The optimization step requires an iterative method.

11.7.3 Generalized Predictive Control

The model used in this approach is the CARIMA (Controlled Auto-Regressive Integrated Moving Average) model

$$A(q^{-1}) = B(q^{-1})u(k - 1) + C(q^{-1})\xi/\Delta \tag{11.43}$$

where Δ is a difference operator ($\Delta = 1 - q^{-1}$) and q^{-1} the backward shift operator $\{q^{-1}y(k) = y(k-1)\}$.

$$A(q^{-1}) = 1 + a_1q^{-1} + \ldots + a_nq^{-n} \quad (11.44)$$

$$B(q^{-1}) = b_0 + b_1q^{-1} + \ldots + b_mq^{-m} \quad (11.45)$$

$$C(q^{-1}) = c_0 + c_1q^{-1} + \ldots + c_{n_c}q^{-n_c} \quad (11.46)$$

ξ is the random disturbances (white noise). The predictor model used is given by

$$y(k+i) = G_j\Delta u(k+j-1) = F_j(k) \quad (11.47)$$

where

$$G_j(q^{-1}) = E_jB \quad (11.48)$$

E_j and F_j are obtained from the identity

$$1 = E_j(d^{-1})A\Delta + q^{-j}F_j(q^{-1}) \quad (11.49)$$

The objective function considered is given by

$$J(N_1, N_2, N_u) = E\sum_{j=N_1}^{N_2}[y(k+j) - W(k+j)]^2 + \sum_{j=1}^{N_u}\lambda(j)[\Delta u(k+j-1)]^2 \quad (11.50)$$

where N_1 is the minimum costing horizon; N_2 the maximum costing horizon; N_u the control horizon, and $\lambda(j)$ is a control-weighting sequence.

The solution of optimization problem is given by

$$\Delta\mathbf{u} = [G^TG + \lambda I]^{-1}G^T(\mathbf{w} - \mathbf{f}) \quad (11.51)$$

The first element of $\Delta\mathbf{u}$ is Δu so that current control $u(k)$ is given by

$$u(k) = u(k-1) + \mathbf{g}^{-T}(\mathbf{w} - \mathbf{f}) \quad (11.52)$$

where $\mathbf{g}^{-T}$ is the first row of $[G^TG + \lambda I]^{-1}G^T$ and $\mathbf{w}$ the set point vector; and

$$\mathbf{f}\,[f(k+1)\,f(k+2)\,f(k+n)]^T \quad (11.53)$$

$$f(k+1) = [G_1(q^{-1}) - g_{10}]\Delta u(k) + F_1y(k) \quad (11.54)$$

etc. $f(k+j)$ is the component of $y(k+j)$ composed of signals which are known at time k. After an interval $N_u < N_2$ projected control increments are assumed to be zero

$$\Delta u(k+j-1) = 0,\ j > N_u \quad (11.55)$$

Clark et al. (1989) discussed the method of selecting parameters N_1, N_2 and N_u. The recommended values are $N_1 = 1$, $N_u = 1$ and N_2 must be equal to the plant rise time.

The GPC is usually able to stabilize and control open-loop unstable processes through judicious choice of tuning parameters N_1, N_2, N_u and λ. The GPC approach is related to the properties of Linear Quadratic Control. The theoretical basis of the GPC algorithm is developed. It has been shown (Clarke and Mothadi, 1987) that for limiting cases of parameter choices, the GPC algorithm is stable and that certain well known controllers (such as deadbeat) are inherent in the GPC structure. The limitations of GPC is that there are no guaranteed stability properties for it except under special conditions. Care must be taken in tuning parameters choices. If $N_u = N_2$ and the control weighting λ is set to zero, then GPC reduces to a minimum variance controller, which is known to be unstable for non minimum phase systems.

A substantial amount of computation is necessary in order to implement a generalized predictive controller. Since the dynamics of most of the processes are described adequately by a delay plus first order transfer function model, Camacho and Bordons (1993) have discussed a simple method of implementing GPC for such processes. Let the process be described by

$$G(s) = k_p \exp(-Ks)/(1 + \tau s) \tag{11.56}$$

The corresponding discrete transfer function model is given by

$$G(q^{-1}) = bq^{-1}/(1 - aq^{-1})z^{-d} \tag{11.57}$$

where $L = dT$ and T is the sampling time. The equation can be transformed into a CARIMA model as

$$y(k + 1) = (1 + a)y(k) - ay(k - 1) + b\Delta u(k - d) + \zeta(k + 1) \tag{11.58}$$

Camacho and Bordons (1993) have considered $N_1 = d + 1$, $N_2 = d + N_u$; N_u being the control horizon. After carrying out the design method of GPC as discussed in Sec. 11.7.3, Camacho and Bordons (1993) have shown that GPC law for the delay plus first order system can be written as

$$\Delta u(k) = l_1 y(k + d) + l_2 y(k + d - 1) + l_3 w(k) \tag{11.59}$$

The predicted values of $y(k + d)$ and $y(k + d - 1)$, from Eq. (1.209) with $\zeta = 0$ to be used in Eq. (1.210). For assumed value of $\lambda = 0.8$ and $Nu = 15$, the fitted expressions for the parameters l_1, l_2 and l_3 are given by (Camacho and Bordons, 1993)

$$l_1 = -0.845 - 0.564[a/(1.05 - a)] \tag{11.60}$$

$$l_2 = 0.128 + 0.459[a/1.045 - a)] \tag{11.61}$$

$$l_3 = -l_1 - l_2 \tag{11.62}$$

These expressions are shown by Camacho and Bordons (1993) to give a very good approximation to the true controller parameters and fit the set of computed data with a maximum error of less than 1 per cent of the nominal values for the range of interest of the open-loop pole. By simulation studies on transfer function models, Camacho and Bordons (1993) have shown that GPC works well even for situations with unknown time delay and unmodeled dynamics. A comparative study of the predictive controller tuning methods was recently given by Rani and Unbehauen (1997).

Summary

Adaptive control strategy is required for controlling time-varying systems. The MRAC, self-tuning control, direct adaptive control and indirect adaptive controller fall under adaptive control techniques. Model predictive control strategies are to be used for controlling uncertain systems with or without constraints.

References

Akaike, H.A., A new look at the statistical model identification, IEEE, AC–19 (1974) 716.

Anand, D.K. and R.B. Zmood, Introduction to control systems, III Ed., Asian Books Pvt Ltd , New Delhi, 1995.

Astrom, K.J. and B. Wittenmark, On self-tuning regulators, Automatica, 9 (1973) 185–199.

Astrom, K.J. and B. Wittenmark, Adaptive Control, Addison Wesley, New York, 1989.

Camacho, E.F. and C. Bordons, Implementation of self-tuning generalized predictive controllers for the process industry, Int. J. Adapt. Signal Proc., 7 (1993) 63–73.

Camacho, E.F. and C. Bordons, Model Predictive Control, Springer-Verlag, (1999).

Chalam, V.V., Adaptive control systems, Marcel Dekker, New York, 1995.

Clarke, D.W., C. Mohtadi and P.S. Tuffs, Generalized predictive control-Part I: The basic algorithm, Automatica, 23 (1987) 137–148.

Garcia, C.E., D.M. Prett and M. Morari, Model Predictive control: Theory and Practice-a survey, Automatica, 15 (1989) 335–348.

Kanjilal, P.P., Adaptive prediction and predictive control, Peter Peregrinus Ltd., London, 1995.

Landau, I.D., R. Lozano and M.M. Saad, Adaptive Control, Springer-Verlag, Berlin, 1998.

Lim, C.M., A self tuning stabiliser for excitation or governor control of power systems, IEEE trans. on energy conversion., 4 (1989) 152–159.

Lim, C.M., Experimental evaluation of a self-tuning controller, IEEE Trans. Ind. Elctr., IE–37 (1990) 193–194.

Martin-Sanchez, J.M. and J. Rodellar, Adaptive Predictive Control. From the concepts to plant optimization, Prentice Hall International, New York, 1996.

Muske, K.R. and J. Rawlings, Model Predictive control with linear models, AIChE J., 39 (1993) 262–287.

Palancer, M.C., J.M. Aragon, J.A. Miguens and J.S. Torrecilla, Application of a model reference adaptive control system to pH control: Effects of lag and delay time, I and EC Res. 35 (1996) 4100–4110.

Rani, K.Y. and H. Unbehauen, Study of predictive controller tuning methods, Automatica, 33 (1997) 2243–2248.

Richalet, Industrial applications of model based predictive control, Automatica, 29 (1993) 1251–1274.

Seborg, D.E., T.F. Edgar and S.L. Shah, Adaptive control strategies for process control: A survey, AIChE J., 32 (1986) 881–913.

Shridhar, R. and D.J. Cooper, A tuning strategy for unconstrained SISO model predictive control, Ind. Eng. Chem. Res., 36 (1997) 729–746.

Slotin, J.E. and W. Li, Applied Nonlinear Control, Prentice-Hall, Englewood Cliffs, New Jersey, (1991).

Review Questions

1. What is meant by adaptive control system?
2. Discuss on gain scheduling controllers. What are the advantages and limitations of such controllers?
3. What is meant by a self-tuning controller?
4. With a block diagram, explain the MRAC systems.
5. Compare the STC and MRAC systems.
6. What is meant by a certainty equivalence principle?
7. Design a suitable Model Reference Controller for the system

$$A(q^{-1})y(k) = q^{-1}Bu(k)$$

$$y(0) = 1$$

where $\quad A(q^{-1}) = 1 + 2q^{-1} + q^{-2}; B(q^{-1}) = (1 + 0.5q^{-1} + 0.5q^{-2})$

The desired reference model of the closed loop system is given by

$$A_m^{-1}(z^{-1})y_r(k) = q^{-1}B_m u(k)$$

where

$$A_m(z^{-1}) = 1 - q^{-1} + 0.25q^{-2}$$

$$B_m(z^{-1}) = 0.5 + 0.15q^{-1}$$

8. Explain the design method of optimal controller.
9. What is meant by a model predictive controller? Name the different MPC algorithms.
10. Explain the basic concepts in designing DMC.
11. Explain the basic concepts in designing MAC.
12. Explain the basic concepts in designing GPC.
13. Explain the tuning method for DMC proposed by Shridhar and Cooper (1997).

12

Introduction to Nonlinear Discrete Models and Control

Chapter 8 discussed the discrete time linear models. Such models are also valid for mildly nonlinear systems. However for highly nonlinear systems, nonlinear models need to be develcped. The nonlinear model in addition to satisfactorily describing the input-output behaviour, should also be linear in unknown parameters and also be suitable for derivation of nonlinear control algorithm. This chapter presents development of nonlinear models and design of nonlinear controllers.

12.1 Block Oriented Nonlinear Models

12.1.1 Hammerstein Model

The first category in nonlinear models are block-oriented models which combine linear dynamic systems with static nonlinearities. Two such models are called Hammerstein model and Wiener model. The Hammerstein model (Fig. 12.1) consists of a static nonlinearity fcllowed by a linear system. One way to represent any member of this class is given by the equation

$$y(k) = \sum_{i=0}^{\infty} h_i f(u(k-i)) \tag{12.1}$$

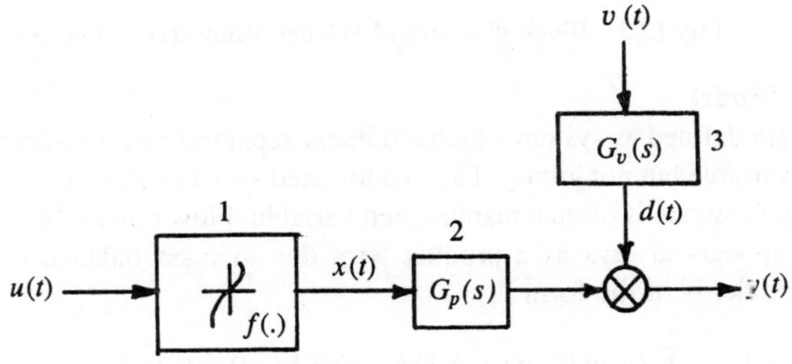

Fig. 12.1 Block diagram of a Hammerstein system

where h_i, $i = 0, 1, \ldots$ are parameters of linear model (finite impulse response model) and $f(.)$ is a static nonlinearity. It is to be noted that the value of $u(k)$ and $y(k)$ are the actual measured values of the process input and output signals and not the deviations from steady-state values. Hammerstein system, with continuous static nonlinearity in the case of bounded input signals, or analytic static nonlinearity in the unbounded case, can be represented arbitrarily well using the equation

$$y(k) = \sum_{i=0}^{M} h_i \sum_{j=0}^{N} g_j u^j (k - i) \tag{12.2}$$

and choosing M and N large enough. The parameters h_i, $i = 0, \ldots, M$ represent the linear system and g_j, $j = 0, \ldots N$ are parameters of the polynomial static nonlinearity.

12.1.2 Wiener Model

Wiener model (Fig. 12.2) consists of a linear system followed by a static nonlinearity. The parameterization is denoted by $W_{N,M}$

$$y(k) = \sum_{j=0}^{N} g_j \left(\sum_{i=0}^{M} h_i u(k - i) \right)^j \tag{12.3}$$

where g_j, $j = 0, 1, \ldots, N$ and h_i, $i = 0, 1, \ldots, M$ as parameters may be used to represent the Wiener model.

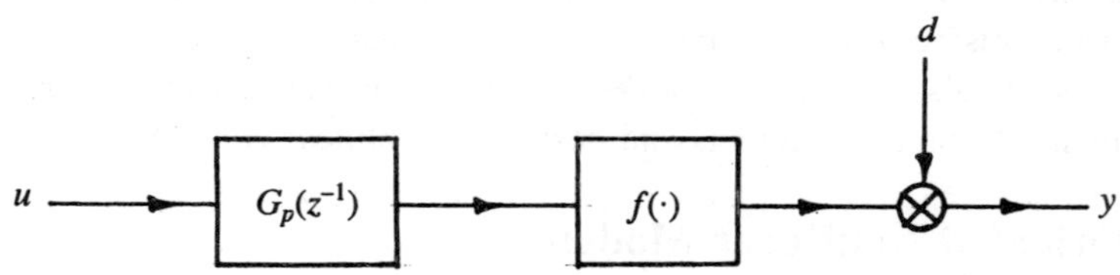

Fig. 12.2 Block diagram of a Wiener system

There are some systems which can be modeled by Wiener-Hammerstein cascade model, in which a linear model is followed by a nonlinear gain subsystem which is in turn followed by another linear model (Fig. 12.3). Usually the identification of Hammerstein system is easier than that of the Wiener system.

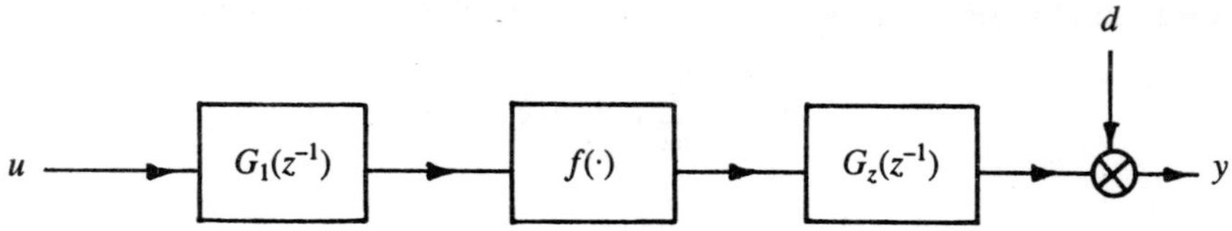

Fig. 12.3 Block diagram of Wiener-Hammerstein system

12.1.3 Bilinear Model

Bilinear systems are defined as system which are linear separately with respect to the output variable and manipulated variable but not jointly. The manipulated variable and the output variable appears as a product. Usually flow rate is often a manipulated variable. Flow rate and concentration or flow rate and temperature appears always as a product term due to mass balance or energy balance. The discrete bilinear model is of the form

$$y(k) = \sum_{i=1}^{N} \{a_i y(k - i) + b_i y(k - i)u(k - i) + c_i u(k - i)\} + d(k) \tag{12.4}$$

In compact vector notation, the model equation is

$$y(k) = \mathbf{p}^T\mathbf{x}(k-1) + d(k) \tag{12.5}$$

where

$$\mathbf{p} = [a_1, \ldots, a_n, b_1, \ldots, b_n, c_1, \ldots, c_n]^T \tag{12.6}$$

$$\mathbf{x}(k-1) = [y(k-1), \ldots, y(k-n), y(k-1)u(k-1), \ldots, y(k-n)u(k-n), u(k-1), \ldots, u(k-n)]^T \tag{12.7}$$

The standard least squares method can be used to get the model parameter vector **p** from the measurement vector $\mathbf{x}(k-1)$.

12.1.4 Volterra Model

The best known class of nonlinear systems that does not possess moving average representations is the Volterra model

$$y(k) = y_0 + \Sigma\, a_j u(k-j) + \sum_{i=0}^{\infty}\sum_{j=0}^{\infty} b_{i,j} u(k-i)u(k-j) + \sum_{l=0}^{\infty}\sum_{i=0}^{\infty}\sum_{j=0}^{\infty} c_{l,i,j} u(k-l)u(k-i)u(k-j) \tag{12.8}$$

where a_j, $b_{i,j}$ etc. are called the Volterra kernels. This class of systems is very broad, but to be practically useful, the summation must be truncated to some finite upper limt M and the number of sums included must also be made finite. A simple example of this model class is the second order Volterra model

$$y(k) = u(k) + bu(k-1)u(k-2) \tag{12.9}$$

For the general Volterra series description of a nonlinear system, consider that in order to obtain a good approximation to the first-order kernels, it is necessary to estimate about 30 points. To obtain the same accuracy for the second order terms requires (30 × 30)/2 = 450 points, taking to account the symmetry. The third order kernels require (30 × 30 × 30)/3 = 9000 points. Consequently, the number of input-output data map required to estimate all the parameters is extremely large. The parameter estimation takes an enormous amount of computational time.

12.1.5 NARX Model

The NARX (Nonlinear Auto-regressive with eXogenous input) model is frequently used with many nonlinear identification methods. The NARX model establishes a relation between the plant input-output data and the predicted output

$$\hat{y}(k+1) = F(y(k), \ldots, y(k-n_y+1), u(k), \ldots, u(k-n_u+1)) \tag{12.10}$$

where k denotes discrete time samples, n_u and n_y are integers related to the system's order, and F denotes a nonlinear model. In the NARX model, the regression vector is a collection of a finite number of past inputs and outputs

$$\mathbf{x}(k) = [\, y(k), \ldots, y(k-n_y+1), u(k), \ldots, u(k-n_u+1)]^T \tag{12.11}$$

The regressand is the predicted output $\hat{y}(k+1)$. Hence, from a set of observed inputs and outputs of an unknown dynamics system, $S\{u(j), y(j), j = 1, 2, \ldots N\}$, the function F (.) in Eq. (12.10) can be approximated by using static nonlinear regression. Pure delays from the input to the output can be directly incorporated in the regression vector

$$\mathbf{x}(k) = [y(k), \ldots, y(k - n_y + 1), u(k - n_d + 1), \ldots, u(k - n_d - n_u + 1)]^T \qquad (12.12)$$

where n_d is the delay in samples. For the sake of brevity, the models will be given for $n_d = 1$. Nonlinear counterparts of other linear block-box structures can be used as well. The Nonlinear Finite Impulse Response (NFIR) model employs only the past values of $u(k)$ as regressors

$$\hat{y}(k + 1) = F(u(k), u(k - 1), \ldots, u(k - n_u + 1)) \qquad (12.13)$$

Also, here the unknown function F (.) can be directly inferred from the data by using statistical regression. Note, however, that n_u may be quite large and is not directly related to the system's order. For parameter estimation problem the RLS algorithm is to be used if the model is linear in its parameters. Otherwise, quasi-Newton or other gradient based methods must be used to solve the resulting nonlinear optimization problem.

One of the main advantages of the difference equation description is the number of parameters required. Generally, less than 30 parameters suffice for a polynomial difference equation. Hence, from practical point of view, the difference equation is more convenient than the Volterra series. From the difference equation model (NARX), it is possible to compute the kernels of the Volterra series (Diaz and Desrochers, 1988).

12.1.6 Nonlinear Output Error model

The Nonlinear Output Error [NOE] model involves past model predictions instead of the data

$$\hat{y}(k + 1) = F(\hat{y}(k), \ldots, \hat{y}(k - n_y + 1), u(k), \ldots, u(k - n_u + 1)) \qquad (12.14)$$

The 'innovations' forms offer additional freedom in describing the effect of disturbances. In the NARMAX model, for instance, the prediction error $e(k) = y(x) - \hat{y}(k)$ and its past values are included in the regression vector as well

$$\hat{y}(k + 1) = F(y(k), \ldots, y(k - n_y + 1), u(k), \ldots, u(k - n_u + 1), e(k), \ldots, e(k - n_e)) \qquad (12.15)$$

The regression vector in NOE and innovation model forms cannot be constructed directly from the data, since it includes the past values of the model output. The corresponding regression problem must be solved iteratively.

It is necessary to impose structural constraints on the NARMAX nonlinearity F (.) in Eq. (12.15), since even in the case of relatively simple polynomial NARMAX models, the quantitative behaviour of the steady state solutions is difficult to classify.

12.1.7 Nonlinear State Space Modelling

The model for the system is assumed as

$$\mathbf{x}(k + 1) = g(\mathbf{x}(k), \mathbf{u}(k)) \qquad (12.16)$$

$$y(k) = h(\mathbf{x}(k)) \qquad (12.17)$$

where state transmission function g maps the current state $\mathbf{x}(k)$ and input $\mathbf{u}(k)$ into a new state $\mathbf{x}(k + 1)$. The output function h maps the states $\mathbf{x}(k)$ into the output $y(k)$.

The nonlinear state-space representation is useful when prior knowledge allows us to model the system from first principles such as mass and energy balance. This approach is called white-box state-space modeling. If the state is directly measured on the system, or can be reconstructed from other measured variables, both g and h can be approximated by using nonlinear regression vectors. An advantage of the state-space modeling approach is that the structure of the model is related to the structure of the real system, and consequently, also the model parameters are often physically relevant. This not usually the case in the input-output models. In addition, the dimension of state-space modeling is often smaller than input-output models, since the state of the vector can be represented with a vector of a lower dimension than the regression (Eq. 12.10). Usually the parameters are also appear nonlinearly in Eqs. (12.16) and (12.17) and nonlinear optimization method is required for their estimation. The nonlinear controller based on Eqs. (12.16) and (12.17) requires most of the state variable measurement (or their estimation).

12.1.8 NARMAX Model

The NARMAX model proposed in Chen and Billings (1989) comprises of a general nonlinear discrete model structure. The model considers all possible permutations of the input and output terms in a linear combination. To introduce the subscript notation, consider an example of a NARMAX model

$$y(k) = c_{01,1}y(k-1) + c_{01,2}y(k-2) + c_{02,1}y(k-1)y(k-1) + c_{10,1}u(k-1) + c_{20,1}u(k-1)u(k-1) + \ldots \quad (12.18)$$

The terms $c_{uy,i}$ are constant coefficients with the following subscripts notation: u refers to the order of the contribution of past input values; y refers to the order of the contribution of past values; and i is the coefficient for a specific regressor structure.

The above model equation can be obtained from the general model equation (Eq. 12.10) by expanding F (.) as a polynomial function. The general NARMAX model allows for many input/output model structure terms to be considered while still maintaining a linear-in-coefficients problem to be solved. Through the use of an orthogonal estimator algorithm (Billings et al. 1988), the key regressor are determined from a statistical analysis.

12.2 Identification Methods

12.2.1 Identification of NARMAX Model

Orthogonal estimator: The first step in developing a NARMAX model is to identify the regressors. The large number of possible permutations between input and output pairings can cause the model order to grow significantly. For example, a cubic NARMAX model using three past states has 99 different combination terms. To identify the significant terms among the 99 possibilities, an orthogonal least squares estimator (Billings et al. 1988) is used to select the terms of the NARMAX model structure. An estimator algorithm is then used to solve the least squares problem once a library of regressors has been orthogonalized. The orthogonalization and manipulation of the least squares problem will yield a term selection criterion referred to as error reduction ratio (ERP). This ERP is a measure of the significance that each orthogonalized regressor has on the variance of the model output. The algorithm uses this criterion to select model regressors until a point of diminishing return is reached.

The NARMAX least squares problem is formulated as

$$y(k) = \sum_{i=1}^{M} p_i(k)\theta_i + e(k) \quad (12.19)$$

where $y(k)$ is the output of the model, $p_i(k)$ the regressor or model structure terms, θ_i the coefficients for each regressor term, and $e(k)$ is the noise model of the system.

To develop a NARMAX model, the noise term $e(k)$ must be minimized such that it is uncorrelated with the input and output signals. The regressors are related to the output of the system as

$$\mathbf{Y} = \mathbf{P}\theta + H \tag{12.20}$$

where **Y** is of size ($N \times 1$), **P** is of size ($N \times M$), θ is of size ($M \times 1$) and H of size ($N \times 1$). The vector θ is solved using linear algebraic manipulations. The orthogonal least squares algorithm solves the standard least square problem (Soderstorm and Stoica, 1989).

To provide insight into the model structure for a system, consider the following **LDU** decomposition

$$\mathbf{P}^T\mathbf{P} = \mathbf{A}^T\mathbf{D}\mathbf{A} \tag{12.21}$$

where **A** and the diagonal matrix **D** are of size $M \times M$. Now define two new matrices from this decomposition as

$$\mathbf{W} = \mathbf{P}\mathbf{A}^{-1} \quad (\text{size of } N \times M) \tag{12.22}$$

$$\mathbf{g} = \mathbf{A}\ \theta \quad \text{of size } (M \times 1) \tag{12.23}$$

where the least squares problem of Eq. (12.19) is now defined as

$$\mathbf{Y} = \mathbf{W}\mathbf{g} + \mathbf{H} \tag{12.24}$$

and **W** is an orthogonal matrix. To develop a measure of significance for a regressor, consider the following operation

$$\mathbf{Y}^T\mathbf{Y} = (\mathbf{W}\mathbf{g})^T\mathbf{W}\mathbf{g} + \mathbf{H}^T\mathbf{H} + \mathbf{g}^T\mathbf{W}^T\mathbf{H} + \mathbf{H}^T\mathbf{W}\mathbf{g} \tag{12.25}$$

Using the assumption that $e(k)$ is uncorrelated to the input and output signal signals, the above equation reduces to

$$\mathbf{Y}^T\mathbf{Y} = (\mathbf{W}\mathbf{g})^T\mathbf{W}\mathbf{g} + \mathbf{H}^T\mathbf{H} \tag{12.26}$$

Dividing above equation by the number of points in the data set yields

$$\sum_{k=1}^{N} \{y^2(k)/N\} = \sum_{i=1}^{M} (g_i^2) \sum_{k=1}^{N} \{w_i^2(k)/N\} + \sum_{k=1}^{N} \{e^2(k)/N\} \tag{12.27}$$

that is,

$$\overline{y}^2 = \Sigma\, g_i^2 \overline{w}_i^2 + \sigma_e^2 \tag{12.28}$$

where $\overline{y}^2$ is the estimate of the variance of the output vector, $\overline{w}^2$ is the estimate of the variance of each regressor, and σ_e^2 is the variance of the zero mean noise signal. Using this result, an error reduction ratio is defined for each regressor by dividing by the standard deviation of the output vector

$$\text{err}_i = \overline{g}_i^2 \overline{w}_i^2 / y^{-2} \tag{12.29}$$

The above equation is a measure of the significance of each transformed regressor vector on the variance of the output time history and is used to identify the key contributing regressors.

12.2.2 Identification of Hammerstein Systems

The system is described by

$$x(z) = f_1 u(z) + f_2 u(z)^2 + \ldots + f_p u(z)^p \tag{12.30}$$

$$A(z^{-1})y(z) = z^{-d_p} B^*(z^{-1})x(z) + d(z) \tag{12.31}$$

where

$$A(z^{-1}) = 1 + \alpha_1 z^{-1} + \alpha_2 z^{-2} + \ldots + \alpha_n z^{-n} \tag{12.32}$$

$$B^*(z^{-1}) = \beta_1^* z^{-1} + \beta_2^* z^{-2} + \ldots + \beta_n^* z^{-n} \tag{12.33}$$

and $d(z)$ is the bias term used to account for unmeasured load disturbances. The time delay d_p and the system order n are known quantities and all other parameters are unknown and may be time-varying. In case the delay d_p is not known exactly, then we use the method of Vogel and Edgar (1980) by assuming

$$z^{-d_p} B^*(z^{-1}) = (\beta_1 z^{-1} + \beta_2 z^{-2} + \ldots + \beta_3 z^{-r})z^{-k_1} \tag{12.34}$$

$$= B(z^{-1})z^{-k_1} \tag{12.35}$$

where k_1 is the minimum expected process dead time (an integer value of sampling time) and r is the order of the polynomial in $B(z^{-1})$.

In the B polynomial, r is selected large enough so that dead time in addition to k_1 is represented by the leading coefficients in B which are zero. The value of r is selected with a knowledge of maximum possible time delays. Thus on-line estimation of the $\alpha's$ and $\beta's$ allows this model to adapt to a process with an unknown or variable dead time without requiring an explicit estimate of the dead time. It is to be noted that k_1 is specified initially and is not estimated on-line. We have from Eqs. (12.31), (12.34) and (12.35)

$$A(z^{-1})y(k) = z^{-k_1} B(z^{-1})\left[u(k) + \sum_{i=2}^{p} f_1 u^i(k) \right] + d(k) \tag{12.36}$$

Let

$$s_i(z^{-1}) = f_1 B(z^{-1}) = s_{i0} + s_{i1}z^{-1} + \ldots + s_{in}z^{-r} \tag{12.37}$$

then

$$s_{ij} = f_i \beta_j, \quad i = 2, 3, \ldots, p, j = 1, 2, \ldots, r \tag{12.38}$$

$$A(z^{-1})y(k) = z^{-k_1} B(z^{-1})u(k) + z^{-k_1} \sum_{i=2}^{p} s_i(z^{-1})u^i(k) + d(k) \tag{12.39}$$

We note here that without loss of generality, f_1 is normalized to unity. The steps involved in the identification method are:

(i) obtain the least square estimates of the parameters set $(\hat{\alpha}_i, \hat{\beta}_i, \hat{s}_{ij}, \hat{d})$ from Eq. (12.39) by RLS method (discussed in Chapter 11) using the data $u^i(k)$ and $y(k)$.

(ii) estimate $\hat{f}_i$ from Eq. (12.38) by means of least squares since $\hat{s}_{ij}$ and $\hat{\beta}_i$ are now known quantities. The solution is given by (Hsia, 1976)

$$\hat{f}_i = \sum_{j=1}^{r} \hat{\beta}_j \hat{s}_{ij} / \sum_{j=1}^{r} (\hat{\beta}_j)^2 \qquad i = 2, 3, \ldots, p \tag{12.40}$$

The equations for the recursive least squares are

$$P_{k/k-1} = P_{k-1} + R \tag{12.41}$$

$$P_k = [P_{k/k-1} - P_{k/k-1}(z_k^T P_{k/k-1} Z_k + \rho)^{-1} Z_k^T P_{k/k-1}]/\rho \tag{12.42}$$

$$\theta = \theta_{k-1} + P_k Z_k[\, y(k) - Z^T \theta_{k-1}] \tag{12.43}$$

where Z_k is the vector of input/output data, θ_k the parameter vector P_k the covariance matrix and r is the forgetting factor. Since the RLS algorithm tries to match the model output (including the bias term, Eqs. (12.39), with the process output by updating model parameter and the bias term, any error in estimating the load disturbances will result in incorrect estimates of all the other parameters, which in turn causes modeling error. Usually, emphasis is placed on the capability of the system to identify unmeasured load disturbances. The covariance setting approach (random walk method) adds a constant positive definite matrix R to the covariance matrix when adaptation is required, thus preventing the covariance matrix from becoming too small. One advantage of the covariance setting is that the individual elements of the covariance matrix can be separately affected by proper choice of R. Since the estimation of load disturbance is of significant importance to the adaptive control system (like adaptive IMC), relative magnitude of the diagonal elements of the random walk matrix need to be selected based on relative variation rates of load disturbance and process parameters. If a sudden load disturbance (fast dynamics) is likely to occur, a larger magnitude of the random walk element corresponding to the bias term is preferred.

When a sudden change occurs in the process, e.g., set-point changes or unmeasured load disturbances, new parameter estimates will be in error for a period of time. As a result, updating the controller based on erroneous estimates for the model parameters may lead to undesirable control performance. Therefore, all the model parameter can be passed through a first-order filter

$$\theta_c = \gamma\theta_c + (1.0 - \gamma)\theta \tag{12.44}$$

where θ is the vector of updated process parameters and θ_c is the vector of filtered process parameters which is used for control calculations. The value of γ is between 0 and 1.

The controller output can be passed through a simple first-order filter

$$u_c(k + 1) = (1.0 - \delta)u_c(k) + \delta u(k + 1)$$

where δ is an adjustable constant, u the controller output, and u_c the filtered controller output which is actually used as a control input to the process. In addition to accounting for expected modeling errors, this filter also serves as a 'detuner' to prevent excessive control action.

Once the structure of the system is assumed, whether Hammerstein or Wiener systems, the parameters of the model can be calculated based on a two step method. First the nonlinear element can be obtained from the steady-state data. The linear portion of the system is then obtained from dynamic data. Instead of the two step method, we can obtain the model parameters using the whole model equations and using dynamic data as discussed in the section on Hammerstein systems. Let us now discuss a method for identifying Wiener systems.

12.2.3 Identification of Wiener Systems

As discussed earlier, this nonlinear system consists of two subsystems: the first is linear dynamics. followed by a nonlinear static sub system. The system output, identical to the nonlinear block is given by

$$y(k) = g_0 + g_1\, x(k) + g_2 x^2(k) + \ldots + g_r x^r(k) \tag{12.45}$$

The linear dynamic system can be described by the difference equation

$$x(k) = \sum_{i=0}^{m} p_i u(k-i) - \sum_{j=1}^{n} s_j x(k-j) \tag{12.46}$$

From the above two equations, we get

$$y(k) = g_0 + \sum_{k=1}^{r} g_k \left[\sum_{i=0}^{m} p_i u(k-i) - \sum_{i=1}^{m} s_j x(k-j) \right]^k + e(k) \tag{12.47}$$

The resulting equation is predominantly nonlinear because of its data and parameter combinations. Moreover the initial vector $x(k)$ is generally not observable. The total number of parameters is $N = (M + n + 1)\, r + 1$. The parameters can be obtained by a suitable nonlinear optimization method (Norquay et al., 1998). The major difficulty in any of the nonlinear optimization method is the knowledge of the initial guess. If the guess values for the parameters are not good then it is difficult to get the convergence.

To overcome the nonlinearity in Eq. (12.47), Voros (1995) had proposed a method. The basic idea in the method is a form of half-substitution. Let $g(x)$ be an analytic nonlinear mapping which can be rewritten into the additive form

$$y(k) = x(k) + G[x(k)] \tag{12.48}$$

which consists of the separated varible $x(k)$, plus the remainder of the original mapping assigned as $G(.)$. We have

$$x(k) = f\,[u(k)] \tag{12.49}$$

If we substitute equation $y = g[x(k)]$, we get a nonlinear equation without any simplification. However, if the above equation is substituted in Eq. (12.48) only for the first term, a new form of the same mapping appears

$$y(k) = f\,[u(k)] + G[x(k)] \tag{12.50}$$

where output depends on both the input $u(k)$ and the internal variable $x(k)$. Hence, the original composite mapping is replaced by Eqs. (12.49) and (12.50). Note the inner mapping appears both explicitly and implicitly in the outer. The method can be applied for Wiener model parameter identification (Voros, 1995).

Assuming that $g_1 = 1$ (this is always possible in this system description), variable $x(k)$ as the key term of mapping G can be separated. Then, after substituting Eq. (12.48) into (12.50) (for the key term only), the system output is

$$y(k) = \sum_{i=0}^{m} p_i u(k-i) - \sum_{j=1}^{n} s_j x(k-j) + g_0 + \sum_{q=2}^{r} g_q x^q(k) + e(k) \tag{12.51}$$

Thus the original complex Eq. (12.47) is replaced by the two less complicated Eqs. (12.46) and (12.51).

Hence, the derived special description of a Wiener type system with the number of parameters $N = m - n + r + 1$, is nonlinear in data, but is linear in all the parameters in Eqs. (12.50) and (12.51). The internal variable $x(k)$ can be estimated by internal variable estimation method (Voros, 1985). The above method for Wiener model is also modified (Voros, 1995) for models with noise term also, i.e. for extended Wiener models.

12.3 Design of Controllers

12.3.1 Design of Controllers for Hammerstein Systems

We have to first design linear controller (say PID or Dahlin's controller) for the subsystem $G_p(z^{-1})$. Now using this controller output as x, we can calculate the nonlinear controller output as the real root of equation $f(u) = x$ (Fig. 12.4). In case more than one real root is obtained for u, then the value which is nearer to the nominal operating value of u is taken for u. This value is implemented in the system. One can easily check that the controller for Hammerstein system will be of the form of a Wiener system (Fig. 12.2).

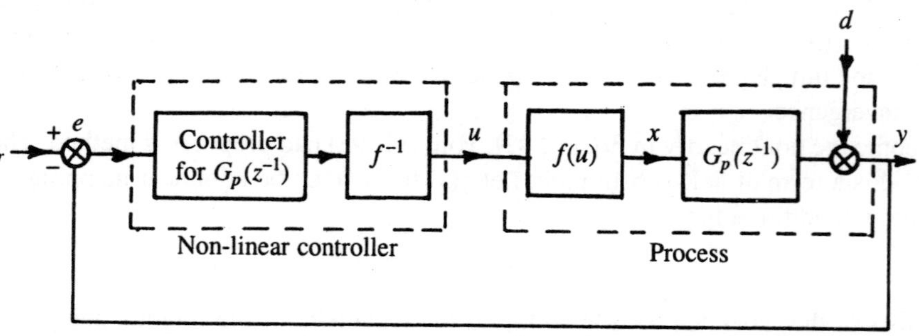

Fig. 12.4 Block diagram of nonlinear controller for Hammerstein system

12.3.2 Design of Nonlinear Controller for Wiener Systems

The block diagram for the control of Wiener type nonlinear system is shown in Fig. 12.5. The nonlinear gain subsystem is considered in the feedback also and the set point value of y enters through a nonlinear gain system (Fig. 12.5). Basically in this figure, we consider controlling x at the desired value given by the corresponding set value of x. The linear controller (Fig. 12.5) is designed for the subsystem $G_p(z^{-1})$. The value of actual manipulated variable u is thus nonlinear in y.

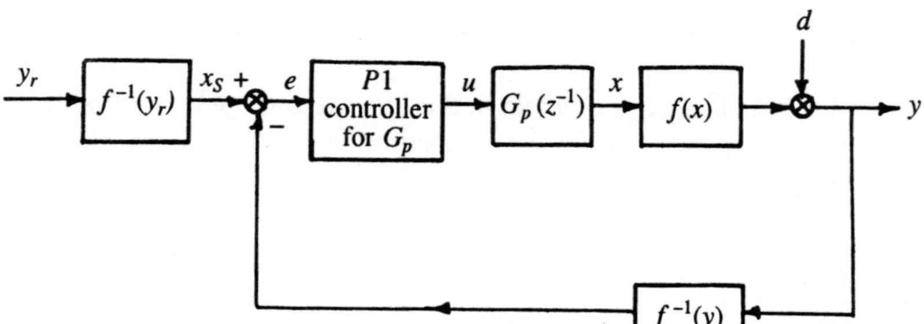

Fig. 12.5 Block diagram of nonlinear controller for Wiener system

12.3.3 Design of Nonlinear State Variables Controllers

The Hammerstein and Wiener models take into account the nonlinear gain variations. However, if the nonlinearity is on the dynamics of the systems (i.e., changing the time constant, time delay), then use of state-space nonlinear models (derived based on basic principles) is required for the design of controllers.

Let us consider a nonlinear system described by the equation

$$\mathbf{x}(k+1) = \mathbf{f}(\mathbf{x}(k)) + \mathbf{g}(\mathbf{x}(k))u(k) \tag{12.52}$$

$$y(k+1) = \mathbf{cx}(k+1)$$

where y and u are scalar output and input, respectively, **f** and **g** are vector containing the state vector **x**, **c** is a row vector. We can define the tracking error in y as

$$e(k+1) = y_r(k+1) - y(k+1) \tag{12.53}$$

where $y_r(k+1)$ is the desired output of the system.

Substituting the expression for $y(k+1)$ from Eq. (12.52) we can write Eq. (12.53) as

$$e(k+1) = y_r(k+1) - \mathbf{cf}(\mathbf{x}(k)) - \mathbf{cg}(\mathbf{x}(\mathbf{k}))u(k) \tag{12.54}$$

Equating right side and left side of the equation to $-\lambda e(k)$, we get the control law as

$$u(k) = [\mathbf{cg}(\mathbf{x}(k))]^{-1}[\, y_r(k+1) + \lambda e(\mathbf{k}) - \mathbf{cf}(\mathbf{x}(k))] \tag{12.55}$$

This control law makes the error dynamics as

$$e(k+1) = -\lambda e(k) \tag{12.56}$$

Thus by specifying the value of λ we can calculate the value for $u(k)$. We can also get an improved robust controller by equating the term on the right side and also separately the left side of Eq. (12.54) to

$$-\lambda_1 e(k) - \lambda_2 e(k-1)$$

We get the error dynamics as

$$e(k+1) + \lambda_1 e(k) + \lambda_2(k-1) = 0 \tag{12.57}$$

The control law becomes

$$u(k) = [\mathbf{cg}(\mathbf{x}(k))]^{-1}[\, y_r(k+1) + \lambda_1 e(k) + \lambda_2\, e(k-1) - \mathbf{cf}(\mathbf{x}(k))] \tag{12.58}$$

Thus, by specifying the error dynamics by a second order system (the values of λ_1 and λ_2 can be selected by specifying the damping coefficient and the settling of the second order error dynamics), we have the expression for the nonlinear controller. The nonlinear control law, Eq. (12.58), will be more robust to parameter uncertainty than that of Eq. (12.55), because of integral action in the control law. The control law requires the measurement of the those state variables appearing in **cx**. Several simulation examples are given for this class of nonlinear controllers for continuous time systems by Chidambaram (1995).

Summary

To model nonlinear systems, the following nonlinear models can be applied in the increasing order of complexity: Bilinear model, Hammerstein model, Wiener model, Wiener-Hammerstein model, Volterra model, NARX model, and NARMAX model. This chapter presented the identification methods and design of controllers for such models.

References

Chen, S. and S.A. Billings, Representations of nonlinear systems: the NARMAX model, Inter. J. Control, 49 (1989) 1013–1032.

Billings, S.A., M.J. Korenberg, and S. Chen, Identification of nonlinear output-affine systems using an orthogonal least-square algorithm, Int. J. Control, 19 (1988) 1559–1568.

Chidambaram, M., Control of Wiener type nonlinear systems with input multiplicities, Int. Conf. Trends in Indust. Measure. and Automation, Jan., Madras, 163–166 (1996).

Chidambaram, M., Applied process control, Allied Pub., New Delhi, (1998).

Chidambaram, M., Nonlinear Process Control, John Wiley, New York (1995).

Dahlin, E.B., Designing and tuning digital controllers, Instr. Contr. Sys., 41, No. 6 (1968) 77–83.

Daiz, H. and A.A. Desrochers, Modeling of nonlinear discrete-time systems from input-output data, Automatica, 24 (1988) 629–641.

Glass, J.H. and M.W.A. Franchek, NARMAX modeling and robust control of internal combustion engines, Int. J. Control, 72 (1999) 289–304

Hsia, T.C., System Identification: Least Squares Methods, Lexington Books, Lexington, P 146 (1977).

Pearson, R.K., Nonlinear input/output modeling, J. Proc. Control, 5 (1995) 197–211.

Soderstrom, T. and P. Stoica, System Identification, Prentice-Hall, 1989.

Vogal, E.F. and T.F. Edgar, An adaptive dead time compensation for process control, Proc. JACC, San Fransisco, CA, Paper TP5–E (1980).

Voros, J., Nonlinear system identification with internal variable estimation, Proc. IFAC Symp. Ident. Sys. Param. Esti., York, (1985) 439–443.

Voros, J., Identification of nonlinear dynamic systems using extended Hammerstein and Wiener models, Control-Theory and Adv., Technology, 10 (1995) 1203–1212.

Welstead, P.E. and P.M. Zanker, Techniques of self-tuning, Opt. Contr. Appl. Mech, # (1982) 305–322.

Review Questions

1. Draw the block diagram of a Hammerstein model for discrete time systems.
2. Draw the block diagram of a Wiener model for discrete time systems.
3. Draw the block diagram of a Hammerstein-Wiener model for discrete time systems.
4. What is a Volterra model?
5. What do you meant by NARX model and NARMAX model?
6. What is an output error model?
7. What is meant by state-space nonlinear model?
8. Discuss a method of designing nonlinear controller for Hammerstein systems.
9. Discuss a method of designing nonlinear controller for Wiener systems.
10. Discuss a method of designing nonlinear controller for Bilinear systems.
11. Eskinath et al. (AIChE J., 1991, V 37, 255–230) have presented a study on the development of Hammerstein model for a Heat exchanger setup. Give a write up on this work.
12. Jyothi and Chidambaram (Bioprocess Engg., 2000, V23, 323-326) have presented a method of identifying a Hammerstein model for a bioreactor exhibiting input multiplicity behaviour. Give a write up about the method.
13. Proll and Karim (AIChE J., 1994, V40, 269–282) have presented a study on identification of NARX model of pH of a waste water system. Give a report on this work.
14. Discuss a method designing nonlinear controller for Hammerstein-Wiener system.

13

Simulation Case Studies on Controller Design

This chapter presents four simulation studies to illustrate the methods involved in the identification of discrete models and design of linear and nonlinear controllers explained in the previous chapters. The case studies include: (i) model identification and control of nuclear reactors, (ii) adaptive IMC of Hammerstein systems, (iii) nonlinear self-tuning regulator for NARMAX models and (iv) design of feed forward control of Wiener system with input multiplicity.

13.1 Model Identification and Control of Nuclear Reactors

The dynamics of a nuclear reactor are nonlinear. An excellent review on the design of controllers for nuclear reactors is given by Kisner (1992). One of the methods for designing a controller is to develop simple linear models based on input-output data for different operating points. The controller can then be designed for each model. However, the model will be applicable only to that operating condition. We can develop linear models for two different operating conditions to design linear controllers to check whether the controller is able to give good response at both the operating points.

A simple functional description of the plant can be stated as follows: Heat generated by nuclear fission in the core is absorbed by liquid sodium that circulates through the reactor. The absorbed heat is transferred in a heat exchanger to a non-radioactive secondary sodium. This forms the primary heat exchanger circuit. The secondary sodium transfers the heat to generate steam in a steam generator. The super heated steam drives a turbine generator to produce electricity. In the present study, only the reactor and the primary heat exchanger loop have been considered. The control rod drive system controls the neutron power. This neutron power is decreased by lowering the rod and increased by raising the rod. The neutron power (y) is to be controlled by manipulating the control rod position (u).

13.1.1 Model Development from Input Output Data

The response of a fast breeder nuclear reactor (FBTR), the neutron power at 4.57 Mwt for a known change in the control rod position is recorded. The input data are given in Table 13.1. The response is shown in Fig. 13.1. Similarly the response in the neutron power at 5.70 Mwt is also recorded. The input data are given in Table 13.1. Figure 13.2 shows that the reactor exhibits an inverse response. Since the response resembles that of a second order system, the discrete model is assumed to be

$$y(k) = -\alpha_1 y(k-1) - \alpha_2 y(k-2) + \beta_1 u(k-1) + \beta_2 u(k-2) \tag{13.1}$$

Since the data shown in Fig. 13.1 contains noise, a smooth curve is first drawn for this data set. The smoothened data is used for identification. The smoothened data sets are given in Table 13.1. The parameters of the model are estimated by linear least square method (Stephanopolous, 1994) and the parameters values are given in Table 13.2 for the two operating points. As can be seen from Table 13.2, the model parameter, particularly those for β_1 and β_2, vary significantly. The response of the model equation for the same type of input (given in Table 13.1) are evaluated using the model parameters meant for operating condition 1 and separately using the model parameters meant for the operating condition 2. The response is good when the model parameters meant for that particular condition are used. Otherwise, the response is not satisfactory as seen from the response shown in Fig. 13.1. Similar results are shown in Fig. 13.2 for the operating condition 2 when the model parameters used are meant for operating condition 1. Hence there is a need to develop a better model suitable for both the operating conditions.

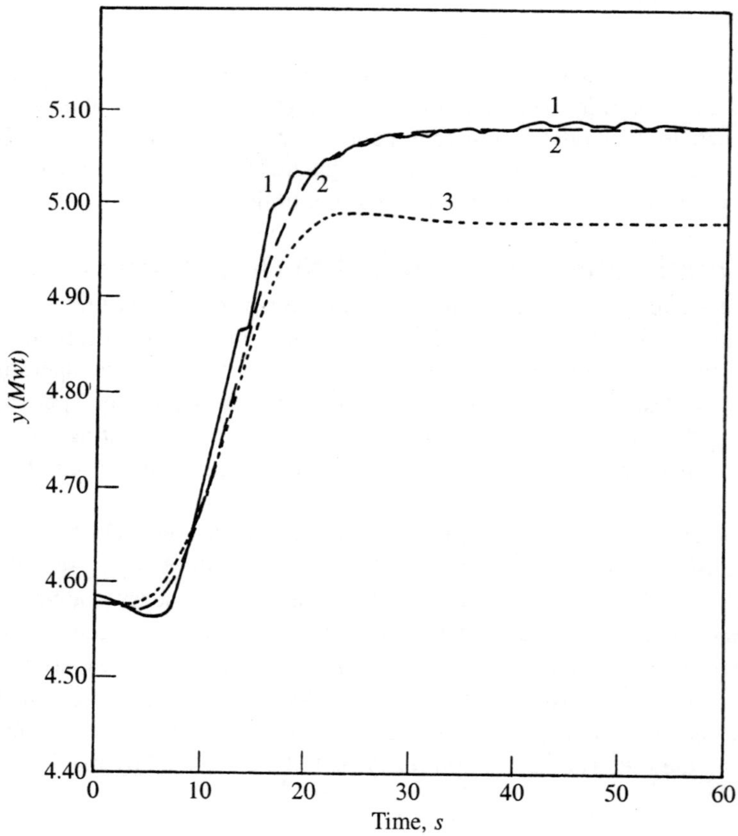

Fig. 13.1 Response at operating point I. 1: plant data; 2: model (Eq 13.1); 3: model [Eq (13.1) with the parameters meant for operating point II]

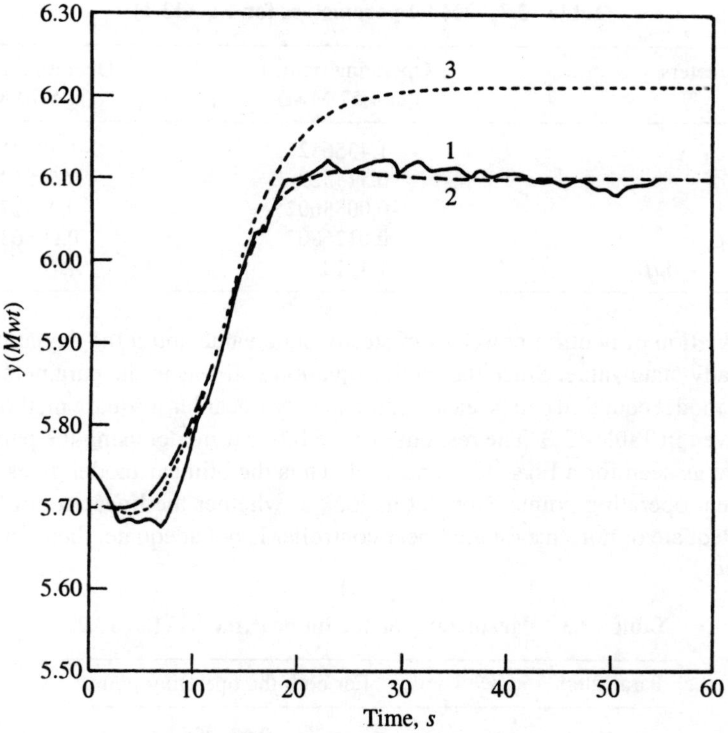

Fig. 13.2 Response at operating point II. 1: plant data; 2: model (Eq 13.1); 3: model [Eq. (13.1) with the parameters meant for operating point I]

The linear model parameters given in Table 13.2 show that parameters β_1 and β_2 vary significantly for the two operating conditions, that is, these parameters depend on the operating condition. Hence, the following bilinear model is developed

$$y(k) = -\alpha_1 y(k-1) - \alpha_2 y(k-2) + [\beta_{10} + \beta_{11} y(k-1)]u(k-1) + [\beta_{20} + \beta_{22} y(k-1)]u(k-2) \quad (13.2)$$

Table 13.1 Deviation in control rod position (mm)

t (*s*)	Operating point I	Operating point II
1	0.0	0.0
2	1.0	1.0
3	2.0	2.0
4	3.0	3.0
5	4.0	4.0
6	5.0	5.0
7	6.0	6.0
8	7.0	7.0
9	8.0	8.0
10	9.0	9.0
11	10.0	9.8
upto 60	10.0 (upto t = 60)	9.8

Table 13.2 Model parameters for Eq. (13.1)

Parameters	Operating point I (at 4.57 Mwt)	Operating point II (at 5.70 Mwt)
α_1	−1.435652	−1.575316
α_2	0.515529	0.646106
β_1	−0.0085602	−0.0037701
β_2	0.0126007	0.0066357
Zero, $z = -\beta_2/\beta_1$	1.4718	1.76

where y is the deviation of neutron power from steady-state value and u the deviation of control rod position from steady-state value. Since the model equation is linear in the parameters β_1 and β_2, the parameters of the model equation can be easily estimated by linear least square method. The estimated parameters are given in Table 13.3. The response of the bilinear model using the parameters given in Table 13.3 is good as seen form Figs. 13.3 and 13.4. Thus the bilinear model gives a good response at both the different operating points. Now let us look at whether the linear controller based on the linear model is adequate or not. In case the linear controller is not adequate, then nonlinear controller has to be designed.

Table 13.3 Parameters for the bilinear model (Eq. 13.2)

Parameters	For both the operating points
α_1	−0.6302064
α_2	0.5232682
β_{10}	−0.0067661
β_{11}	0.1968557
β_{20}	0.0105670
β_{22}	−0.1151290

13.1.2 Controller Design for Linear Models

For discrete time systems there are several controller design methods available. These methods include deadbeat method, Dahlin's algorithm (Stephanopolous, 1984) and Vogel-Edgar method (1980). Vogel-Edgar method (1980) which is based on the IMC (Internal Model Control) principle is one of the best methods for controller design. The block diagram for the Vogel-Edgar controller is shown in Fig. 13.5. The controller (G_I) is a stable inverse of process model G_M. Controller G_I includes a fine tuning filter for robustness. It can be seen from Table 13.2, the root of the numerator (i.e., zero of $y(z)/u(z)$ is greater than the one lying outside unit circle, $z = (-\beta_1/\beta_2) > 1$). The controller, which is an inverse of the process model, will then be unstable. The system has delay of one sampling time. The delay term on inversion for designing of controller leads to a prediction term for the error. To avoid these problems due to numerator positive zero and delay in the model, numerator steady-state gain alone is considered in the controller design as suggested in Vogel-Edgar method {i.e., $1/[(\beta_1 + \beta_2 z^{-1})z^{-1}]$ is approximated by $1/(\beta_1 + \beta_2)$ on substituting $z = 1$}. Hence controller

$$G_I(z^{-1}) = (1 + \alpha_1 z^{-1} + \alpha_2 z^{-2})(1 - \varepsilon)/[(1 - \varepsilon z^{-1})(\beta_1 + \beta_2)] \tag{13.3}$$

where $\varepsilon (0 < \varepsilon < 1)$ is a tuning parameter. This parameter ε takes care of the nonlinearity and parameter variations for the time is varying process.

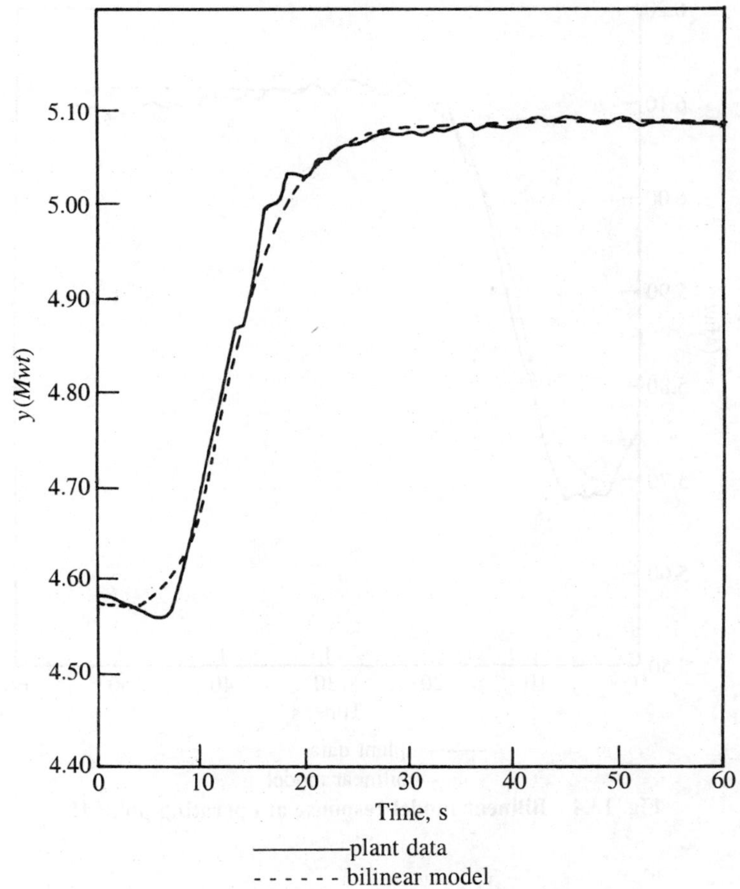

Fig. 13.3 Bilinear model response at operating point I

13.1.3 Simulation of Closed Loop System

The process is assumed to be a linear model meant for the first operating point and the same parameters are also used for a controller design. The response of the closed loop system for a set point change (Table 13.1) is evaluated and is shown in Fig. 13.6. Next the model parameters (α_1, α_2, β_1 and β_2) meant for second operating condition are used for the controller design G_I and for G_M. The closed loop response is shown in Fig. 13.6. The response is satisfactory even when the model parameters meant for the second operating condition are used for controller design for the process (G_p) at the first operating point. Similar response is also obtained (Fig. 13.7) if the process (G_p) is operating at the second operating condition and the controller (G_c and G_M) is designed based on first model parameters. That is, the closed loop response of the controller based on the first model parameters work well for both the operating points. As stated earlier, the open loop response is not acceptable on interchange of the model parameters (Figs. 13.1 and 13.2). However, the closed loop system response is satisfactory if we use the model parameter of the other operating point for the controller design. Since the linear discrete time controller gives good responses for both the operating points, the design of nonlinear controller for bilinear model is not required. The bilinear model may be useful for open loop prediction of the reactor.

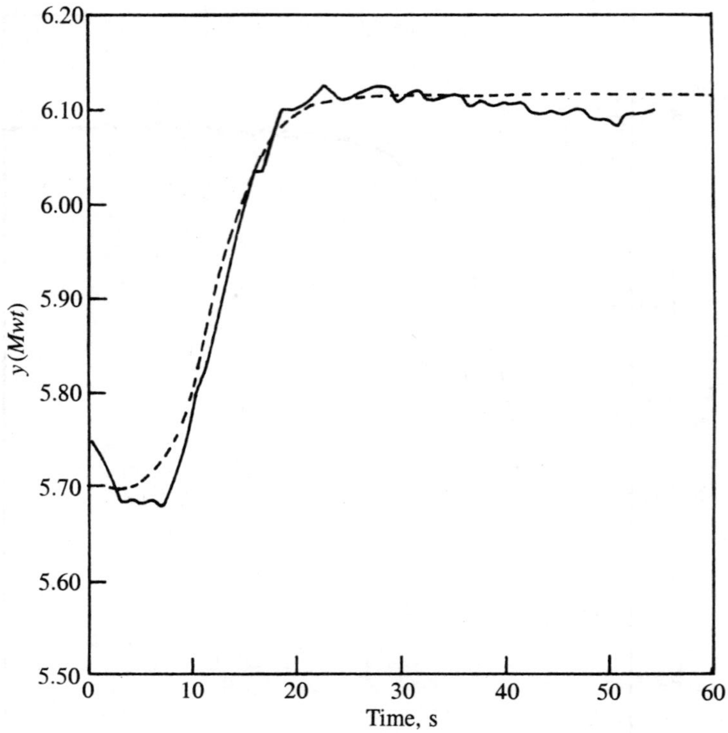

Fig. 13.4 Bilinear model response at operating point II

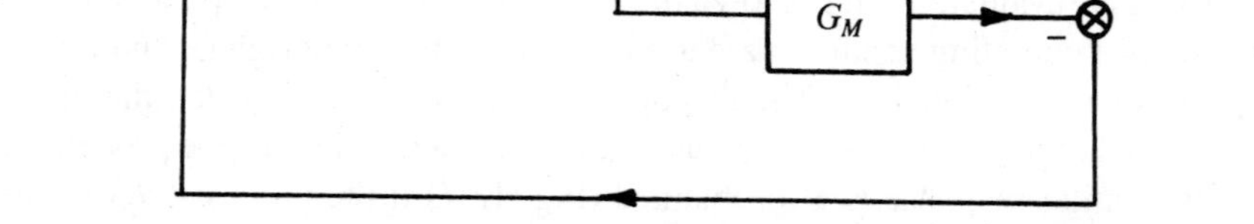

G_I: controller; G_M: transfer function of the model; G_p: transfer function of the process

Fig. 13.5 Block diagram of closed loop system with Vogel-Edgar control scheme

As stated earlier, performance of the controller is evaluated on the model developed. Since the original data is noise corrupted, the effect of noise needs to be considered by adding a noise term to the model developed and then the performance of the controller is to be evaluated. This is left as an exercise for the readers.

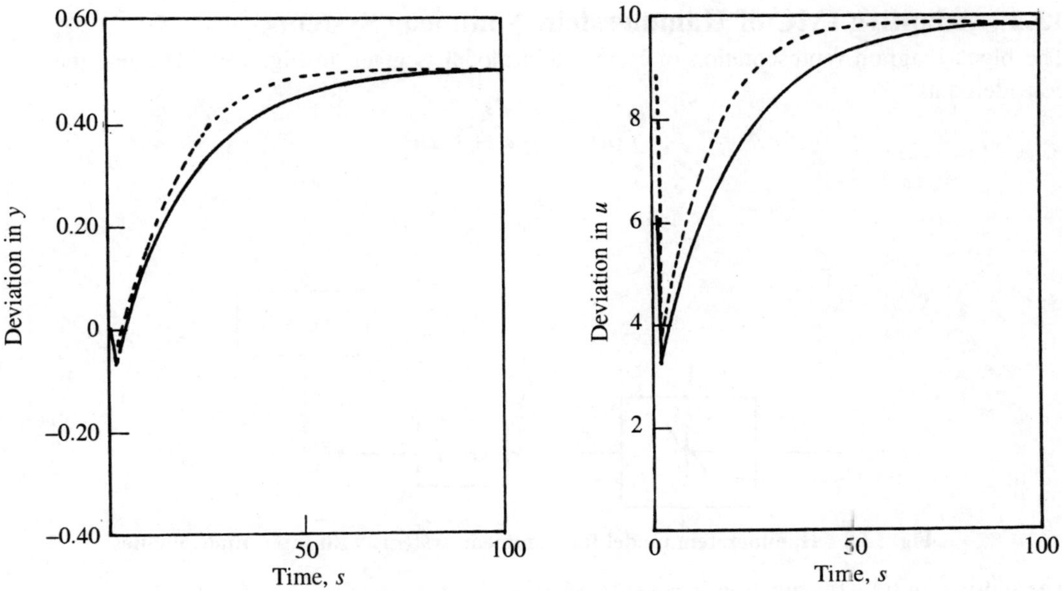

——— controller designed on model I parameters

- - - controller designed on model II parameters ($\varepsilon = 0.95$)

Fig. 13.6 Closed loop servo response of model I

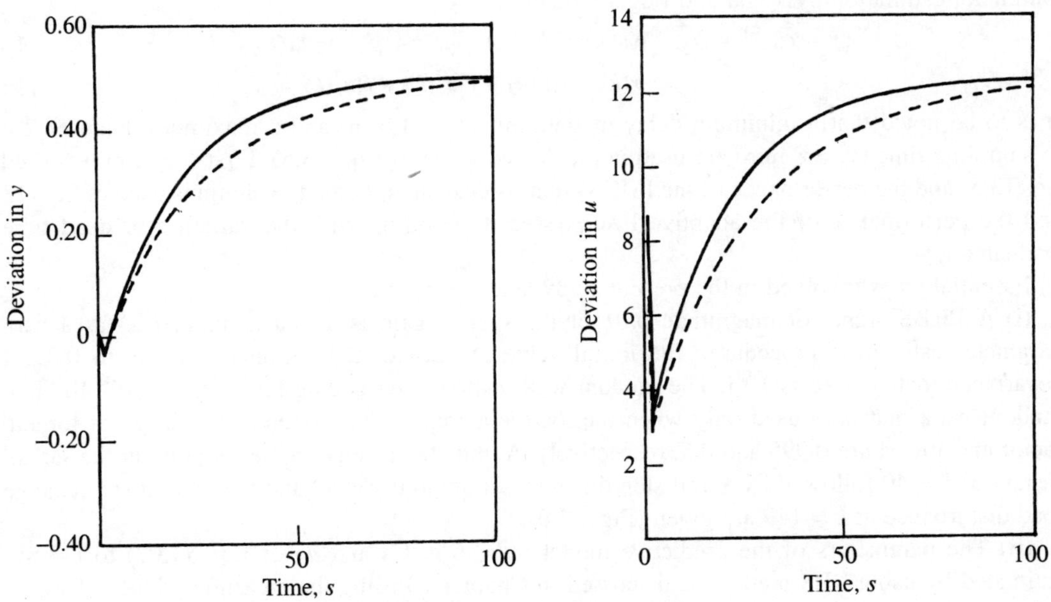

——— controller designed on model II parameter

- - - controller designed on model I parameter ($\varepsilon = 0.95$)

Fig. 13.7 Closed loop servo response of model II

13.2 Adaptive IMC of Hammerstein Nonlinear Systems

The block diagram representation of Hamerstein model is given in Fig. 13.8. The nonlinearity is considered as

$$f(u) = u - u^2 + 0.2u^3 \tag{13.4}$$

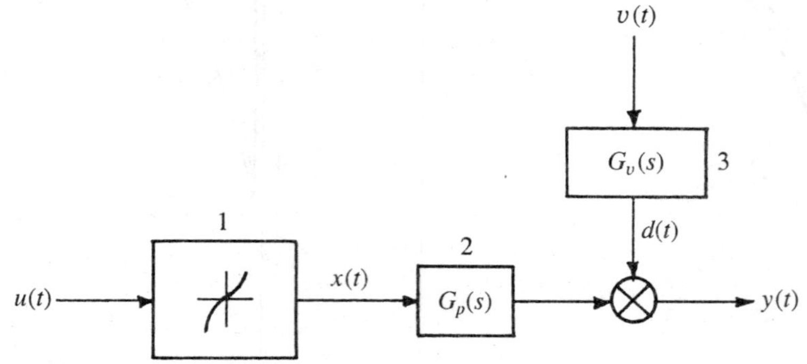

Fig. 13.8 Hammerstein model for nonlinear systems with input multiplicities

The unity gain transfer function is considered as

$$G_p(s) = (1 - 4s)\exp(-4s)/[(4s + 1)(10s + 1)] \tag{13.5}$$

The disturbance transfer function is assumed to be:

$$G_v(s) = 1/[(4s + 1)(10s + 1) \tag{13.6}$$

The system is simulated by using a sampling time of $T = 4$ min. The predictive model considered for parameter estimation by Vogal and Edgar's method is

$$(1 + \alpha_1 z^{-1} + \alpha_2 z^{-2})y(k) = (\beta_1 z^{-1} + \beta_2 z^{-2} + \beta z^{-3})z^{-1}x(k) + d(k) \tag{13.7}$$

$$x(k) = f_1 u(k) + f_2 u^2(k) + f_3 u^3(k) \tag{13.8}$$

It is to be noted that a minimum delay of one unit ($\tau_d = 4$ min) and a maximum delay of 2 unit of sampling time ($\tau_d = 8$ min) are used in the development of Eq. (13.5). First $\tau_d = 4$ min is used in Eq.(13.5) and the performance of the IMC system is evaluated. Later $\tau_d = 8$ min is used in Eq. (13.5) and the performance of the adaptive IMC system to cope up with the variation of dead time is evaluated.

Essential steps involved in the present study are:

(i) A PRBS signal of magnitude 0.01 of the system gain is added to input $u(k)$ as a part of parameter estimation procedure. The initial value of each of the parameters is set to 0.01. The covariance matrix used is 10^8I. The random walk matrix used is diag $[10^8\ 10^8 \ldots 10^8\ 10^{10}]$. This random walk matrix is used only when the load is acting on the system. The values of forgetting factor and γ used are 0.995 and 0.5, respectively. A unit step change in the set point at the sampling instant of $k = 40$ followed by a unit step down of set point at $k = 80$ and then a unit step change in load disturbance at $k = 140$ are given (Fig. 13.9).

(ii) The parameters of the predictive model (α's, β's, f's and d) of Eqs. (13.7) to (13.8) are estimated by using RLS method as discussed in Chapter 10 using the measured values of $y(k)$ and $u(k)$. The stable inverse of the model with a filter for IMC of linear system is calculated from

$$G_I = \frac{(1 + \alpha_1 z^{-1} + \alpha_2 z^{-2})(1 - \varepsilon)z^{-1}}{\left[\sum_{i=1}^{3} \beta_i)z^{-1}\right](1 - \varepsilon z^{-1})} \tag{13.9}$$

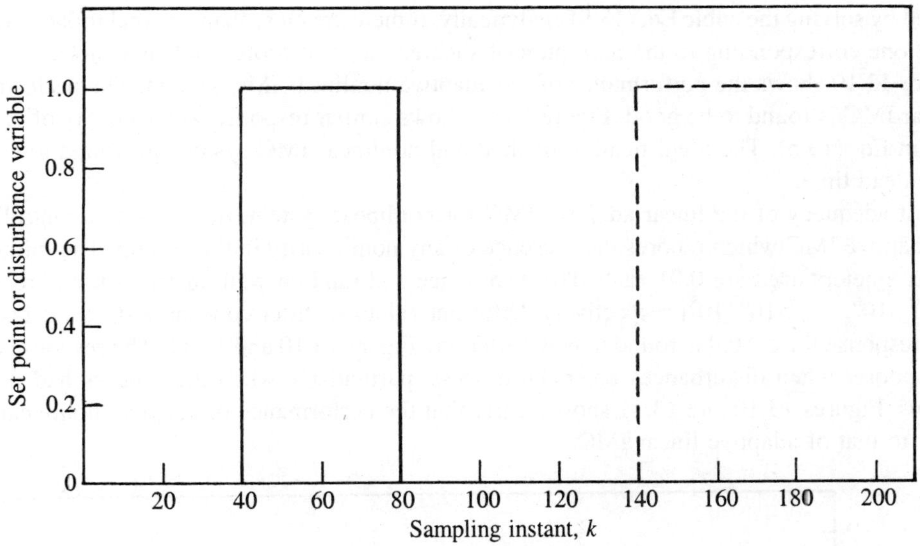

—— set point; - - - disturbance variable

Fig. 13.9 Types of changes in set point and the disturbance variable

Different values of filter constant ε ($0 < \varepsilon < 1$) are used and $\varepsilon = 0.6$ is found to give a better response (Fig. 13.10).

(iii) Using the error signal $e(k)$ and then $G_I(k)$ from Eq (13.9) the values of $x(k)$ are calculated [refer to Fig. 13.5 for evaluation of $e(k)$]. Using the calculated $x(k)$ and estimated f's values, $u(k)$ is

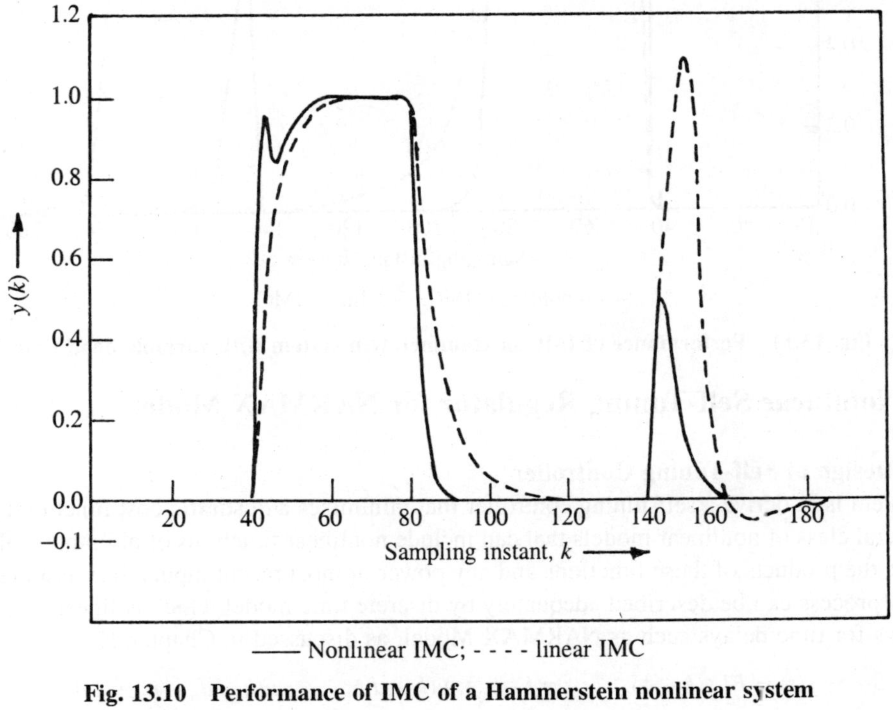

—— Nonlinear IMC; - - - - linear IMC

Fig. 13.10 Performance of IMC of a Hammerstein nonlinear system

evaluated by solving the cubic Eq. (13.8) analytically. If there are more than one real root of Eq. (13.8), then the one corresponding to the minimum of squared value of roots is taken as $u(k)$.

Figure 13.10 shows the performance of the adaptive nonlinear IMC system. The performance of nonlinear IMC is found to be good. Figure 13.11 shows similar response when a delay of $\tau_d = 8$ min is used in Eq. (13.5). The identification method and nonlinear IMC system are found to cope with variable dead time.

To test adequacy of the linear adaptive IMC for nonlinear system, the system is controlled by a linear adaptive IMC which ignores the presence of any nonlinearity in the system. The initial values of the parameters used are 0.01 each. The covariance and random walk matrices used are $10^6 I$ and diag $[10^6, 10^6, \ldots, 10^6, 10^8]$ respectively. Different values of filter constant ε $(0 < \varepsilon < 1)$ are used and the response for $\varepsilon = 0.9$ is found to be a better one Figure 13.10 and 13.11. The response of linear IMC is poorer when disturbances enter the process, particularly when the process had dead time variations. Figures 13.10 and 13.11 show clearly that the performance of adaptive nonlinear IMC is superior to that of adaptive linear IMC.

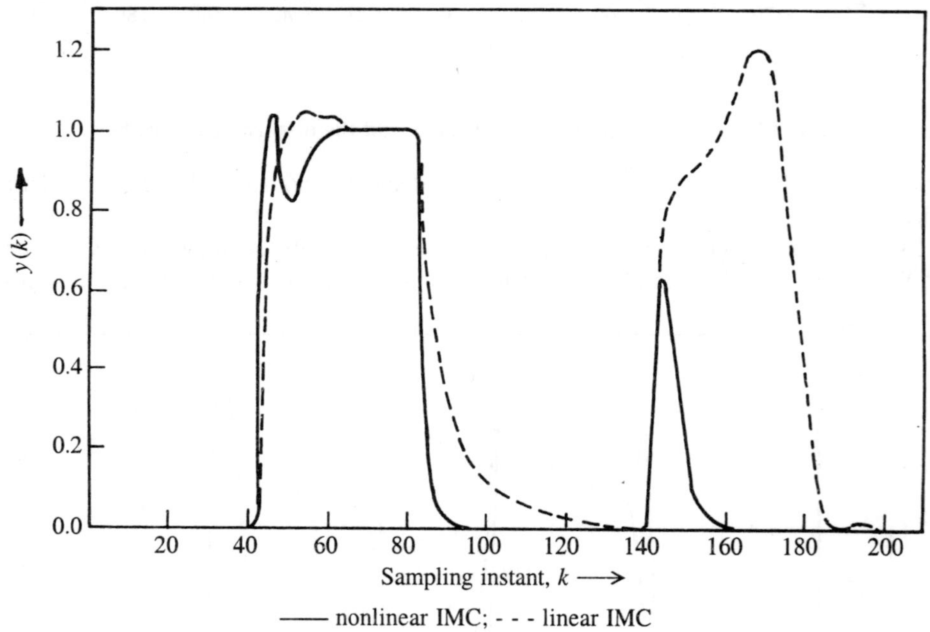

Fig. 13.11 Performance of IMC of Hammerstein system with variable dead time

13.3 Nonlinear Self-Tuning Regulator for NARMAX Models

13.3.1 Design of Self-Tuning Controller

The problem is to derive a self-tuning controller that minimizes a quadratic cost function based on a very general class of nonlinear models that can include nonlinear functions of old inputs, old outputs as well as the products of these functions and any power of most recent inputs. It is assumed that the nonlinear process can be described adequately by discrete time model which is linear in parameters and allows for time delays such as NARMAX Model, as discussed in Chapter 12.

$$y(t) = F[y(k-1), \ldots, y(t-n_y), u(k-d_u), \ldots, u(k-d_u-n_u+1)] \tag{13.10}$$

where n_y and n_u are the orders of $y(k)$ and $u(k)$, respectively and $F(.)$ is some nonlinear function of u and y. Process measurements and sampling errors are stochastic in nature. Stochastic disturbances arise from random variations occuring continuously in many proceses. In tune with reality it is assumed that the output is corrupted by the most basic stochastic process, a discrete white noise sequence $\{e(k);\ k = 1, 2 \ldots\}$, where $e(k)$ are independent, identically distributed, random variables with zero mean and variance σ^2

$$x(k) = y(k) + e(k) \tag{13.11}$$

As the superposition does not apply to nonlinear systems a more realistic formulation is by substituting Eq. (13.11) into (13.12) to get

$$z(k) = F^l[x(k-1), \ldots, x(k-n_y); u(k-n_d), \ldots, u(k-n_d-n_u+1); e(k-1), \ldots, e(k-n_e)] \tag{13.12}$$

where l is the degree of nonlinearity. Since NARMAX model is linear in parameters it lends itself easily to the well established and easily implementable recursive estimation scheme (Goodwin and Payne, 1977) and provides a natural basis upon which to develop nonlinear self-tuning controllers.

The basic generalized minimum variance (GMV) controller includes weighting of the controlled input within the cost function. This is modified to include the NARMAX model terms in the following manner

$$J(k+n_d) = E\{[(P(z^{-1})x(k+n_d) - R(z^{-1})w(k))^2 + Q(z^{-1})f(u(t))^2]\} \tag{13.13}$$

where P, Q and R are polynomials in the backward shift operator z^{-1} and $f(u(k))$ is a function that includes all the $u(k)$ terms that are in the model

$$\begin{aligned} f(u(t)) = F^l[&x(k+n_d-1), \ldots, x(k+n_d-n_u); u(k), \ldots, u(k-n_u+1); \\ &e(k+n_d-1), \ldots, e(k+n_d-n_e)] - F^l[x(k+n_d-1), \ldots, x(k+n_d-n_u); \\ &u(k-1), \ldots, u(k-n_u-n_e); e(k+n_d-1), \ldots, e(k+n_d-n_e)] \end{aligned} \tag{13.14}$$

The weighting has to be placed on a function like $f(u(k))$ because since it is only on a linear function of $u(k)$ the estimation does not produce the parameter estimates expected for minimum variance control. Clarke and Gawthrop (1979) showed that the minimization of GMV cost function is equivalent to minimizing the variance of the auxiliary output function, $\phi(k)$. The same can be shown in the NARMAX case as

$$\phi(k+n_d) = P(z-1)x(k+n_d) - R(z^{-1})w(k) + Q(z^{-1})f(u(k)) \tag{13.15}$$

However, in the NARMAX case $f(u(k))$ is not extracted as a unique function although it is feasible to minimize the auxiliary output, it can not be directly measured from the system. The algorithm can be made self-tuning by employing an extended least square (ELS) estimator (Goodwin and Payne, 1977). The data and parameter vectors can be constructed as

$$\phi(k/k-n_d) = \phi^*(k/k-n_d) + e(k) \tag{13.16}$$

where

$$\begin{aligned} \phi(k/k-n_d) &= F^l[x(k-1), \ldots, x(k-n_y); u(k-n_d), \ldots, u(k-n_d-n_u); \\ &\quad e(k-1), \ldots, e(k-n_e)] - Rw(k) \qquad (13.17) \\ &= \psi^T(K)\theta(k) \qquad (13.18) \end{aligned}$$

Here we use the expression for the function F as given by Eq. (12.10), where ψ^T is data vector and $\theta(k)$ the parameter estimate vector at sampling instant k

$$\psi^T(k) = [x(k-1), \ldots, x(k-n_y); u(k-1), \ldots, u(k-n_d), \ldots, u(k-n_d-n_u+1);$$
$$w(k), \ldots, w(k-n_r); e(k-1), \ldots, e(k-n_y); w(k), \ldots, w(k-n_r);$$
$$x^2(k), \ldots, x(k-1)u(k-1), \ldots, x(k-1)e(k-1) + \text{higher order terms}] \quad (13.19)$$

and

$$\theta = [\hat{\theta}_1, \ldots, \hat{\theta}_{ny}; \theta_{ny+1} \ldots \theta_{ny+nu+ne}, r_0 \ldots r_n, r_{11} \ldots \theta_{1(ny+1)} \ldots \theta_{1(ny+nu+1)}$$
$$\ldots + \text{higher order terms}] \quad (13.20)$$

With linear systems, the noise is often disregarded if the output to noise ratio is high. However, with nonlinear systems where the noise is internally additive and it may contribute to more significant part to the output signal, in general it should be estimated and included in the control law wherever possible. The noise may be estimated as

$$\hat{e}(k) = \phi(k) - \psi^T(k-n_d)\,\hat{\theta}(k-n_d) \quad (13.21)$$

and the control law is

$$F^l[x(k+n_d-1), \ldots, x(k+n_d-n_1); u(k), \ldots, u(k-n_2);$$
$$e(k+n_d-1), \ldots, e(k+n_d-n_3)] - Rw(k) = 0 \quad (13.22)$$

where $n_1 = n_y + n_p$; $n_2 = n_u + n_p - 1$ or $n_u + n_q - 1$; $n_3 = n_e + n_p$, and n_p and n_q are degrees of P and Q polynomial.

The control law, Eq. (13.22), contains future unknown output terms in $x(k + n_d - 1)$ that occur if $k > 1$. These terms can be proxied by their predicted values. The predictions can be found by employing a sequential algorithm based on setting the predictions of the auxiliary output to zero. Let us apply this method by simulation to a continuous bioreactor.

13.3.2 Continuous Fermentor Model Equations

The unsteady-state model for a continuous fermentor is given by

$$dX/dt = -DX + \mu X \quad (13.23)$$

$$dS/dt = D(S_f - S) - (\mu X/y) \quad (13.24)$$

$$dp/dt = -DP + (\alpha\mu + \beta)X \quad (13.25)$$

where D is the dilution rate, S_f the feed substrate concentration, X the cell-mass concentration, S is the substrate concentration and P the product concentration. Here μ is the specific growth rate, y the cell mass yield, and α and β are kinetic parameters for the product. The specific growth rate is given by the model

$$\mu = \mu_m[1 - (P/P_m)]S/[K_m + S + (S^2/K_i)] \quad (13.26)$$

This model contains four parameters: μ_m, the maximum specific growth rate; P_m the product saturation constant; K_m the substrate saturation constant and K_i the substrate inhibition constant. The nominal values for the parameters and operating conditions are given in Table 13.4. The controlled variable is the cell concentration and the manipulated variable is the dilution rate.

Table 13.4 Parameters values and Operating conditions for continuous bioreactor

$X = 6$ g/l	$S = 5$ g/l
$P = 19.14$ g/l	$D = 0.202\ h^{-1}$
$S_f = 20$ g/l	$y = 0.4$ g/g
$\alpha = 2.2$ g/g	$\beta = 0.2\ h^{-1}$
$\mu_m = 0.48\ h^{-1}$	$P_m = 50$ g/l
$K_m = 1.2$ g/l	$K_I = 22$ g/l

13.3.3 Simulation Results

Simulation studies are carried out to evaluate the performance of the nonlinear self tuning controller. The prediction model used in the self-tuning calculations are with $n_y = n_e = 2$ and $n_u = 3$. The observed process dead time in the simulation study is nearly zero. Parameter estimation method used in the present study is ELS with variable forgetting factor. The nonlinear model expressed earlier has Hammerstein and bilinear terms. Thus, control input is a polynomial equation which has to be solved on-line. A polynomial roots solving procedure is used and the smallest real root nearer to the nominal value is taken as the control signal, $u(t)$. For the sake of simplicity, zero initial parameter estimates are assumed for NSTC. The performance of NSTC is compared with a PI controller. The PI controller parameters are based on IMC tunings (Henson and Seborg, 1991). A sampling time of 30 min is employed for cell concentration measurement. Constraints on the manipulated variable are assumed to be ±75% on the nominal steady-state value.

Figure 13.12 shows the performance of the nonlinear STC and PI controllers for a set point change (– 1 *g/l* in ΔX) is made. It can be seen (Fig. 13.12) that the PI controller gives sluggish response and takes a longer time to settle. The performance of NSTC is better. When an uncertainty in the maximum specific growth rate parameter is introduced (– 12.5% in μ_m from its nominal value) one could notice, from Fig. 13.13, that the performance of PI controller is deteriorated whereas the performance of NSTC is robust. The performance of NSTC and PI algorithms are shown in Fig. 13.14 when a

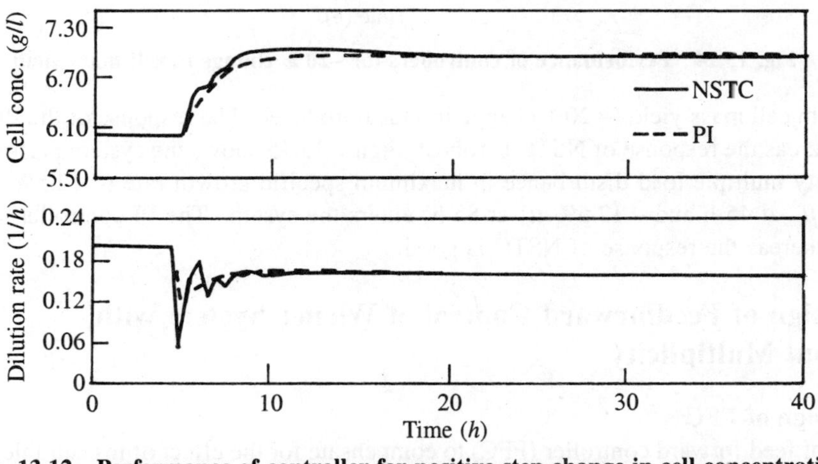

Fig. 13.12 Performance of controller for positive step change in cell concentration

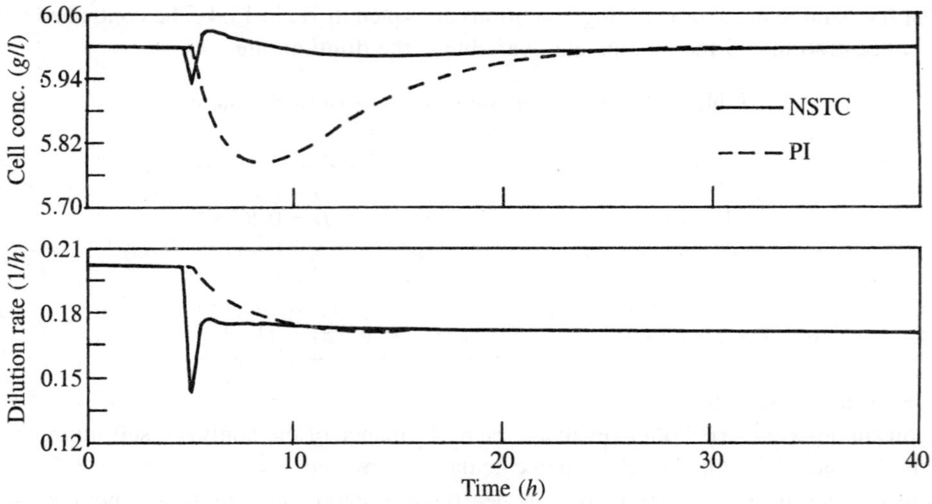

Fig. 13.13 Performance of controllers for –12.5% change in maximum specific growth rate

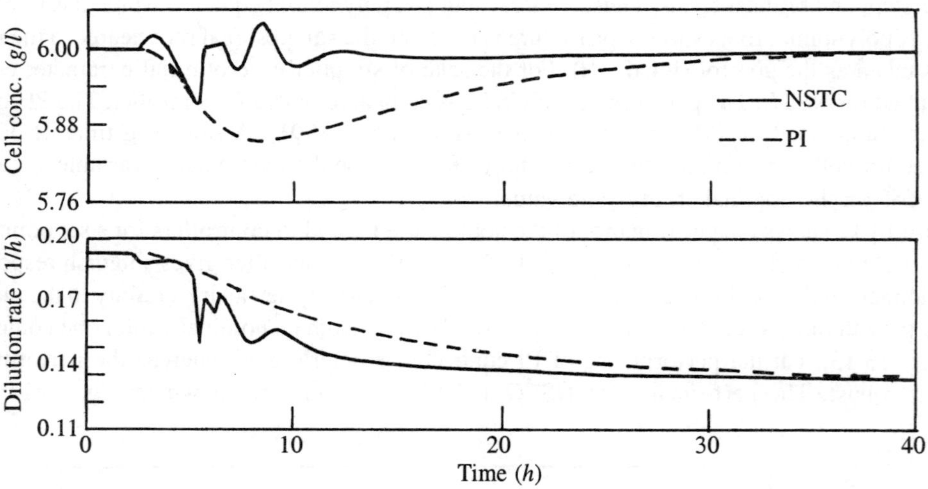

Fig. 13.14 Performance of controllers for –20% change in cell mass yield

disturbance in cell mass yield (– 20% change in y) is introduced. The response of the PI controller is sluggish whereas the response of NSTC is robust. Figure 13.15 shows the system performance when an uncertainty multiple load disturbance in maximum specific growth rate (– 12.5% μ_m change at 5 h, + 25% μ_m at 45 h and – 12.5% μ_m at 85 h) enters the system. The PI controller shows a large overshoot whereas the response of NSTC is good.

13.4 Design of Feedforward Control of Wiener System with Input Multiplicity

13.4.1 Design of FFC

The design of feed forward controller (FFC) to compensate for the effect of measurable disturbances of linear systems is discussed in Chapter 9. The performance of the linear controller is adequate for

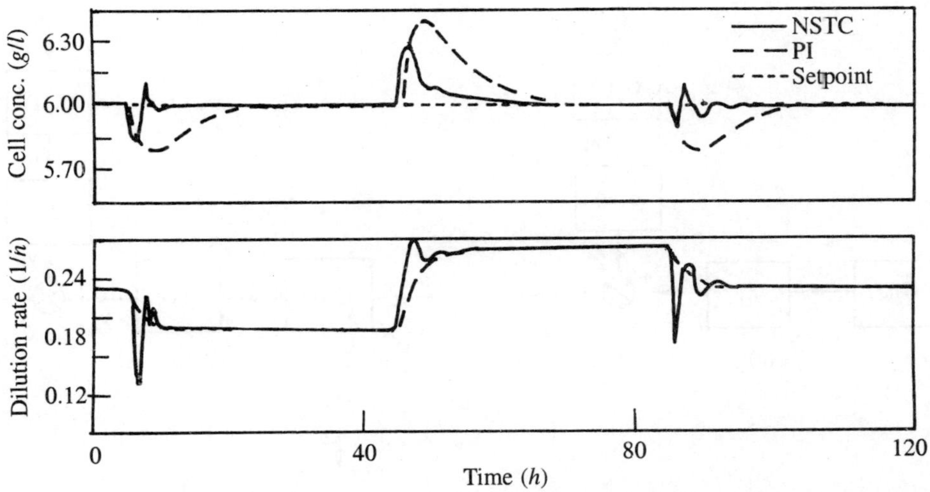

Fig. 13.15 Performance of controllers for multiple changes in maximum specific growth rate

linear or mildly nonlinear systems. However, for highly nonlinear systems, the performance of the linear controller will not be satisfactory. For several nonlinear systems, the gain varies significantly. For some of the nonlinear systems like biochemical reactors and chemical reactors, for a given value of output, there is more than one value of the input variable (manipulated variable). For each values of the manipulated variable, the gain changes significantly in magnitude and in sign also. Such systems can be described by Wiener model (Fig. 13.16). In the present work we shall see the performance of linear FFC and the nonlinear FFC. Here u is the manipulated variable and v the disturbance variable. If we linearize the nonlinear gain for u alone, then the overall linear system is described by Fig. 13.17a. Similarly, the linearized model for the nonlinear gain for v alone is shown in Fig. 13.17b. If the linearization is carried out for both the nonlinear gains of u and v, then the total linear model is as shown in Fig. 13.17c. From Fig. 13.17, we have

$$A(z^{-1})x(z) = z^{-dp}B(z^{-1})u(z) \tag{13.27}$$

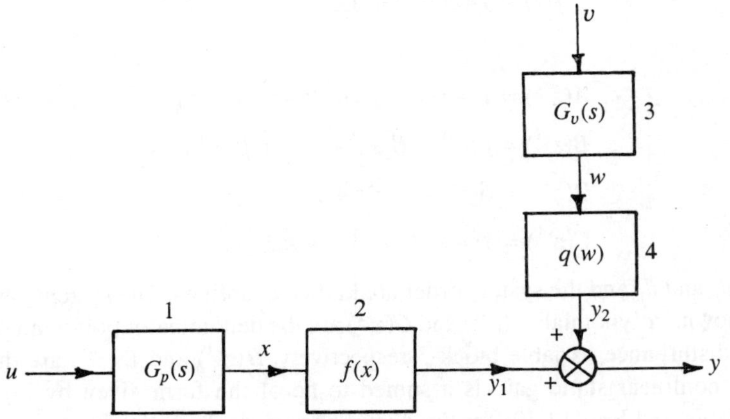

Fig. 13.16 The Wiener type nonlinear system

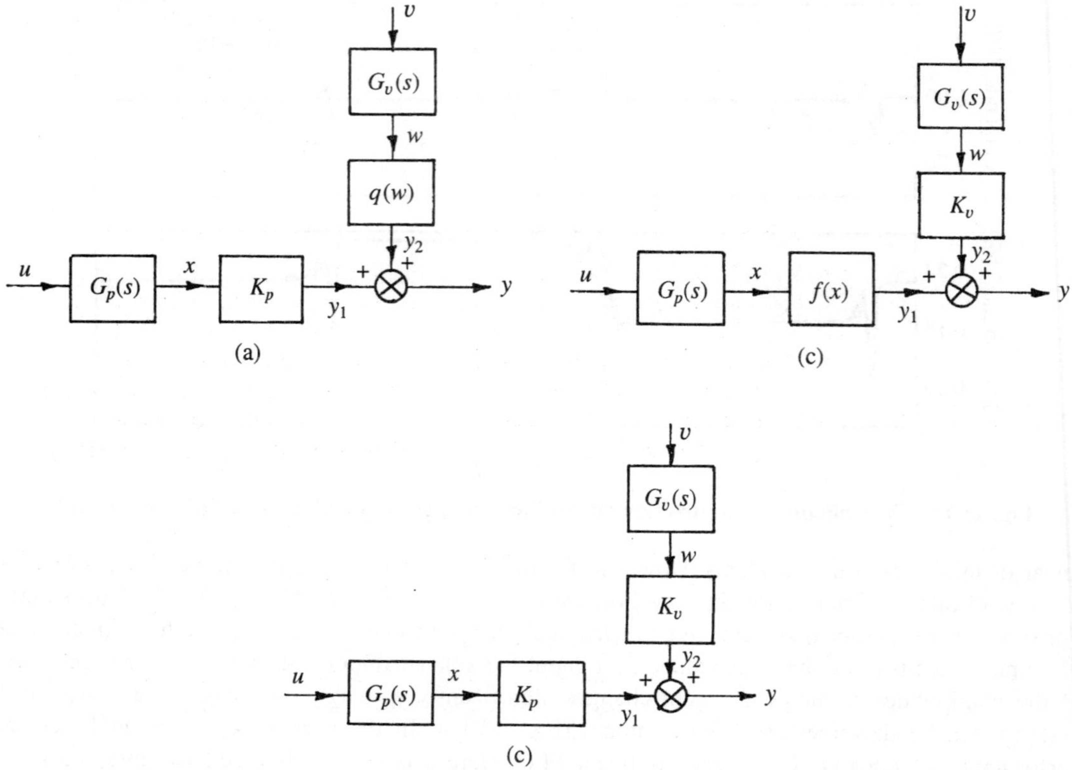

Fig. 13.17 The system model used for design of FFC: (a) linearized model for *u* and nonlinear model for *v*; (b) nonlinear model for *u* and linearized model for *v* and (c) linearized model for both *u* and *v*

$$y_1(z) = f_1[x(z)] + f_2[x(z)]^2 + \ldots + f_p[x(z)]^p \qquad (13.28)$$

$$C(z^{-1})w(z) = z^{-dp}D(z^{-1})v(z) \qquad (13.29)$$

$$y_2(z) = q_1w(z) + q_2[w(z)]^2 + \ldots + q_P[w(z)]^P \qquad (13.30)$$

$$y(z) = y_1(z) + y_2(z) \qquad (13.31)$$

where

$$A(z^{-1}) = 1 + \alpha_1 z^{-1} + \alpha_2 z^{-2} + \ldots + \alpha_n z^{-n} \qquad (13.32)$$

$$B(z^{-1}) = \beta_1 z^{-1} + \beta_2 z^{-2} + \ldots + \beta_n z^{-n} \qquad (13.33)$$

$$D(z^{-1}) = \delta_1 z^{-1} + \delta_2 z^{-2} + \ldots + \delta_n z^{-n} \qquad (13.34)$$

$$C(z^{-1}) = 1 + \varepsilon_1 z^{-1} + \ldots + \varepsilon_n z^{-n} \qquad (13.35)$$

The time delays d_p and d_v and the system order are known quantities. The system parameters are also assumed to be known. Polynomials $A(z^{-1})$ and $C(z^{-1})$ are the denominator polynomials for the process block-1 and the disturbance variable block-3 respectively. $B(z^{-1})$ and $D(z^{-1})$ are the corresponding numerators. The nonlinear static gain is assumed to be of the form given by Eq. (13.28) for the manipulated variable and Eq. (13.30) for the disturbance variable. Referring to Fig. 13.16, we have for perfect regulation (i.e., for $y(k) = 0$]

$$y_1(z) = -\,y_2(z) \tag{13.36}$$

then $x(k)$ can be obtained from the roots of the equation

$$f_1 x(z) + f_2[x(z)]^2 + \ldots + f_p[x(z)]^p = y_1(z) \tag{13.37}$$

In case of multiple real roots for $x(k)$ from Eq. (13.37), the real root which is nearer to the nominal operating point is to be taken for implementation since this requires lesser control effort. The value of $u(k)$ can be obtained from applying static linear FFC law [between $u(k)$ and $x(k)$] as

$$u(k) = [A(1)/B(1)]x(k) \tag{13.38}$$

13.4.2 Simulation Results

In the present work, the nonlinear Wiener system is considered with a third order polynomial for the static nonlinear gain (for relating y and u) with $f_1 = 1$, $f_2 = -1$ and $f_3 = 0.2$. Similarly, the static nonlinear gain (for relating y and v) is assumed as $q_1 = 1$, $q_2 = -1.0$ and $q_3 = 0.2$. At steady-state $u = 0$ and $y = 0$. The system has three multiple values for x as 0, 1.4 and 3.6. The nominal steady-state value of x is assumed to be 0. Similarly, the nominal steady-state value for w is assumed to be zero. The process and the disturbance transfer functions are assumed to be

$$G_p(s) = 1/[(7s + 1)(5s + 1)] \tag{13.39}$$

$$G_v(s) = 1/(6s + 1) \tag{13.40}$$

Discretizing with a sampling time of $T = 4s$, yields the following coefficients for the polynomial $A(z^{-1})$, $B(z^{-1})$, $C(z^{-1})$ and $D(z^{-1})$ as (Isermann, 1982)

$$\alpha_1 = -1.036, \qquad \alpha_2 = 0.2636$$

$$\beta_1 = 0.1387, \qquad \beta_2 = 0.0889$$

$$\delta_1 = 0.48658, \qquad \varepsilon_1 = -0.51342 \tag{13.41}$$

First let us design a linear FFC for the overall linear system (Fig. 13.17c). The linearized gain is given by $K_p = 1$ (by taking the partial derivative of Eq (13.28) with respect to x and substituting the nominal steady-state value of $x = 0.0$). Similarly, we get the linear gain in v as $K_v = 1.0$ Hence, the complete linear static FFC is given by

$$u(k) = -\,\{[D(1)A(1)K_v]/[C(1)B(1)K_p]\}v(k) \tag{13.42}$$

The performance of linear FFC is evaluated on the nonlinear Wiener model. A unit step change is given in v. The response is shown in Fig. 13.18. The response of the system without any FFC is also evaluated and is shown in Fig. 13.18. The response of the linear FFC is worse than that of system without any controller. The manipulated variable is calculated as $u = -1.0$. Performance of the nonlinear FFC is also evaluated. The response is shown in Fig. 13.18. The nonlinear FFC shows an excellent regulatory response. Since we have considered the static nonlinear FFC, the manipulated variable is calculated as $u = -0.17$ (Table 13.6). The steady-state offset in the response y is shown in Fig. 13.18. The response for a unit step down change in v (i.e., $v = -1.0$ at $t = 0$) is shown in Fig. 13.19.

The response of the controlled system using linearized model for relating y to u and nonlinear model for relating y to v is also evaluated (Fig. 13.17a). The response is shown in Fig. 13.19. The responses of the controlled system with linearized model for relating y to v and nonlinear model for relating y to u is also evaluated (Fig. 13.17b). The response is also shown in Fig. 13.19. Though the

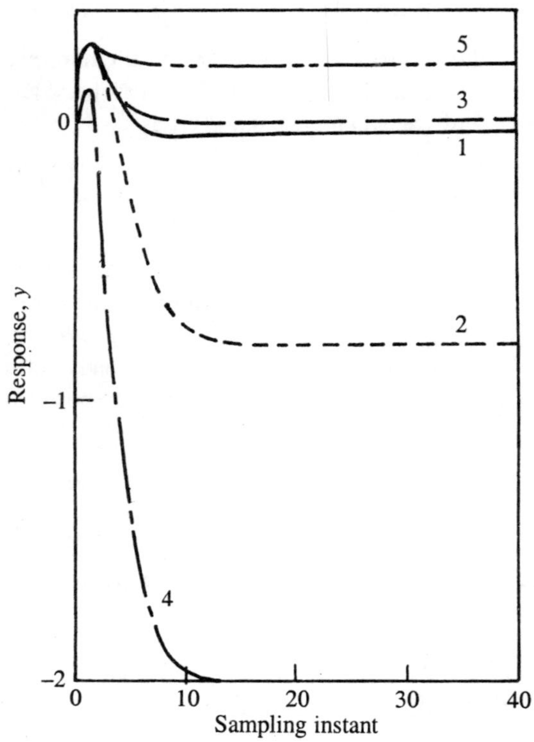

1: Linear for *u* and Nonlinear for *v*
2: Nonlinear for *u* and linear for *v*
3: Nonlinear for *u* and nonlinear for *v*
4: Linear for *u* and linear for *v*
5: No controller (*u* = 0)

Fig. 13.18 The performance of various FFCs on nonlinear system for step input in *v* as *v* = 1

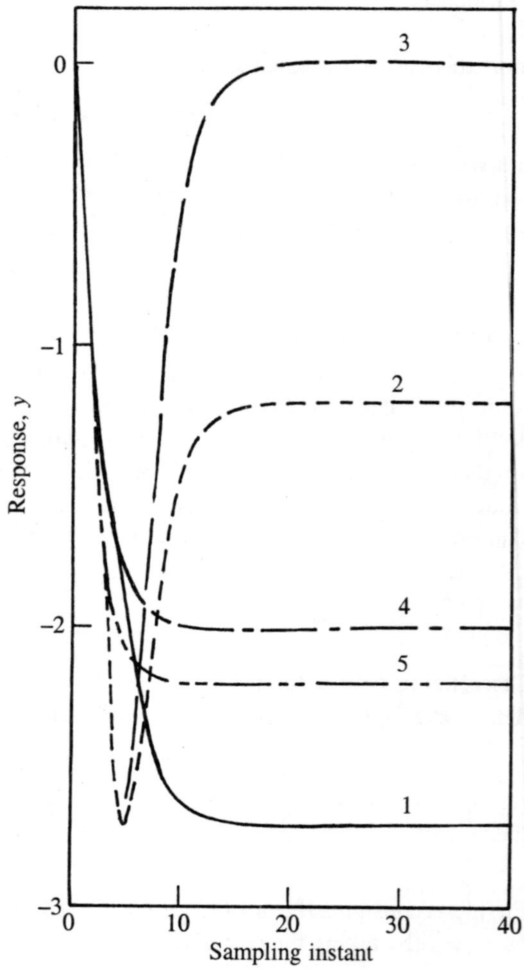

Legend: as in Fig. 13.18

Fig. 13.19 The performance of various FFCs on nonlinear system for step input in *v* as *v* = − 1

Table 13.6 The steady state value of *u* and offset in *y* for various FFC (simulation results on nonlinear model)

No.	Model used in FFC design		Value of *u* for step input in		Offset in *y* for step input in	
	for *u*	for *v*	*v* = 1	*v* = 1	*v* = 1	*v* = −1
1	L	NL	−1.0	1.0	−2.0	−2.0
2	L	NL	−0.2	2.2	0.0416	−2.71
3	NL	L	−0.60	4.074	−0.8	−1.20
4	NL	NL	−0.17	4.432	0.0	0.0
5	(without any FFC)		0.0	0.0	0.20	−2.20

performance of the controlled system considering the nonlinear only in y to v relation or nonlinear only in y to u relation is better than that of the totally linearized system; the performance of the completely linearized system is not adequate when compared to that of the system without any controller.

Summary

In this chapter, simulation case studies are given: (i) to illustrate the identification of a bilinear model for a nuclear reactor (ii) to design of linear and nonlinear IMC for Hammerstein nonlinear systems; (iii) to develop a NARMAX model for a bioreactor and design of controllers for NARMAX model; and (iv) to design a linear and nonlinear FFC for a Wiener model. In all these studies the need for nonlinear model develpment and nonlinear controller is brought out.

References

Agarwal, P., G. Koskhy and M. Ramseier, An algorithm for operating a fed-batch fermentor at optimum specific growth rate, Biotech. Bioeng., 30(1984) 61–90.

Benitez-Read, J.S., M. Jamshidi and R.A. Kisner, Advanced control design for nuclear reactors, Control Theory and Adv. Tech. (C-TAT), 8 (1992) 447–464.

Bhat, J., M. Chidambaram and K.P. Madhavan, Adaptive internal model control of Hammerstein nonlinear systems, Chem. Eng. Commun., 89 (1990) 177–186.

Chidambaram, M. and S.N. Jyothi, Feedforward control of Wiener type nonlinear system, Indian. Chem. Engr., 41 (2) (1999) 191–192.

Chidambaram, M. and S.N. Jyothi, Feedforward control of Hammerstein nonlinear system, Proc. NSC-97 on Integrated systems with multiple technologies, Allied Publishers Ltd., New Delhi, 144–148.

Chidambaram, M. and S.N. Jyothi, Feedforward control of Hammerstein nonlinear systems with input multiplicities, Trends in Measurement and automation (TIMA)-99, Allied Publishers Ltd., New Delhi, 280–283.

Chidambaram, M. and S.N. Jyothi, Feedforward control of Wiener type nonlinear systems, Proceedings of Chem. Engg. Congress (CHEMCON), Chandigarh, 18–21 Dec (1999).

Chen, S., and S.A. Billings, Representation of nonlinear systems; The NARMAX approach, AIChE J., 40 (1994) 269–282.

Clark, D.W. and P.J. Gawthrop, Self-tuning control, Proc. IEE, 126 (1979) 633–640.

Goodwin, G.C. and R.L. Payne, Dynamic system identification experiment design and data analysis, Academic Press, NY (1977).

Henson, M.A. and D.E. Seborg, An internal model control strategy for nonlinear systems, AIChE J., 37 (1991) 1065–1081.

Hsia, H.C., System identification: Least squares methods, Lexington Books, Lexington, (1977) 146–148.

Isermann, R., Digital Control Systems, Berlin, Springer-Verlag, (1982).

Kisner, R.A., A framework for selecting appropriate control technologies for nuclear power plant systems. Control-Theory and Adv. Tech. (C-TAT), 8 (1992) 405–418.

Leontaritis, I.J., and S.A. Billings, Input-output parametric model, Int. Jl. Contr., 41 (1985) 303–344.

Proll, T. and M.N. Karim, Model predictive control using NARX approach, AIChE J., 40 (1994) 269–282.

Radhakrishnan, T.K., S. Sundaram and M. Chidambaram, Nonlinear control of continuous bioreactors, Bioprocess Engg., 20 (1999) 173–178.

Reddy, G.P and M. Chidambaram, Nonlinear control of bioreactors with input ultiplicities, Bioprocess Engg., 11 (1994) 97–100.

Reddy, G.P., and M. Chidambaram, Nonlinear control of bioreactors with input multiplicities in dilution rate, Bioprocess Engg., 12 (1995) 151–155.

Shen, G.C. and W.K. Lee, Adaptive inferential control for chemical processes with perfect measurements, Ind. Engg. Chem. Res., 27 (1988) 71–80.

Stephanopolous, G., Chemical Process Control: An Introduction to Theory and Practice. Prentice-Hall of India Pvt. Ltd., New Delhi (1994).

Subramaniam, M., A. Rajakumar and M. Chidambaram, Nonlinear control of nuclear reactors, Control Theory and Adv. Tech. (C-TAT), 10 (1995) 1531–1540.

Subramaniam, M., A. Rajakumar and M. Chidambaram, Evaluation of control loops ineraction in liquid metal cooled nuclear reactors, Indian J. Chem. Tech., 2 (1995) 301–306.

Subramaniam, M., A. Rajkumar, B.C. Reddy and M. Chidambaram, Multivariable PI controllers for a nuclear reactor, Proc. Inter. Conf. Adv. Chem. Eng., I.I.T Madras, (1996).

Subramaniam, M. and M. Chidambaram, Model identification and control of a nuclear reactor, Proceedings of Automation in process engineering and manufacture, Kalpakkam, November 26–28 (1997) 213.

Vogel, E.F. and T.F. Edgar, An adaptive dead time compensation for process control, Proc. JACC, San Francisco, Paper TPS-E (1980).

Zafiriou, E. and M. Morari, Digital controllers for SISO systems: a review and a new algorithm, Int. J. Contr. 42 (1985) 855–876.

Review Questions

1. Herbst et al. (Powder technoloy, 1992, V69, 21–32) have presented an experimental work on supervisory control of ball mill grinding plant. Give a brief write up about this work.
2. Rohani and Bourne (Chem. Engg. Sci., 1990, V45, 3457–3466) presented an experimental application of self tuning control of crystallizer. Give a brief write up about this study.
3. Norquay et al. (Chem. Engg. Sci., 1998, V53, 75–84) have presented an experimental study on model predictive control of pH process based on Wiener models. Give a write up about the identification of the Wiener model and design of MPC for the model and comparison of the performance with that of a PID controller.
4. Polihronakis et al. (Compu. and Chem. Engg., 1993, V17, 1167–1179) have reported experimental study on the adaptive control techniques for anaerobic digesters. Give a write upon the model developed, controller design method and the performance obtained.
5. Machado and Bolzan (The Chem. Engg. Jl., 1998, V70, 1–8) have presented the implementation of self tuning PID control of a batch suspension polymerization reactor. Give a write up on the development of discrete time model and on the design of digital controller.
6. Najim et al. (Powder Techn., 1995, V82, 59–68) have presented a long range predictive control of a simulated grinding circuit. Give a summary of the discrete model development and on the design of the predictive controller.
7. Goochee et al. (Biotech. and Bioengg., 1989, V33, 282–292) have presented an experimental study on DMC of a competitive mixed culture system. Give a write up on this work.
8. Marco et al. (Ind. Eng. Chem. Res., 1996, V36, 1708–1716) have given a comparative study on different nonlinear MPC. Give a write-up on this study.

14

Fuzzy Logic Control

14.1 Need for Fuzzy Logic Control

Most industrial applications of modeling and control rely on intuition, sound engineering judgment, past experience, and heuristic procedures. Let us discuss the problem in modeling and control of polymerization reactors. There are considerable difficulties in modeling of emulsion polymerization reactors and the non availability of robust on-line measurement techniques for critical properties of the product. Quality control is interpreted as proper recipe formulation, optimum initial conditions selection, and ad hoc operation based on experience.

Mathematical models available presently relate the operating and measured variables to molecular properties such as composition or molecular weight or particle size of the polymer and not directly to the end use properties (like mechanical strength, melt viscosity, impact strength, drawability etc.). The on-line control problem is to control one or more of the molecular properties. These molecular properties are also difficult to measure on-line. Temperatures, pressures and flow rates of the process are considered as controlled variables for the closed loop control system design. Vague control and measurement objectives, difficult sampling, discrete sensor output, large dead times and sampling intervals, off-line analysers, large measurement errors and high noise levels are the reasons for the difficulty in controlling polymerization reactions (Shork, 1994).

An operator of polymerization reactors is able to control the processes based on his/her experience, without a mathematical model. Hence, one way to handle the control problem is to develop a controller based on operators' knowledge. The process operator thinks and acts according to linguistic quantities like 'high', 'medium', 'low', etc. The theory of fuzzy sets developed by Zadeh enables us to mathematically describe and handle these linguistic expressions. A fuzzy logic controller (FLC) based on fuzzy logic provides a means of converting a linguistic control strategy based on an operators' knowledge into an automatic control strategy. Excellent review of work done on the theory of fuzzy logic control is given by Lee (1990) and Driankov et al. (1993).

The advantages of FLC are: (i) increased robustness; (ii) a few rules encompassing great complexity; (iii) FLC is rule-oriented, the order of rules is arbitrary; (iv) FLC is tolerant to redundant or contradictory rules; (v) the method can be easily extended to multi-input multi-output systems. The limitations or drawbacks of FLC are: (i) the response of FLC is difficult to predict with analytical tools and simulation is often the only course of evaluation and (ii) it is difficult to program in machine language.

Zadeh's definition of fuzzy versus conventional control systems is given in Table 14.1.

Table 14.1 Zadeh's definition of fuzzy and crisp control systems

No.	Crisp control	Fuzzy control
1	Differential equations based	Rule based
2	Start with system equations, goals and constraints	Start with human solution
3	Provides certain tools	Adds to those tools

14.2 Classification of FLC

Hang and Tyan (1994) have classified FLCs into five types.

Type 1: FLC for input selection: The purpose is to select the right input from a set of input functions or a specific value out of a finite choice of parameters.

Type 2: FLC for feedback error/output control (Fuzzy linguistic control): To generate a control input based upon the error/output information resulting from feedback.

Type 3: FLC for controlling parameters of dynamic systems.

Type 4: FLC for choosing the best compensator: It is the generalization of Type 3 where the decision is applied to the compensator or subsystem. The idea of adaptation and learning was quiet popular during 1960s and 1970s. Now the trend is to apply FLC for this purposes.

Type 5: Fuzzy linguistic rules as mathematical model for unknown or complex system dynamic.

14.3 Design of Fuzzy Logic Controllers

The fuzzy controller design procedure (Fig. 14.1) consists of the following steps (Sugano, 1985):

1. Specifying the ranges of manipulated and controlled variables.
2. Deducing the linguistic labels that can be used to describe these variables and specifying the appropriate ranges of applicability of each label, alongwith the fuzzy sets.
3. Selecting the type of membership function and the parameter in that function for the manipulated variable and the output variables.
4. Determining the rules that will relate the linguistic labels of controlled variables to that of the manipulated variables. The execution of these rules gives the fuzzy output.

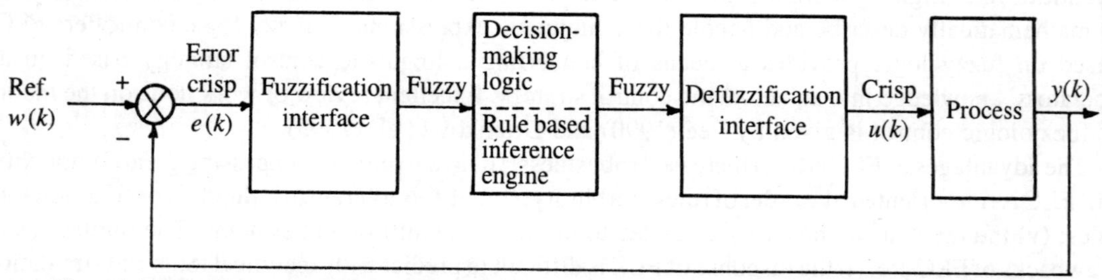

Fig. 14.1 Basic configuration of a closed loop system involving a FLC in which the fuzzy and crisp data flow is identified

5. Selecting a suitable defuzzification algorithm to calculate the non fuzzy (crisp) value for the manipulated variable to be implemented in the process.

All these steps are to be based on the operators' experience. Kobayashi et al. (1993) have presented a method of inducing fuzzy rules and membership functions using the plant operating data. Let us discuss the generation of membership function for fuzzy numbers. A standard form of function for all the relevant numbers can be chosen from the programing point of view. A triangular membership function is the widely used membership function. The fuzzy numbers associated with this type of function are referred to as triangular fuzzy numbers (Fig. 14.2). The main advantage of this type of fuzzy number is that it can be completely defined by three values (a, b, c). 'a' represents the extreme left of the fuzzy number range while its grade of membership is zero. Value c represents the extreme right of the range and its grade of membership is again zero. Point b is the apex in the triangular membership where the grade of membership is unity. All the points in ranges a-b and b-c have grades of membership (γ) defined by the relations:

$$\gamma = 0 \quad x \le a$$

$$\gamma = (x - a)/(b - a) \quad a \le x \le b$$

$$\gamma = (c - x)/(c - b) \quad b \le x \le c$$

$$\gamma = 0 \quad x \ge c \tag{14.1}$$

where x is a non-fuzzy numerical value and γ is the grade of membership of the fuzzy set. Turksen (1991) has discussed membership functions and their measurements in more detail.

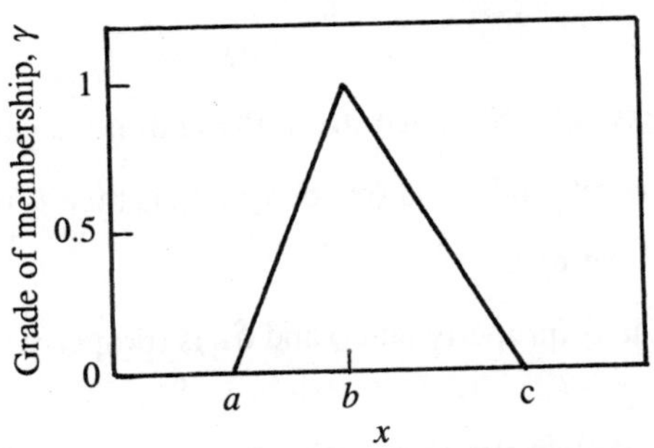

Fig. 14.2 Triangular membership function

Once the fuzzification of the process measurement data is carried out, then the rule base is to be developed, based on the operators' experience. The form of the rules used depends on the process under control and the heuristics employed. The rule will have the format.

If (condition) Then (Result)

where condition is the antecedent part of the rule and the action is the consequent part. Conditions can be combined by AND or OR combinations and negated by the logic NOT. The techniques to perform the logical combination are

cond_1 AND cond_2 AND cond_3 . . . AND cond_n

$$\gamma_{\text{output}} = \min(\gamma_1, \gamma_2, \gamma_3, \ldots \gamma_n) \tag{14.2}$$

cond_1 OR cond_2 OR cond_3 . . . OR cond_n

$$\gamma_{output} = \max(\gamma_1, \gamma_2, \gamma_3, \ldots \gamma_n) \tag{14.3}$$

NOT (condition)

$$\gamma_{output} = 1 - \gamma \tag{14.4}$$

where $\gamma_1, \gamma_2, \gamma_3, \ldots \gamma_n$ refer to the grade of membership associated with $cond_1, cond_2, \ldots, cond_n$ respectively.

FLC implementing proportional control action

The symbolic representation of a rule for a proportional-like FLC is given by

If e is ⟨property label⟩ then u is ⟨property label⟩

where e is the error and u the manipulated variable. Similarly for PI-like FLC we have

If e is ⟨property label⟩ and Δe is ⟨property label⟩ then Δu is ⟨property label⟩

where Δu is the incremental output of a controller. The above rule is arrived from the conventional PI control law

$$u = K_c e + K_I \int e dt \tag{14.5}$$

i.e.,

$$\dot{u} = K_c \dot{e} + K_I e \tag{14.6}$$

Using

$$\dot{u} = (u_k - u_{k-1})/T \tag{14.7}$$

and

$$\dot{e} = (e_k - e_{k-1})/T \tag{14.8}$$

we get the corresponding fuzzy rule. Similarly for a PD action we can write

If e is ⟨property label⟩ and Δe is ⟨property label⟩ then u is ⟨property label⟩

Similarly, for a PID-like FLC we have

If e is ⟨property label⟩ and Δe is ⟨property label⟩ and δe is ⟨property label⟩ then u is ⟨property label⟩

where δe is the sum of errors.

One can also use the process state variables rather than error, rate of error and sum of error. The symbolic expression for a fuzzy rule for a scalar output case is given by

If x_1 is ⟨property label⟩ and x_2 is ⟨property label⟩ . . .

x_n is ⟨property label⟩ then u is ⟨property label⟩

Once all the rules have been scanned, then the result will be a set of degree of truth. The purpose of defuzzification step is to extract a crisp value (non-fuzzy value) from the fuzzy output. The defuzzified value will be applied to the process. There are different methods of defuzzifying the output: (i) maximum membership method, (ii) first of maximum method, (iii) last of maximum method and (iv) fuzzy-mean method. One criteria for the choice of a suitable defuzzifier is the computational simplicity.

In recent years, there has been growing interest in using fuzzy logic control (FLC) systems. Ying et al. (1990) have shown that the fuzzy PI controller is better than the conventional one for nonlinear systems. This is not a drawback of the conventional PI controller, since is designed for linear systems and this objective is achieved. The fuzzy PI controller is driven by a set of control rules rather than by two constants namely proportional and integral gains. However, there are simple rules together with few simple formulas. Because of these rules and formulas, which change the controller gains automatically and continuously with the outputs, the controller becomes self tuning and suitable for nonlinear systems. In general the sources of nonlinearity of a fuzzy controller include: the fuzzification algorithm, fuzzy control rules, inference mechanisms and the defuzzification algorithm used for the controller output.

14.4 A Detailed Example

The five stages in designing a FLC are illustrated by an example of continuous co-polymerization reactor control problem. Associated with each input is a linguistic variable. Each linguistic variable may take a number of possible values. Each value is represented by a fuzzy set. In the continuous stirred tank co-polymerization reactor, we have to control the mole fraction of monomer in the co-polymer (C) and reactor temperature (T) using the flow rate of feed monomer (F). The ranges are assumed to be 0.56-0.64 for C 76-80°C for T and 16-20 kg/h for F respectively. We have the labels 'CLow', 'CMedium' and 'CLarge' associated with the linguistic variable (C) and the fuzzy set 'Tlow', 'TMedium' and 'THigh' associated with the linguistic variable temperature and the fuzzy set 'Flow', Fmedium' and 'FHigh' with the flow rate of active monomer.

14.4.1 Membership Function

Triangular membership function for each fuzzy set is assumed here. The values of *a*, *b* and *c* of the membership for each fuzzy set is assumed to be as follows (refer to Fig. 14.3 for C)

For composition:

CLOW	(0.56, 0.58, 0.60)
CMEDIUM	(0.58, 0.60, 0.62)
CHIGH	(0.60, 0.62, 0.64)

(14.9)

For Temperature:

TLOW	(76.0, 77.0, 78.0)
TMEDIUM	(77.0, 78.0, 79.0)
THIGH	(78.0, 79.0, 80.0)

(14.10)

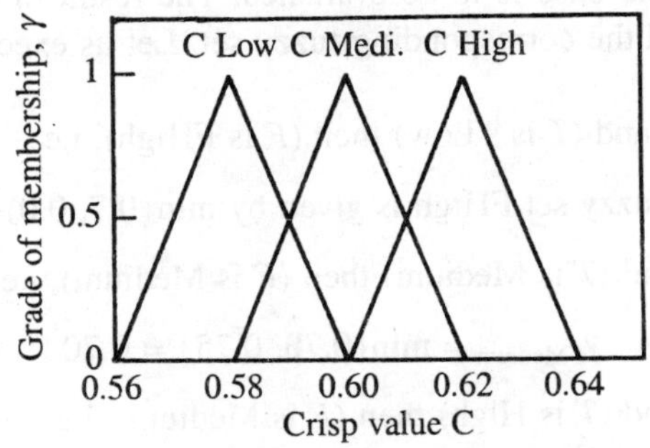

Fig. 14.3 Fuzzy relation for composition C

For Flow rate:

FLOW	(16.0, 17.0, 18.0)
FMEDIUM	(17.0, 18.0, 19.0)
FHIGH	(18.0, 19.0, 20.0)

(14.11)

The following fuzzy rules, relating fuzzy sets C and T to fuzzy set F, are considered

Rule No 1—If (C is CLow) and (T is TLow) Then (F is FHigh)
Rule No 2—If (C is CLow) and (T is TMedium) Then (F is FMedium)
Rule No 3—If (C is CLow) and (T is THigh) Then (F is FMedium)
Rule No 5—If (C is CMedium) and (T is TMedium) Then (F is FMedium)
Rule No 6—If (C is CMedium) and (T is THigh) Then (F is FLow)
Rule No 7—If (C is CHigh) or (T is THigh) Then (F is FLow)

(14.12)

Input acquisition: Let at time t, the composition C and temperature (T) are read from the sensors as

$$\text{Composition C} = 0.584$$

$$\text{Temperature} = 81.5°\text{C}$$

Fuzzification: Irrespective of the inference and defuzzification method used, the first stage evaluates the grade of membership for each of the terms. From Eq (14.1) and using the corresponding a, b and c values, we get the grade of membership for the composition C

$$\gamma_{\text{CLow}}(0.584) = 0.70$$

$$\gamma_{\text{CMedium}}(0.584) = 0.20$$

$$\gamma_{\text{CHigh}}(0.584) = 0.0 \tag{14.13}$$

Similarly, we get the grade of membership for the temperature T as

$$\gamma_{\text{TLow}}(78.5) = 0.0$$

$$\gamma_{\text{TMedium}}(78.5) = 0.75$$

$$\gamma_{\text{THigh}}(78.5) = 0.25 \tag{14.14}$$

14.4.2 Rule Evaluation

Each of the 7 rules of the rule base is to be evaluated. The result of each rule gives the grade of membership for output F and the corresponding fuzzy set. Let us execute these rules.

Rule No. 1 If (C is CLow) and (T is TLow) then (F is FHigh), i.e.

$$\gamma \text{ for fuzzy set FHigh is given by } \min\{0.7, 0.0) = 0.0$$

Rule No. 2 If (C is Low) and (T is Medium) then (F is Medium), i.e.

$$\gamma_{\text{FMedium}} = \min\{0.70, 0.75\} = 0.70$$

Rule No. 3 If (C is Low) and (T is High) then (F is Medium), i.e.

$$\gamma_{\text{FMedium}} = \min\{0.7, 0.25\} = 0.25$$

Rule No. 4 If (C is Medium) and (T is Low) then (F is Medium), i.e.

$$\gamma_{F\text{Medium}} = \min\{0.20, 0.0\} = 0.0$$

Rule No. 5 If (C is Medium) and (T is Medium) Then (F is Medium), i.e.

$$\gamma_{F\text{Medium}} = \min\{0.20, 0.75\} = 0.20$$

Rule No. 6 If (C is Medium) and (T is High) Then (F is Low), i.e.

$$\gamma_{F\text{Low}} = \min\{0.20, 0.25\} = 0.2$$

Rule No. 7 If (C is High) or (T is High) Then (F is Low), i.e.

$$\gamma_{F\text{Low}} = \max\{0.0, 0.25\} = 0.25 \quad (14.15)$$

A quick review of the rule membership reveal that Rule No. 2 has the highest membership grade (0.7) and Rules No. 1 and 4 have the lowest membership grade (0.0).

14.4.3 Inferences and Defuzzification

In this example two Rule No. 6 and 7, prescribe the 'FLow' output. There are four Rules No. 2, 3, 4 and 5, which prescribe the output to be 'Medium' and only Rule No. 1 which prescribes the output to be 'High'. Since the rules are considered to be combined in the rule base by OR operation, the membership grade associated with each output term is obtained by identifying the rule which prescribes that output term with the maximum membership grade. The membership grade for each output terms is computed below:

Table 14.2 Inference of fuzzy rules

Output term	Rules (R_i)	Membership grade for (F)
Low	6, 7	max{0.2, 0.25) = 0.25
Medium	2, 3, 4, 5	max{0.7, 0.25, 0.0, 0.2} = 0.7
High	1	0

The defuzzified output for F is obtained by using fuzzy mean method as

$$\text{crisp output of } F = \Sigma\ [(F)_i \gamma_i]/\Sigma\ \gamma_i \quad (14.16)$$

where $(F)_i = [(a + c)/2.0]$ {'a' and 'c' of that particular fuzzy set given in Table 14.2 and Eq (14.9) to (14.11)} and γ_i is the grade of membership given in the Table 14.2. Let us discuss the method calculating the defuzzified output for this example. We have

$$F = \frac{(0.25 \times [(16 + 18)/2] + 0.70 \times [(17 + 19)/2] + 0.0 \times [(18 + 20)/2]}{(0.0 + 0.70 + 0.25)}$$

$$= 17.74 \quad (14.17)$$

The above defuzzification method (mean defuzzification method) is applicable for only for symmetric fuzzy membership functions. For non-symmetrical membership functions, there are other methods for defuzzification, namely first of maximum and last of maximum methods. Table 14.2 shows that the maximum value for γ is for the fuzzy set FMedium. The values of F for a grade of membership $\gamma_{F\text{Medium}} = 0.7$ are obtained from Eq. (14.1) with a, b and c given by Eq. (14.11) as 17.6 and 18.4. By the defuzzification method of first of maximum, we get the output F as 17.7. By the last of maximum method we get F = 18.3. This value of F is applied to the process and then the process status is again measured. The feedback control calculations by FLC is then repeated. If the control performance is not adequate then the fuzzy rules and/or membership functions for the inputs and outputs are to be modified.

Though a single output control problem is explained here, extension to multi-output problems can be easily carried out. The design methods of such multivariable controllers for general systems are given by Ray and Majumder (1985) and Chen and Chen (1991). Let us review the reported work on the fuzzy logic control of polymerization reactors.

14.5 FLC of Semi-Batch Polymerization Reactors

14.5.1 Control of Reactors

As stated earlier, the precise control of polymer properties such as composition, molecular weight etc., has become one of the major issues in industrial co-polymerization reactions. Many co-polymerization reactions are carried out in semi-batch reactors. Generally, the operation of semi-batch co-polymerization reactors involves the addition of a more active monomer to a batch (containing a lesser reactive monomer and a solvent) during the reaction with a time-varying or constant feed rate, thus minimizing the concentration drift which results from the different reactivities of the co-monomer (Odian, 1991). The addition of a pre-cooled monomer during polymerization has the beneficial effect of providing extra cooling to the reaction mixture, thus permitting better temperature control and higher productivity. The reactor model equations are highly nonlinear. The reaction kinetics involve the determination of several kinetic parameters. Application of the conventional PI controller requires several tuning adjustments, particularly for the control of semi-batch reactors.

One way of solving the control problem is to develop a controller based on the operators' knowledge. Mahesh et al. (1993) have applied FLC for the control of the semi-batch co-polymerization reaction of styrene and acrylonitrile with xyline as a solvent. To acquire an operators' knowledge of the process, Mahesh et al. (1993) have considered the reported mathematical model of a semi-batch reactor. In practice, the fuzzy control rule will be developed on the basis of operators's knowledge on a real plant.

For a semi-batch co-polymerization reactor, the output variable is the copolymer composition (C) and the manipulated variable is the feed flow rate (F) of the more active monomer (styrene). The composition error (C1E) and the flow rate (FR) are considered as linguistic variables. The choice also follows from the desire to make the rules as simple as possible, both for development and implementation. The number of necessary fuzzy sets, their ranges and the corresponding membership functions are determined after the selection of the control range. The number of fuzzy sets and their ranges depends on the finess of the control rules and the desired control action. For development and simulation of FLC, Mahesh et al. (1992) have considered the reported mathematical model for a semi-batch co-polymerization reactor. The system considered is styrene (M_1) -acrylonitrile (M_2) solution copolymerization in a semi-batch reactor where the styrene (M_1) feed rate is continuously varied under isothermal conditions. The equations representing the above system are given by Kravaris et al. (1989).

In the work of Mahesh et al. (1993), the selected allowable ranges span across a set of nine fuzzy sets for the linguistic variable composition error (C1E) and a set of eight for the flow rate (FR). The selected labels and fuzzy sets are as follows (Figs. 14.4 and 14.5).

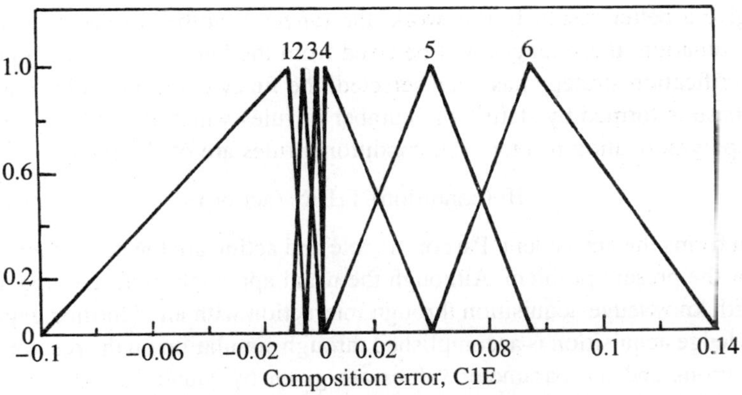

Fig. 14.4 Fuzzy relation for composition error. 1: C1NOKE; 2: C1NSE; 3: C1ZE; 4 C1PSE; 5: C1POKE; 6: C1PME

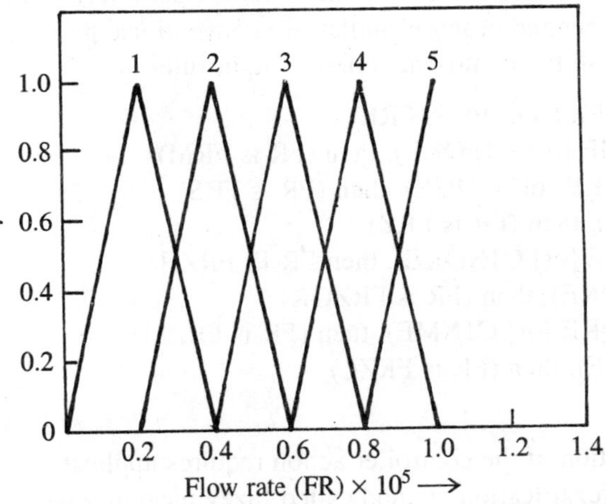

Fig. 14.5 Fuzzy relation for flow rate. 1: FRZL; 2: FRZML; 3: FRZOK; 4: FRZH; 5: FRZ

Composition error (C1E)		Flow rate (FR)	
C1NLE	(−0.40, −0.10)	FRZL	$(0.0, 2\times10^{-6}, 4\times10^{-6})$
C1NME	(−0.40, −0.10, −0.01)	FRZML	$(2\times10^{-6}, 4\times10^{-6}, 6\times10^{-6})$
C1NOKE	(−0.10, −0.01, −0.005)	FRZOK	$(4\times10^{-6}, 6\times10^{-6}, 8\times10^{-6})$
C1NSE	(−0.01, −0.005, 0.0)	FRZH	$(6\times10^{-6}, 8\times10^{-6}, 1\times10^{-5})$
C1ZE	(−0.005, 0.0, 0.002)	FRZ	$(8\times10^{-6}, 1\times10^{-5}, 4\times10^{-5})$
C1PSE	(0.0, 0.002, 0.04)	FRS	$(1\times10^{-5}, 4\times10^{-5}, 0.006)$
C1POKE	(0.002, 0.04, 0.075)	FRM	$(4\times10^{-5}, 0.006, 0.012)$
C1PME	(0.04, 0.075, 0.14)	FRL	(0.006, 0.012, 0.014)
C1PLE	().75, 0.14, 1.0)		

(14.18)

The fuzzy labels C1NLE, C1NME, . . . , C1PLE denote C1E being negative large error, negative medium error, negative okay error, negative small error, zero error, positive small error, etc. The fuzzy labels for FR can be interpreted similarly. Triangle membership functions, as discussed earlier, are chosen from the programming viewpoint. At present there is no systematic way of selecting fuzzy ranges which give a better result. In this work, the ranges have been selected from the simulation study. In a real situation, these ranges will be fixed from the knowledge of the operator.

Once the fuzzification strategy has been selected, the fuzzy controller rules may be constructed. An initial rule base is formed by stating the number of rules which the operator deems suitable for the semi-batch polymerization reactor. The conditional rules are of the type

IF (condition) THEN (action)

where condition forms the antecedent. Part of the rule and action are the consequent which represents the flow rate for the present problem. Although the usual approach in developing a knowledge base system starts with knowledge acquisition through interaction with an identified expert in the field, in this work, knowledge acquisition is accomplished through simulation on the reactor model equations. The model equations and the parameter values are given by Mahesh et al. (1993). The rules are written down and investigated as to their physical correctness, and the rule base is formed after an iterative process in which new rules are added and some previous ones are deleted or changes are made to the existing ones. After several simulation runs, a set of eight rules is finalized. These rules are simple, of small number and in sense similar to a conventional proportional controller in how and what it tries to accomplish the control action. The eight rules are:

1. If (C1E is C1PLE), then (FR is FRL)
2. If (C1E is C1PME [or] C1POKE), then (FR is FRM)
3. If (C1E is C1POKE [or] C1PSE), then (FR is FRS)
4. If (C1E is C1ZE), then (FR is FRZ)
5. If (C1E is C1NSE [or] C1NOKE), then FR IS FRZH)
6. If (C1E is C1NOKE), then (FR is FRZOK)
7. If (C1E is C1NOKE [or] C1NME), then (FR is FRZM)
8. If (C1E is C1NLE), then (FR is FRZL)

(14.19)

As stated earlier, generation of the controller action requires application of inference mechanism to combine the rules. A defuzzification method called fuzzy mean has been selected. The procedure is carried out as follows.

At every sampling interval C1E was calculated (actually obtained from the numerical solution of the model equations). Since the calculated value of C1E can be a member of more than one fuzzy set, the grades of membership of C1E in every fuzzy set were determined through the triangular membership functions. In this way, the grades of membership of the antecedent part of each and every rule are developed. The action of each rule is calculated by the corresponding grade of membership and the fuzzy set for the flow rate. From this, the crisp value (F_r) for each rule is calculated. The final control action is calculated by weighting the actions of each rule and hence the value of F_r is given by

$$F_r = \sum_{i=1}^{n} [(F_r)_i \gamma_i] / \sum_{i=1}^{n} \gamma_i$$

14.5.2 Simulation Results

Figure 14.6 shows the response of C of the controlled system and it indicates that a good response is obtained. The fuzzy rules and grade of membership function chosen are thus adequate. The manipulated variable vs time behaviour is shown in Fig. 14.7. The effect of sampling time on the C response is shown in Fig. 14.8. A sampling time of 60 sec is found to be adequate.

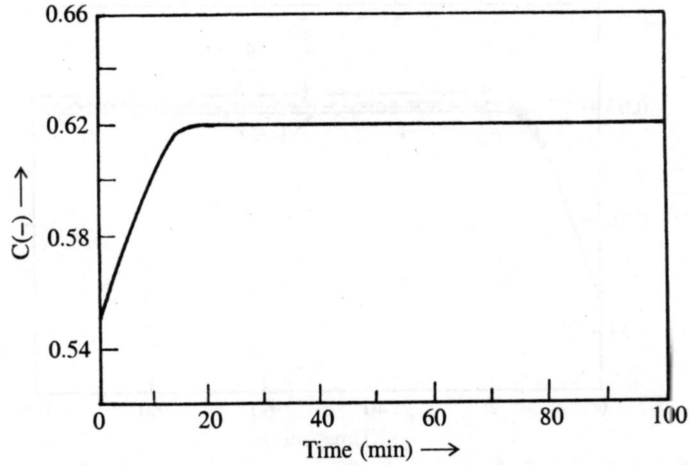

Fig. 14.6 Performance of the fuzzy controllers. Fuzzy range given by Eq. (14.19). Set point value = 0.62

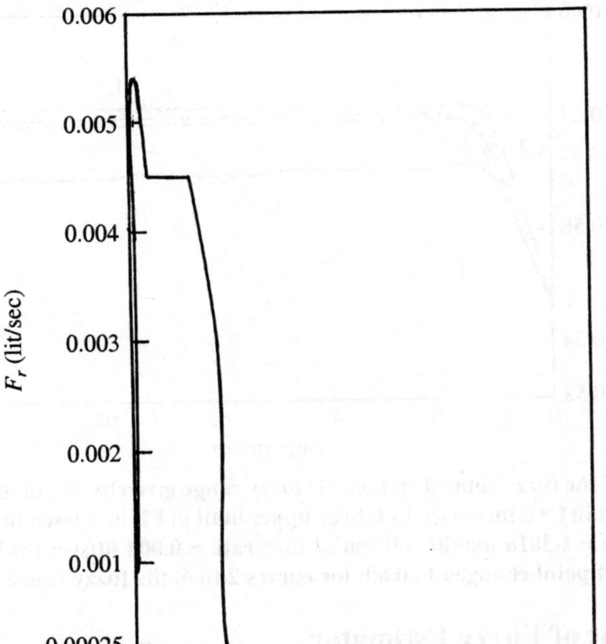

Fig. 14.7 Manipulated variable vs time for Fig. 14.6

To test the robustness of the control system, the operating point is changed to 0.60. The response is shown in Fig. 14.9. A similar response is obtained when the initial concentration is changed from 1.256 to 1.3816 mol/lit. The robust response for a change in the initial flow rate from 0.001 to 0.004 lit/sec is shown in Fig. 14.9. Thus the fuzzy control rules given by Eqs. (14.19) work well for other operating conditions.

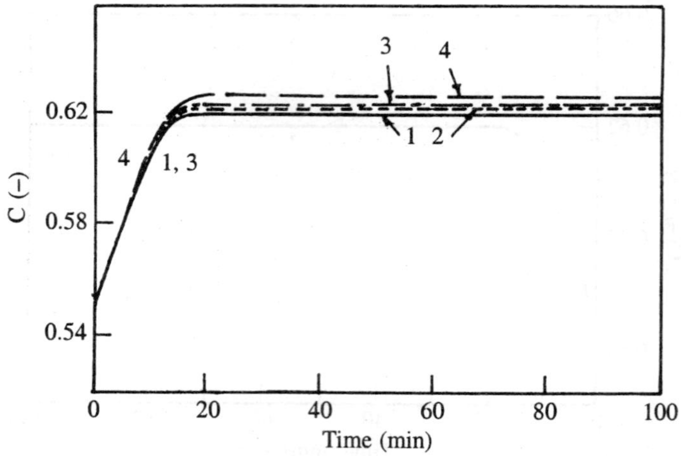

Fig. 14.8 Effect of sampling time on the performance of the control systems. 1, 40 sec; (2), 50 sec; 60 sec; (4), 90 sec

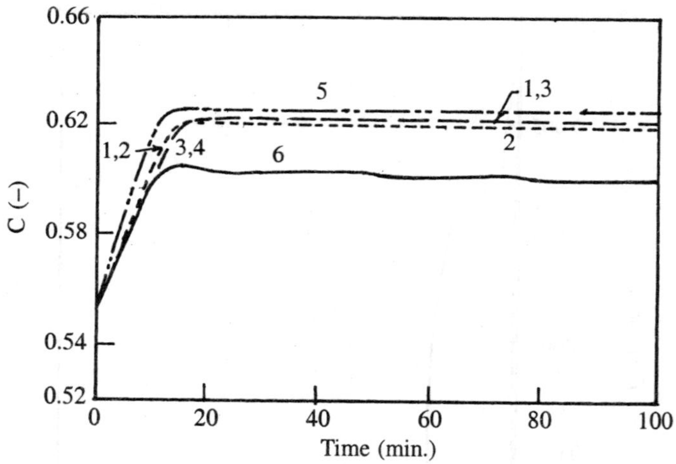

Fig. 14.9 Robustness of the fuzzy control system. (1) fuzzy range given by Eq. (6) of Mahesh et al. (1993); (2) upper limit in C1E increased to 1.20 or upper limit of FR increased to 0.02; (3) initial styrene concentration = 1.3816 mol/lit; (4) initial flow rate = 0.004 lit/sec; (5) initial reactor volume: 5.25 lit; (6) set point changed to 0.60; for curves 2 to 6, the fuzzy range is given by Eq. (14.18)

14.6 Development of Fuzzy Estimator

As stated earlier, inadequate on-line measurement of polymer properties such as composition or molar mass is one of the major difficulties in achieving good control of polymerization reactors. Many attempts have been made to develop mathematical model based estimators for various polymerization systems (Dimitratos et al. 1994). So far, these estimators have not been applied by the industry. This may be due to the complexity of the estimator and the need for wider knowledge of modeling and on-line estimation techniques. Also many polymerization reactions are sensitive to impurities which, even in very small and often undetectable quantities, can change the reaction mechanism considerably. In addition, laboratory measurements of co-polymer properties take considerable time. One way to overcome this problem is to develop an estimator based on the operators' experience. Mahesh and Chidambaram (1992) have reported a simulation work on the fuzzy estimator for a co-polymerization reactor. Let us briefly discuss their method.

The first step in the development of the estimator is assignment of numerical values of each of the fuzzy ranges for both input and output variables. Here, the input variables are: M_1, the concentration of monomer 1 in the reacting mixture; ζ_1, mole fraction of monomer 1 in the reacting mixture; and V, volume of the reactor. The output variable is C (mole fraction of monomer 1 in the co-polymer).

The equation for C depends on ζ_1 which in turn, depends on M_1. In this work, only ζ_1 is considered as an input variable. In real systems, ζ_1 is a fuzzy variable. However, in the present simulation, the non-fuzzy value of ζ_1 is obtained from the model equations. Using its grade of membership, it is fuzzified. The fuzzification of reactor volume is unnecessary since volume can be easily measurable.

The next step involves the generation of membership functions for fuzzy numbers. A standard form of triangle function for all relevant numbers are chosen from the programming view point. Four fuzzy numbers are defined for ζ_1 and C. The qualitative descriptions of the fuzzy numbers for each of the variables are ZERO, SMALL, MEDIUM, and LARGE. The actual numerical ranges for the fuzzy variables are:

for ζ_1

ZERO	(0.40, 0.55)
SMALL	(0.40, 0.55, 0.85)
MEDIUM	(0.55, 0.85, 0.97)
LARGE	(0.85, 0.97, 1.0)

(14.20)

for C

ZERO	(0.35, 0.65)
SMALL	(0.35, 0.65, 0.85)
MEDIUM	(0.65, 0.85, 0.90)
LARGE	(0.85, 0.90, 1.0)

(14.21)

The ranges used for fuzzy numbers may affect the sensitivity of the subsequent rule base. Unfortunately, there is at present no systematic way of selecting appropriate fuzzy ranges which would give better results (Xu and Zai, 1987). In this work, the ranges are fixed after some time has been spent on working with the simulation in order to develop a 'feel' for the correct qualitative ranges. Ranges are selected according to the physical limits. Simulation studies indicated that the starting values of C and ζ_1 are 0.35 and 0.40, respectively, and that ζ_1 and C cannot exceed unity. To ensure sensitivity, the four fuzzy numbers which qualitatively describe C and ζ_1 are skewed towards the larger end of the possible range.

The input to the simulator is ζ_1 and it is obtained from the simulation. The first step consists of converting the non-fuzzy values of these quantities into a fuzzy format. This is achieved very simply by applying the grade of measurement relationships over various ranges for the variable ζ_1. Now the fuzzy rules have to be developed. An initial rule base is developed by specifying the rules that are assumed by the operator to apply to a semi-batch reactor. The rules used in the estimator are as follows:

S.No.	Rule
1	(If ζ_1 is ZERO), then (C is ZERO)
2	(If ζ_1 is SMALL), then (C is SMALL)
3	(If ζ_1 is MEDIUM), then (C is MEDIUM)
4	(If ζ_1 is LARGE), then (C is LARGE)

(14.22)

Rules are developed by observing the simulation behaviour each time the estimated value of C is compared with the actual value of C obtained by solving the model equations (Mahesh and Chidambaram, 1992). In an actual situation, the rules would be developed on the basis of the experience.

Once all the rules have been scanned, the result is a set of degree of truth for each of the possible C. To make use of this information say in conventional control system it has to be defuzzified into a crisp value. Fuzzy mean type is used in this work.

Figure 14.10 shows the estimated value of C compared with the actual values of C_1 obtained by solving the model equations. The results are satisfactory. The fuzzy rules and the grade of membership function chosen are adequate. To test the robustness of the estimator, the simulation study is carried out for f_i = 0.015 lit/sec, i.e. +50% variation from the nominal value used for standardizing the estimator. The results are shown in Fig. 14.11.

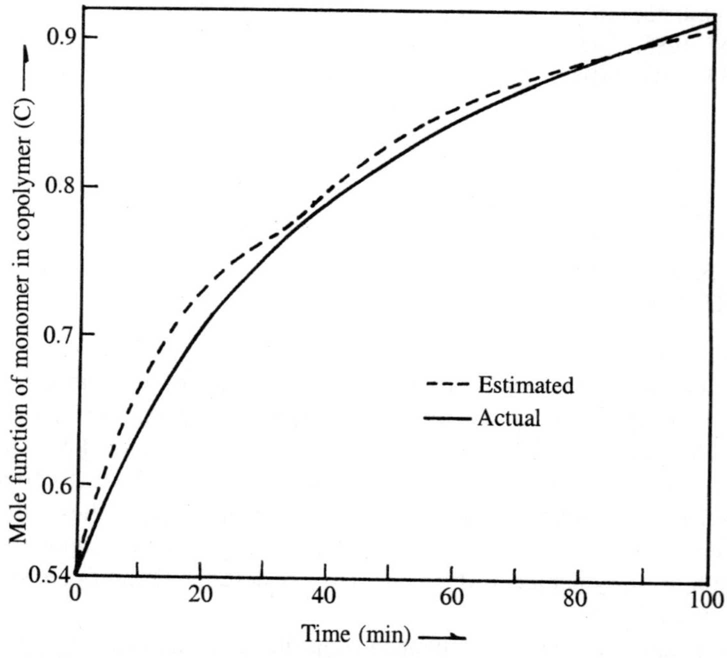

Fig. 14.10 Comparison of estimated and actual values for C_1 = 0.01 lit/sec; fuzzy ranges for the variables as given by Eqs. (14.20) and (14.21)

Simulations studies are performed by including both $[M_1]$ and ζ_1 in the rule base. The rules considered are:

1 If ($[M_1]$ is ZERO or ζ_1 is ZERO), then (C_1 is ZERO)
2 If ($[M_1]$ is SMALL or ζ_1 is SMALL), then (C_1 is SMALL)

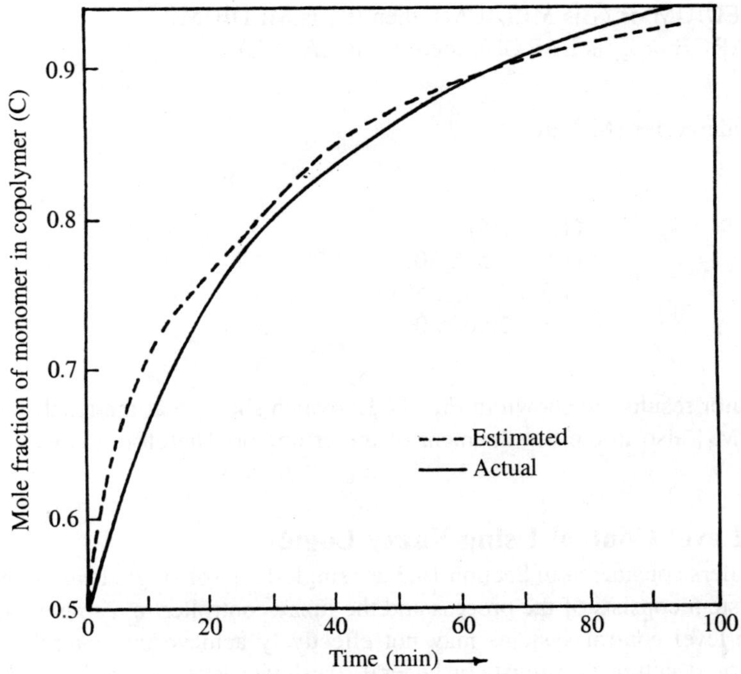

Fig. 14.11 Comparison of estimated and actual values for F_i = 0.015 lit/sec

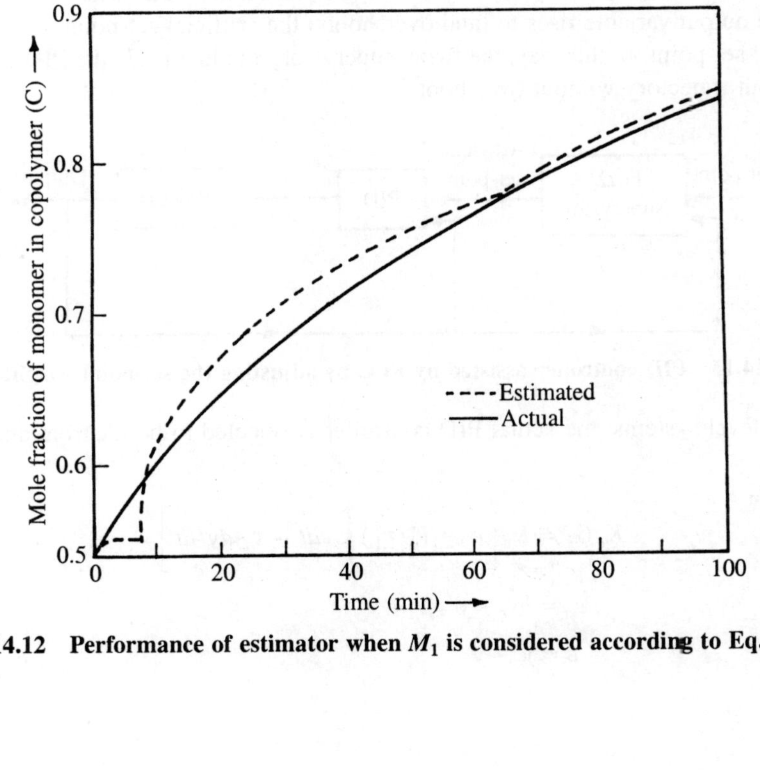

Fig. 14.12 Performance of estimator when M_1 is considered according to Eq. (14.26)

3 If ([M_1] is MEDIUM or ζ_1 is MEDIUM), then (C_1 is MEDIUM)
4 If ([M_1] is LARGE or ζ_1 is LARGE), then (C_1 is LARGE)

(14.23)

The ranges considered for [M_1] are:

for [M_1]

ZERO	(1.0, 1.55)
SMALL	(1.0, 1.55, 2.30)
MEDIUM	(1.55, 2.30, 2.50)
LARGE	(2.30, 2.50, 5.00)

(14.24)

The fuzzy estimator results are shown in Fig. 14.12, which shows that the results are not improved by considering [M_1] also in the development of the estimator. Therefore the estimator using ζ_1 is sufficient.

14.7 Multi-Level Control Using Fuzzy Logic

The fuzzy controllers considered in Section 14.3 are single-loop (or single level) controllers, i.e., the whole control system consists of the process and the fuzzy controller in a single loop. For complex processes, single-level control systems may not effectively achieve the control objectives, and a multi-level control structure is a must. In general, the lower-level controllers perform low-speed supervision. Usually one of the levels is constructed using fuzzy systems. The various possibilities of using fuzzy systems in two-level control system are shown in Figs. 14.13-14.17. In Fig. 14.13 fuzzy logic is used to determine artificial set-points that are fed to a conventional PID controller. The PID controller usually gives high overshoot. The fuzzy supervisory module detects impending overshoot, it 'fools' the PID controller to aim for the output variable value that is somewhat lower than the actual set-point. As the output variable rises to (and overshoots) the artificial set-point, the fuzzy set-point lowers the actual set-point. In this way, the fuzzy supervisory module leads the PID controller along a reference output trajectory without overshoot.

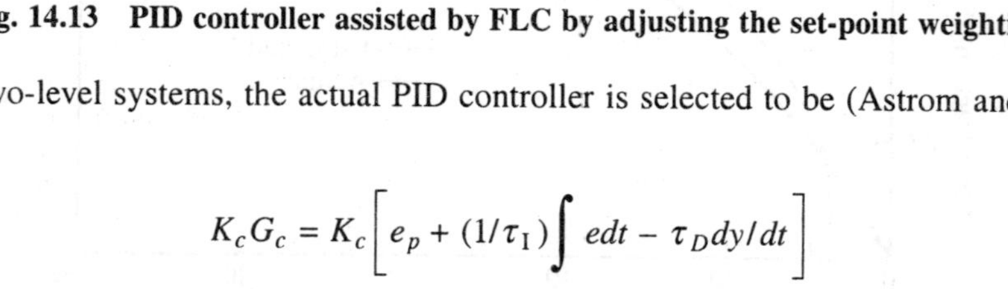

Fig. 14.13 PID controller assisted by FLC by adjusting the set-point weighting

In some two-level systems, the actual PID controller is selected to be (Astrom and Wittenmark, 1989)

$$K_c G_c = K_c\left[e_p + (1/\tau_I)\int edt - \tau_D dy/dt\right] \quad (14.25)$$

Here
$$e_p = by_r - y \quad (14.26a)$$

$$e = y_r - y \quad (14.26b)$$

The value of b is less than 1 and range is 0 to 1. The error for the proportional control action is lesser than the actual value. However, for integral action the correct error value is used. This guarantees zero offset. The derivative action is taken on the process measurement and not on the reference value (y_r). This is to reduce the excessive control action for a step change in the set-point. Fuzzy supervisory control can select appropriate value of b depending on the operating conditions (Visioli, 1999).

Similarly a fuzzy supervisory (Fig. 14.14) can tune a conventional PID controller according to certain heuristic rules. Simulation studies of this type of two-level systems are given by He et al. (1993), Zhao et al. (1993) and Ramkumar and Chidambaram (1995) for stable systems and Satishkumar and Chidambaram (1999) for unstable systems.

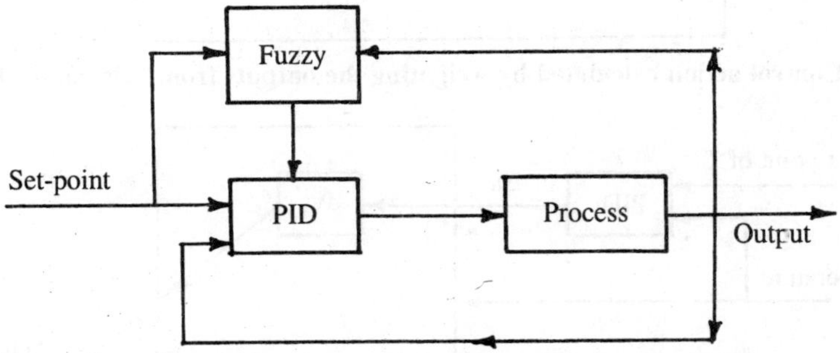

Fig. 14.14 PID controller settings adjustment by FLC

In another type of two-level system (Fig. 14.15), the fuzzy controller is designed in the first stage without considering stability and the supervisory crisp controller is used in the second stage to deal with stability related problems. Hence much freedom is given in selecting the fuzzy controller parameters and its performance is improved.

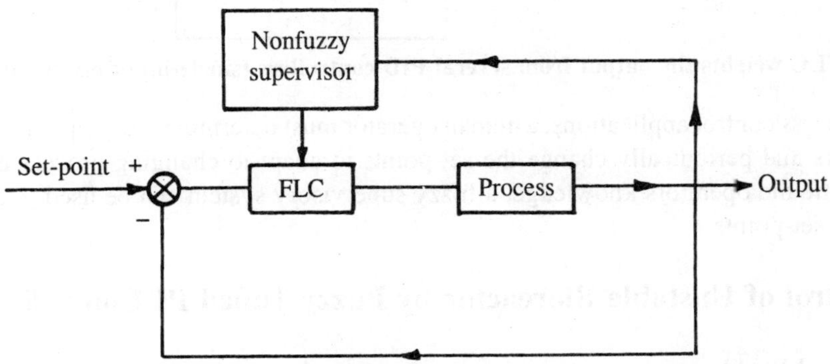

Fig. 14.15 Tuning of FLC by a nonfuzzy controller

In some applications, fuzzy controller is supplied parallel to a PID controller (Fig. 14.16). The combined weighted outputs is applied to the process.

In some applications, a fuzzy supervisory module can be used to combine the output of several PID controllers. For example in polymerization reactor control problems, two PID controllers can be used to regulate the temperature and composition of the product. However there is only one manipulated variable (the feed flow rate of more active monomer). The fuzzy supervisory module assigns weights to the different PID controller outputs (Fig. 14.17).

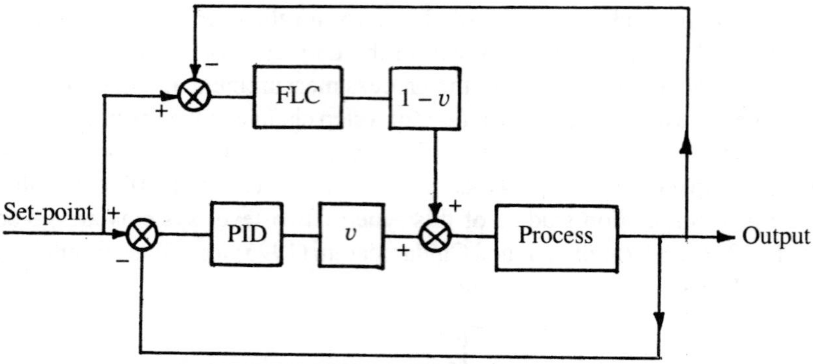

Fig. 14.16 Control action calculated by weighting the outputs from PID controller and FLC

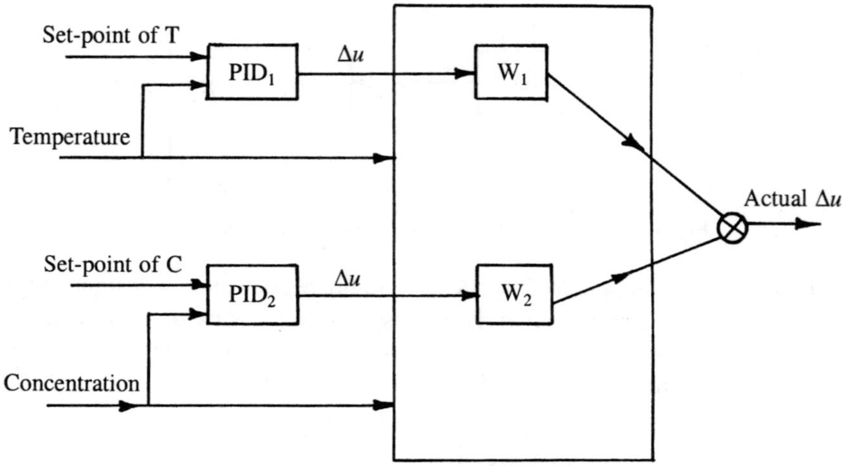

Fig. 14.17 FLC weights the output from several PID controllers (single-input multi-output systems)

In most process control applications, a human operator must determine the set-points for numerous PID controllers and periodically change the set-points to adapt to changing process conditions. In order to emulate the operators knowledge, a fuzzy supervisory system can be used to automatically determine the set-points.

14.8 Control of Unstable Bioreactor by Fuzzy Tuned PI Controller

14.8.1 Control Problems

Bioreactors have nonlinear behaviour and exhibit output multiplicities for certain parameter values. Some of the steady-states may also be unstable such that, even a small perturbation in the process conditions will make the system to move away from the steady-state and hence a feedback controller is required to operate it in the unstable steady-state. Conventional PI controllers do not give satisfactory response for nonlinear processes like bioreactor because of the change in the process gain and time constant with the operating conditions. Further, there is a significant delay in the output measurement which deteriorates the performance of the closed loop system. So, a controller tuned at one operating condition may even destabilize the system for other operating conditions. A review of the work done on control of biorectors is given by Agarwal and Lim (1984) and Shimizu (1984). Chen and Chang

(1984) have analyzed the difficulties in controlling the unstable biorecator using a proportional controller.

Controller design for open loop unstable system is more difficult than that for stable systems. Some of the common performance specifications for stable systems may be impossible to achieve for unstable systems. DePaor and O'Malley (1989) have proposed Ziegler-Nichols type tuning formulae for P/PI/PID controllers for unstable first order plus time delay systems. Chidambaram (1997) has given a review of the work done in control of unstable systems.

As the operating conditions change, the PI/PID controller with fixed settings may not maintain a satisfactory response. Frequent tuning of the controller is carried out by the operator based on his experience to get a good response. Hence, it is desirable to translate the operators experience into an automatic scheme. The fuzzy logic developed by Zadeh (1965) can be used for this purpose. Satish Kumar and Chidambaram (1999) have applied fuzzy logic for tuning PI controller for an unstable bioreactor. In this section their work is reported.

14.8.2 PI Controller Design

The model equations of a continuous bioreactor exhibiting substrate inhibition, proposed by Agrawal and Lim (1984) are used for simulation studies. For some operating parameters the system exhibits output multiplicity. The model equations are:

$$dX/dt = (\mu - D)X \tag{14.27}$$

$$dS/dt = (S_f - S)D - (\mu X/\gamma) \tag{14.28}$$

where

$$\mu = \mu_m S/(k_m + S + k_1 S^2) \tag{14.29}$$

It is desirable to control the cell mass concentration X of the reactor by manipulating the dilution rate D. Feed substrate concentration S_f is considered as disturbance. This control structure is termed to be turbidostat. The system parameters are

$$\gamma = 0.4 \text{ g/g},\ S_f = 4 \text{ g/l};\ \mu_m = 0.53 \text{ h}^{-1}$$

$$k_m = 0.12 \text{ g/l};\ k_1 = 0.4545 \text{ g/l};\ D = 0.3 \text{ l/h} \tag{14.30}$$

Analytical solution of the steady-state model equations [with $dX/dt = 0$ and $dS/dt = 0$) in Eqs. (14.22) to (14.29)] with these parameters gives the three steady solutions. On linearizing the dynamic equations around the set-points and analyzing the characteristic roots gives the following steady-state nature of the bioreactor

$$\begin{array}{ll} [X, S] = [0.4] & \text{(wash-out conditions)} \\ [X, S] = [0.9951, 1.5122] & \text{(unstable)} \\ [X, S] = [1.5301, 0.1746] & \text{(stable)} \end{array} \tag{14.31}$$

Of these steady-states, in the wash-out condition, there is no conversion at all, and the bioreactor is not operated in this steady-state. In the stable steady-state, the conversion rate is too high, which may lead to product inhibition on specific growth rate. Hence, as stated, it is desirable to operate the reactor in the unstable steady-state. Linearizing the model equations around this unstable steady-state condition (0.9951, 1.5122) gives the transfer function relating deviation in cell mass concentration (from the steady-state) and the deviation in dilution rate

$$\Delta\hat{X}/\Delta\hat{D} = K_p(\tau_0 s + 1)[(\tau_0 s + 1)(\tau s - 1)] \quad (14.32)$$

where ^ denotes the Laplace transform of the deviation variables and

$$\tau_0 = 1/\overline{D};\ \tau = \gamma/(\overline{X\varepsilon});\ K_p = -\gamma/\varepsilon;\ \varepsilon = (\overline{\partial\mu/\partial S}) \quad (14.33)$$

The parameters τ_0, τK_p and ε are calculated based on steady-state conditions. Considering a measurement delay of L units and canceling the zero and the stable pole, transfer function becomes

$$\Delta\hat{X}/\Delta\hat{D} = K_P \exp(-Ls)/(\tau s - 1) \quad (14.34)$$

DePaor and O'Malley (1989) have proposed a graphical method for calculating PI controller settings for an unstable FOPTD system. For the chosen operating conditions and parameters value, we get $\tau = 5.888$, $K_p = -5.859$. A measurement delay of $L = 1$ is considered. The PI controller settings are calculated as

$$K_c = -0.7356,\ \tau_I = 5.452$$

Performances of the closed loop system are evaluated by simulating the nonlinear model equations with the controller parameters given above. The system being initially in the stable steady-state 3 (i.e., at $X = 1.5301$, $S = 0.1746$) is forced to move to the unstable steady-state 2 (i.e., at $X = 0.9951$, $S = 1.5122$). The response shows a large overshoot and it takes a long time to settle (Fig. 14.18). Response of the system for a 30% step change in the set-point of cell mass concentration (X) is shown in Fig. 14.19. The response is much oscillatory. Since there may be uncertainties in the kinetics of bioreactions, the robustness of the PI controller is studied for uncertainties in measurement delay and the cell mass yield γ. For a 20% uncertainty in measurement delay L, the PI controller with the settings as given is unable to stabilize the system for a 30% step change in the set-point from

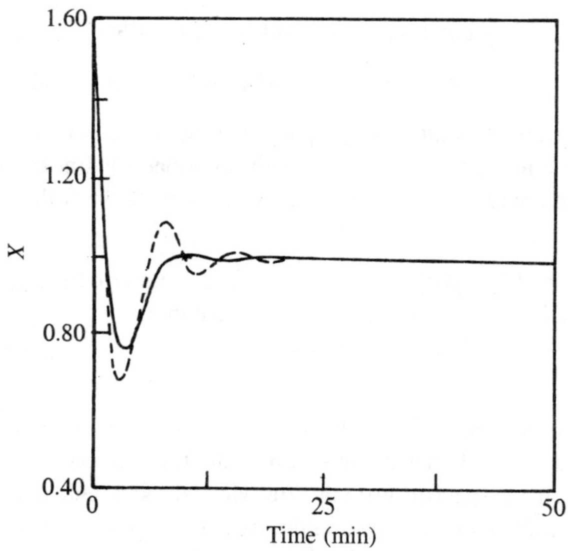

Fig. 14.18 Response of the bioreactor for step change in set-point of X from stable steady-state (1.5301) to unstable steady-state (0.9951). —— fuzzy tuned PI; - - - fixed setting PI

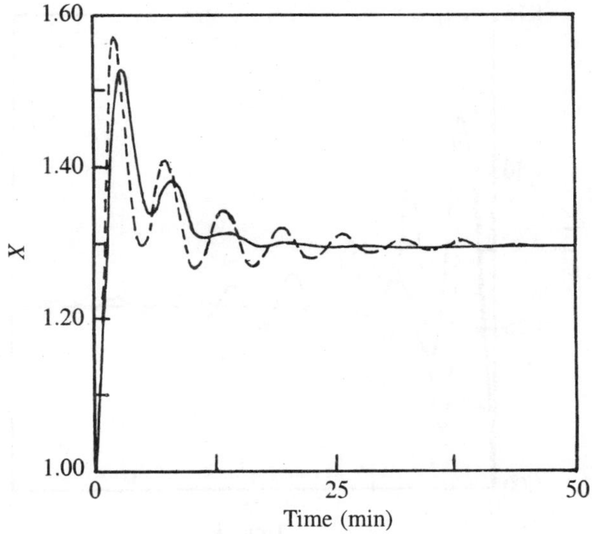

Fig. 14.19 Response of the bioreactor for 30% change in set-point from the unstable steady-state (X = 0.9951 to X = 1.294). Legend: as in Fig. 14.18

X = 0.9951. The system exhibits continuous oscillations as given by Jacob and Chidambaram (1996) and the response is shown in Fig. 14.20. For a 25% uncertainty in the yield factor γ, the system shows a highly oscillatory behaviour and it takes a long time to settle. The response is shown in Fig. 14.21.

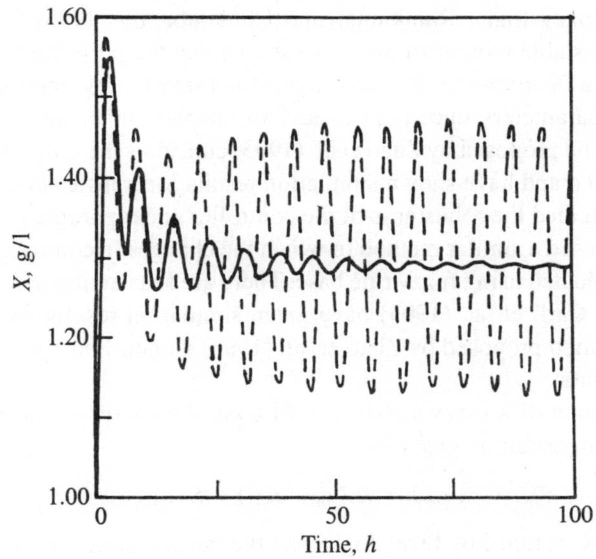

Fig. 14.20 Response for 30% step change in set-point from X = 0.9951 with 20% uncertainty in measurement delay. Legend: as in Fig. 14.18.

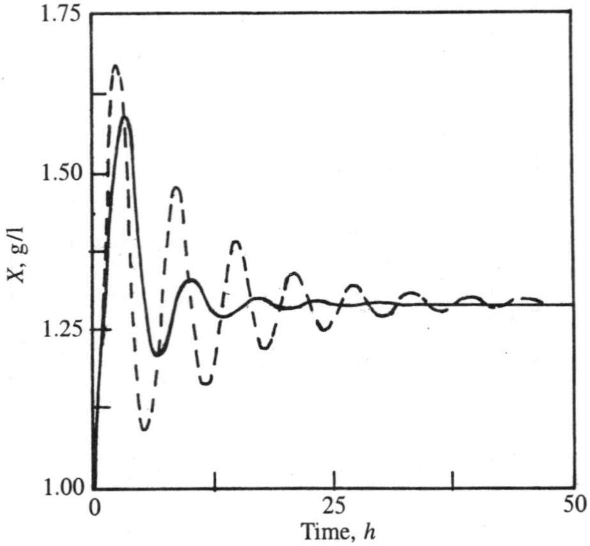

Fig. 14.21 Response for 30% step change in set-point from $X = 0.9951$ with 25% uncertainty in yield factor γ. Legend: as in Fig. 14.18.

14.8.3 Fuzzy Logic Application for Tuning PI Controller

By tuning conventional PI controllers using fuzzy logic, good improvement in the response is shown in the literature for stable systems. He et. al. (1993) have parameterized the Ziegler-Nichols tuning rule by a single parameter and then tuned that parameter using fuzzy logic and have applied it for stable transfer function models. Ramkumar and Chidambaram (1995) have modified the He et al., (1993) method by parametrizing the Ziegler-Nichols formulae by two parameters and then tuning these parameters by fuzzy logic. Ramkumar and Chidambaram (1995) have applied this tuning scheme to a non-linear stable bioreactor and have shown that the performance is improved over that of the fixed settings of a PI controller. It is shown that the fuzzy tuning improves the performance and is robust to process parameters uncertainties and to changes in magnitude and direction of the disturbances. The method proposed by Zhao et al. (1993) considers the range over which the controller parameters can be varied and have shown simulation results for transfer function models with stable poles. They have not studied the robustness of the controller under parametric uncertainties. De Carli et al. (1994) have proposed a similar method in which the changes in controller parameters from their nominal values are evaluated using fuzzy rule base. There are four tuning parameters and also the rule base is very large. De Carli et. al. (1994) have given simulation results for a second order system without delay. The method proposed by Zhao et. al. (1993) is modified here for tuning PI controllers for an unstable bioreactor.

The schematic diagram of a fuzzy self-tuning PI controller is shown in Fig. 14.22. The velocity form of PI controller algorithm is given by

$$\Delta u = K_c(e_n - e_{n-1}) + K_I e_n \tag{14.35}$$

The proportional gain K_c is tuned by fuzzy logic, and the integral gain constant. K_I ($K_I = K_c/\tau_I$) is kept constant so as to reduce the number of rules. This means that the integral time τ_I is varied in such a way that the integral gain K_I is maintained constant. The proportional gain K_c is normalized between 0 and 1 by the linear transformation

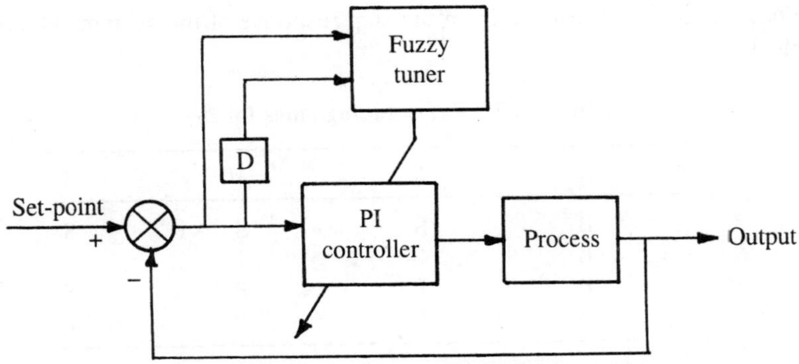

Fig. 14.22 Schematic diagram of fuzzy self-tuning PI controller (D: derivative)

$$\delta_{Kc} = (K_c - K_{c,\text{mina}})/(K_{c,\text{maxa}} - K_{c,\text{mina}}) \quad (14.36)$$

where the subscripts mina and maxa denote the allowable minimum and maximum controller gains. These values are to be selected from the ultimate values of $K_{c,\min}$ and $K_{c,\max}$ (obtained from stability analysis of the characteristic equation of the closed loop linear system). The error and the change in the error are normalized with respect to their maximum values so far obtained and the resulting absolute value are taken. The fuzzy inference mechanism needs this e and Δe as the input and evaluate δ_{Kc} using its rule base. Four fuzzy sets (Z, S, M, L) are considered for the output δ_{Kc}. The membership functions for e, Δe and δ_{Kc} are to be specified. Triangular membership function is chosen for the inputs e and Δe and logarithmic membership function is used for the output δ_{Kc}. For the unstable bioreactor, these membership functions are shown in Figs. 14.23 (a) and (b). The grade of membership functions ψ for the crisp output variable denoted by x (= δ_{Kc}) is given by

$$\psi_{\text{small}}(x) = -[\ln(x)]/4 \quad \text{(or)} \quad x_{\text{small}}(\psi) = \exp(-4\psi) \quad (14.37)$$

$$\psi_{\text{big}}(x) = -[\ln(1-x)]/4 \quad \text{(or)} \quad x_{\text{big}}(\psi) = 1 - \exp(-4\psi) \quad (14.38)$$

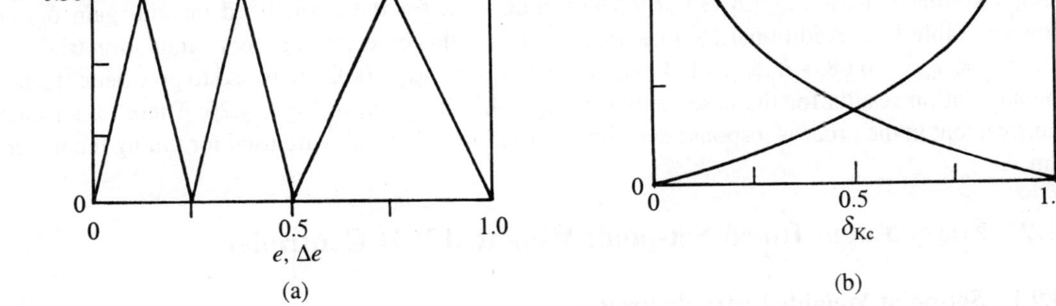

Fig. 14.23 (a) Membership function for e and Δe; (b) Logarithmic membership function for output δ_{Kc}

The fuzzy rule base is to be constructed from the step response of the system. The rule base for δ_{Kc} is given in Table 14.3.

Table 14.3 Fuzzy tuning rules for δ_{Kc}

e \ Δe	Z	S	M	L
Z	B	S	S	S
S	B	B	S	S
M	B	B	B	S
L	B	B	B	B

From Table 14.3, for example, the first fuzzy rule can be interpreted as: if e is Z and Δe is S then δ_{Kc} is B. Similar interpretation can be made for other entries in the table. The centroid method of defuzzification is used to get the crisp value of δ_{Kc} from the fuzzy variables. After determining δ_{Kc}, the controller parameter is calculated using Eq (14.36). By stability analysis of Eq. (14.34), the minimum and maximum values of controller gain are: $-K_{c,\max} = 1.47$ and $-K_{c,\min} = 0.17$. The negative sign indicates that if error increases the manipulated variable has to decrease. The maximum and minimum allowable values of the controller parameter K_c is chosen around the design values calculated earlier for PI controller. After some simulation trail values, $-K_{c,\,\mathrm{maxa}} = 0.68$, $-K_{c,\mathrm{mina}} = 0.3$ with integral gain $-K_I = 0.1$ are found to give a good performance for the present problem.

14.8.4 Performance Evaluation

The performance of the closed loop system when forced to move from the stable steady-state 3 to an unstable steady-state 2 is shown in Fig. 14.18. The overshoot is reduced and the response settles faster. For a 30% step change in the set-point, the oscillations are reduced (Fig. 14.19). With 20% uncertainty in measurement delay L, for a 30% step change in the set-point from $X = 0.9951$, fuzzy tuning helps to stabilize the bioreactor whereas, as stated earlier, the PI controller does not stabilize the bioreactor. The response is shown in Fig. 14.20. For a 20% uncertainty in γ, oscillations are reduced and the response settles faster (Fig. 14.21). The variations of proportional gain for the cases shown in Figs. 14.19 and 14.20 are shown in Figs. 14.24 and 14.25. Since the integral gain (K_I) is kept constant, τ_I will vary similarly but with different scale. Fuzzy tuning improves the robustness of the controller under process parameter uncertainties which is a major requirement for nonlinear processes such as bioreactors with uncertainties in reaction kinetics.

Tuning both the proportional and integral gain K_I, individually, has also been tried for this unstable bioreactor control problem. The integral gain also has been normalized between 0 and 1 using a similar equation as Eq (14.36). The tuning rules for the normalized integral gain δ_{Kc} are shown in Table 14.3. Additional 16 rules are required for this case. After several simulation trials, the values: $-K_{c,\mathrm{maxa}} = 0.68$, $-K_{c,\mathrm{mina}} = 0.4$, $-K_{I,\mathrm{maxa}} = 0.1$, $-K_{I,\mathrm{mina}} = 0.02$ are found to give better results. The simulation results for the case shown in Fig. 14.18 is given in Fig. 14.26. There is not much improvement in the process response even through additional 16 rules are used for tuning the integral gain.

14.9 Fuzzy Logic Tuned Set-point Weighted PID Controller

14.9.1 Set-point Weighted PID Controllers

As stated in Chapter 8, the dynamics of many processes are described adequately by first order plus

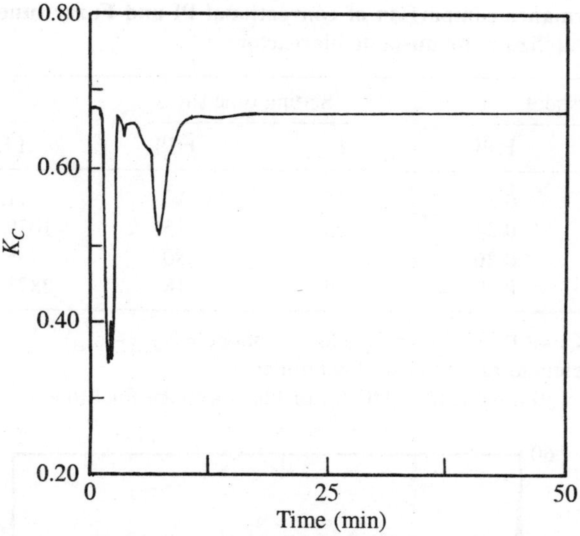

Fig. 14.24 Controller gain variation for the response shown in Fig. 14.18

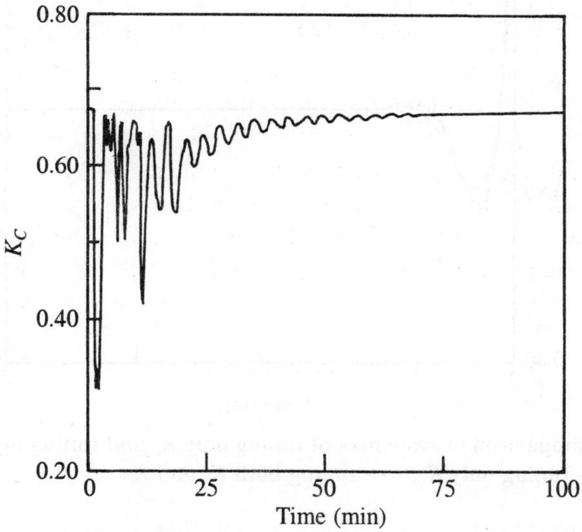

Fig. 14.25 Controller gain variation for the response shown in Fig. 14.20

time delay transfer function models. Methods for tuning PI/PID controllers for stable first order plus time delay systems are available. Controller design for unstable systems with time delay is more difficult than that of the stable systems with time delay. For example, a PI controller cannot be designed for ε (ratio of time delay to time constant) > 0.67 for unstable first order plus time delay systems. Most of the methods for designing PI/PID controllers give excessive overshoot for servo problems. For stable systems, Astrom and Hagglund (1988) have shown that the use of set-point weighting in PI/PID controllers reduce the overshoot for the servo problem. The PID controller law is then given by Eqs. (14.25) and (14.26). Actual error is used for integral calculation in order to ensure zero offset at steady-state. Here b is the set-point weighting parameter. However, no systematic method is available for the selection of the set-point weighting parameter.

Table 14.3 Performance comparison of conventional PI and Fuzzy tuned PI controller (for stabilizing the unstable bioreactor)

Fig. No	Overshoot		Setting time (h)		ITAE	
	PI	FtPI	PI	FtPI	PI	FtPI
14.18	0.31	0.23	15	10	711.05	418.93
14.19	0.27	0.24	25	15	1059.11	584.59
14.20	*	0.26	*	80	*	2025.91
14.21	0.38	0.30	35	18	2873.03	613.75

PI controller settings are: $K_c = -0.7356$, $C_I = 5.452$h; overshoot = $(y_{peak} - y_{set})$
*Unstable response; FtPI refers to fuzzy tuned PI controller
ITAE calculated upto time = 50 h with $\Delta T = 0.01$ h (for Fig. 14.20 the final time is 100 h)

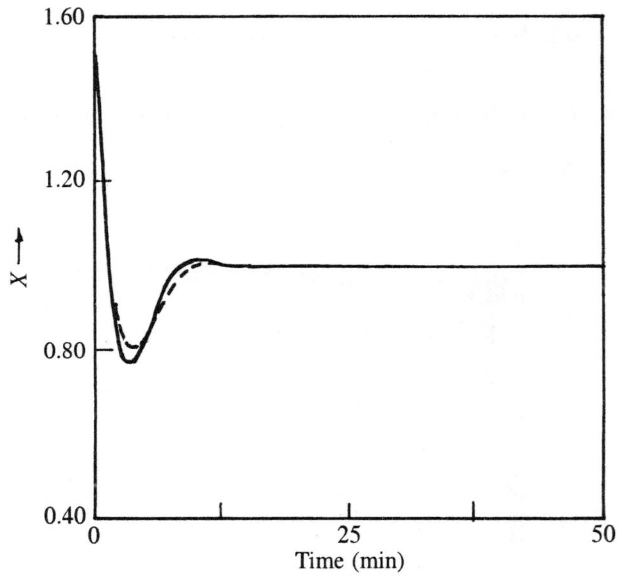

Fig. 14.26 Comparison of responses of tuning only K_c and tuning both K_c and K_I. — tuning only K_c; - - - tuning both K_c and K_I

Hang et al. (1991) have used a numerical optimization method to get the value of parameter b for stable systems. Leva and Colombo (1999) have used two weighting parameters, one in the error term used for the proportional control calculation as $e_p = by_r - y$ and second parameter in the error term used for the for derivative calculations as $cy_r - y$. A numerical optimization method is used for the calculation of the weighting parameters b and c. Chidambaram (2000) has derived an expression, by an analytical optimization method, for the calculation of the set-point weighting parameter b. A simple equation is given for the calculation of b as a function of $\varepsilon (= L/\tau)$. The method is used only for stable systems. Chidambaram (2000) has proposed a simple equation for b for unstable first order plus time delay systems. Though the use of a fixed set-point weighting approach reduces the overshoot significantly, the method in general, however, increases the rise time. Hang and Cao (1996) have shown that variable weighting parameter b improves the performance (by reducing both the overshoot and the rise time).

In industries, the operator changes the set-point weighting based on his experience. The translation of the operator knowledge into an automatic strategy can be made by applying fuzzy logic. Visioli (1999) has applied fuzzy logic for tuning parameter b. Genetic algorithms are used for automatic adjustment of the membership functions. Simulation results are given for delay plus stable systems and integrating systems. The rise time and the overshoot is reduced significantly when compared to that obtained with a constant value of b. Al-Sunni and Al-Nemer (1997) have used a fuzzy tuner for combined tuning of the set-point weighting parameter and the integral time of the PID controller for stable plus time delay systems. This method reduces the overshoot and the rise time significantly. Simulation results are given for delay plus stable systems.

Since the variable weighting parameter gives an improved performance of the controller and b is a function of ε (ratio of time delay to the dominant time constant), a method based on fuzzy logic is proposed by Bharadwaj and Chidambaram (2000) for calculation of b. Since the overshoot is large for unstable plus time delay systems than that in stable systems (Chidambaram, 1998), the fuzzy logic tuner is designed by Bhardwaj and Chidambaram (2000) for the selection of b for PID control of unstable systems. Let us discuss this method in detail.

14.9.2 Fuzzy Logic Tuner

The block diagram of the fuzzy tuned set-point weighting parameter is shown in Fig. 14.27. The input to the fuzzy logic system is the error and rate of change of error. Fuzzy rules are developed relating these inputs to the set point weighting parameter b. The basic computations involved in a fuzzy tuner are: (i) fuzzification of the measured system error and rate of change of error, (ii) fuzzy control rule base, (iii) inference engine and defuzzification.

Three fuzzy sets are considered for error (e): NEGATIVE, ZERO and POSITIVE. Similarly three fuzzy sets NEGATIVE, ZERO and POSITIVE are considered for rate of change of error (Δe). Triangle memberships are considered. The membership functions are shown in Fig. 14.28. Three

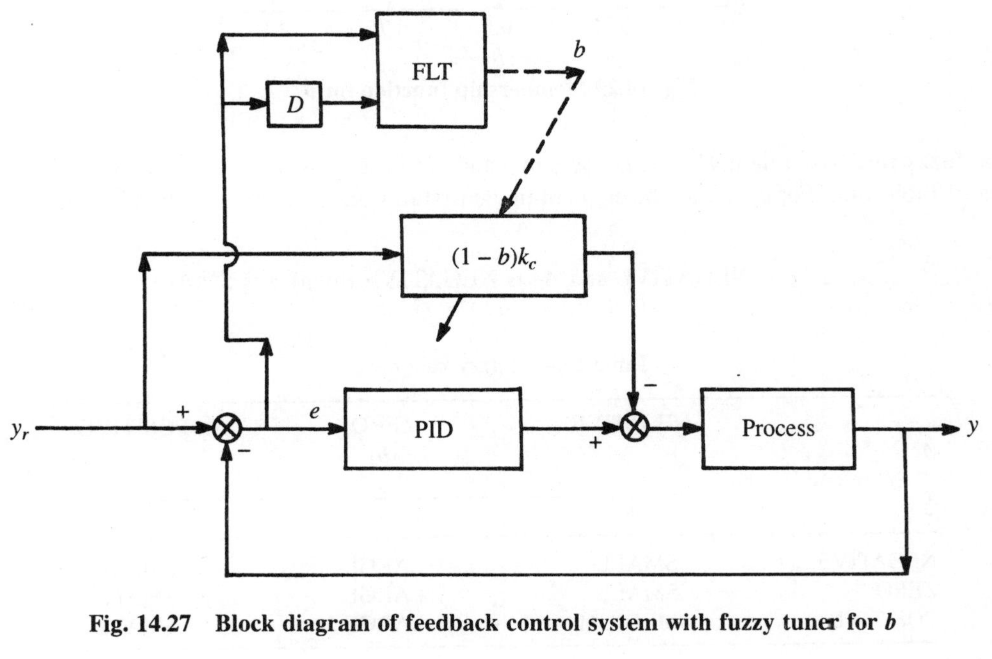

Fig. 14.27 Block diagram of feedback control system with fuzzy tuner for b

fuzzy sets are considered for the weighting parameter b as SMALL, MEDIUM and LARGE. These membership functions for the fuzzy sets in b are shown in Fig. 14.29.

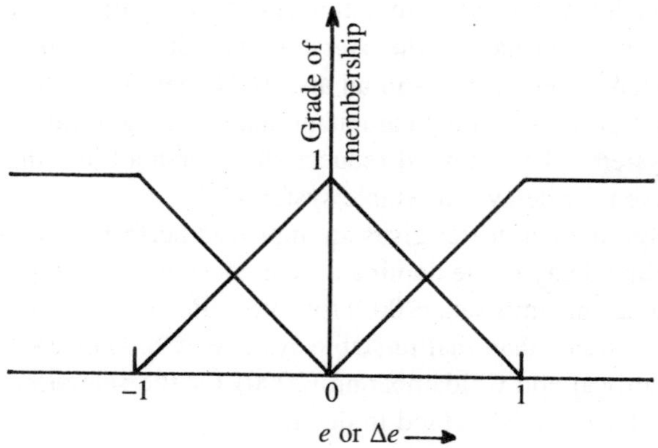

Fig. 14.28 Membership functions for e and Δe

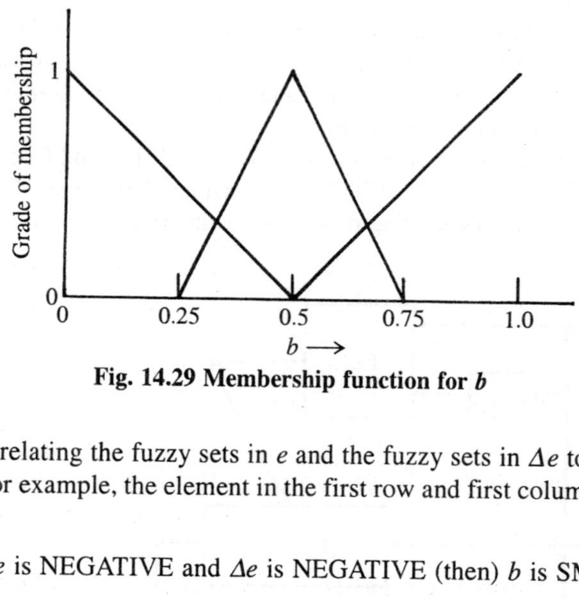

Fig. 14.29 Membership function for b

The fuzzy rule base relating the fuzzy sets in e and the fuzzy sets in Δe to the fuzzy sets in b are shown in Table 14.4. For example, the element in the first row and first column should be interpreted as

IF e is NEGATIVE and Δe is NEGATIVE (then) b is SMALL

Table 14.4 Fuzzy rule base

$e \rightarrow$ / $\Delta e \downarrow$	NEGATIVE	ZERO (b)	POSITIVE
NEGATIVE	SMALL	LARGE	SMALL
ZERO	SMALL	LARGE	SMALL
POSITIVE	MEDIUM	LARGE	SMALL

Similarly, the other entries in the Table 14.4 should be interpreted. The fuzzy mean method is used for the defuzzification step.

14.9.3 Simulation Results

Let us consider a transfer function model $K_p \exp(-Ls)/(\tau s - 1)$ with $K_p = 1.4$, $\tau = 1.0$ and $L = 0.50$. The PID settings are calculated based on the equations given by Chidambaram (1998) as $K_c = 1.4$, $\tau_I = 5.688$ and $\tau_D = 0.341$. The fixed weighting parameter is calculated from the equation suggested by Chidambaram (2000) as 0.1072. Servo responses for the system without and with the set-point weighting parameter are shown in Fig. 14.30. Overshoot is reduced significantly from 1.70 to 0.35 by using the constant weighting parameter. However, rise time is increased from 1.1 to 3.5.

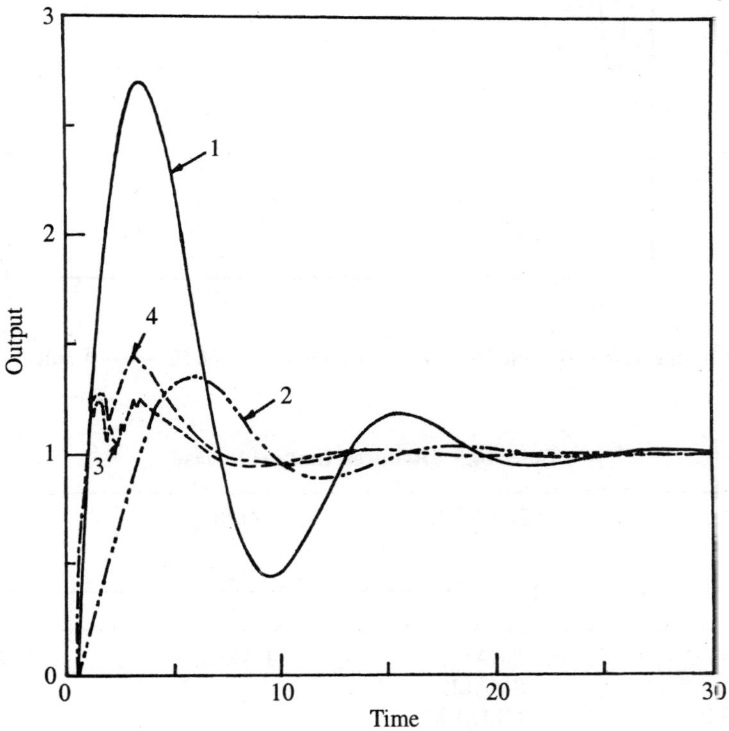

Fig. 14.30 The response *y* versus time. 1: without weighting parameter *b* = 1; 2: with fixed (optimum) value *b* = 0.107; 3: with FL1-tuned *b*; 4: with FL2-tuned *b*

Now the weighting parameter is tuned by fuzzy logic discussed in Sec. 14.9.2 The performance is also shown in Fig. 14.30. The rise time is now decreased from 3.5 to 1.1 andovershoot is reduced to 0.24. The behaviour of b is shown in Fig. 14.29. The fuzzy rule is reduced to 5 (as shown in Table 14.5) and the response is shown in Fig. 14.30. The behaviour of b vs time is also shown in Fig. 14.31. The overshoot with 5 rule base is 0.45 whereas the overshoot with 9 rules is 0.24. The rise time is not changed. Hence, the use of 9 rules is recommended. The results are summarized in Table 14.6.

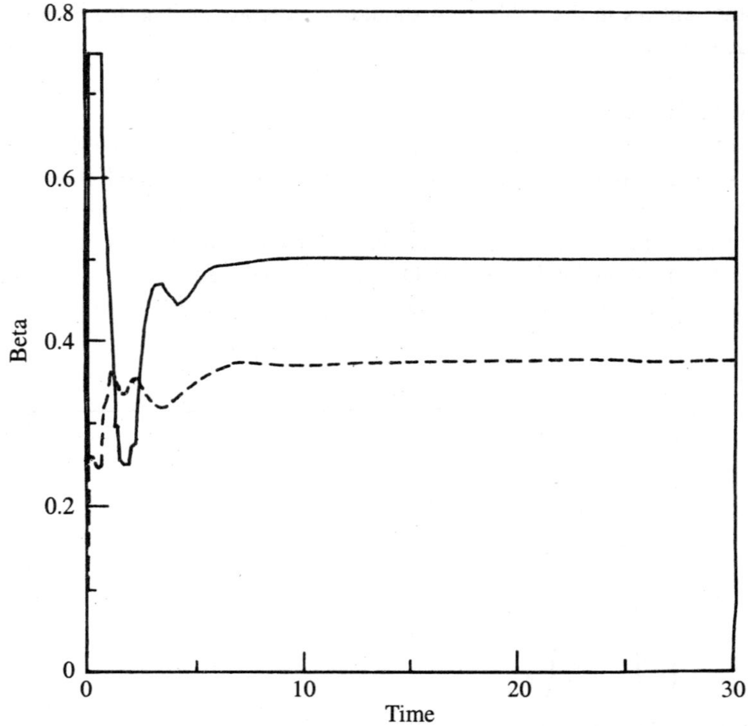

Fig. 14.31 Change of b with time for the conditions in Fig. 14.30. —— 5 rules; - - - - 9 rules

Table 14.5 Reduced fuzzy rule base

$e \rightarrow$ Δe ↓	NEGATIVE ↓	ZERO (b) ↓	POSITIVE ↓
NEGATIVE	SMALL	LARGE	SMALL
ZERO	SMALL	—	—
POSITIVE	MEDIUM	—	—

Table 14.6 Performance of set-point weighting parameter for the example considered

S. No.	b	Overshoot	Rise time
1	1.0	1.70	1.1
2	0.107	0.35	3.5
3	FLT-1	0.24	1.1
4	FLT-2	0.38	1.1

FLT-1: fuzzy logic tuner with 9 rules
FLT-2: fuzzy logic tuner with 5 rules

Summary

The fuzzy logic controller (FLC) provides a means for converting a linguistic control strategy based on operators experience into an automatic control strategy. A fuzzy logic controller is complimentary to the conventional PI controller. The essential steps involved in the design of FLC are fuzzification, fuzzy control rule base and rule inference and defuzzification. Because of these steps, FLC is essentially a nonlinear controller. For effectively controlling complex processes, a multi-level control strategy involving conventional PI controller at one stage and the use of FLC in the second level can be used. In this chapter, several simulation studies were given to illustrate the method of designing FLC systems.

References

Agarwal, P., H.C. Lim, Analysis of various control schemes for continuous bioreactors, Adv. Biochem. Biotechnol., 30 (1989) 61–90.

Al-Sunni, F., T. Al-Nemer, A smith predictor FL-based controller for processes with long dead time, Int. J. System Sci., V 28 (1997) 1251–1258.

Astrom, K.J. and T. Hagglund, PID controllers, Theory, design and tuning, 2nd Ed. ISA Pub., North Carolina, (1995).

Bhardwaj, A.N. and M. Chidambaram, Fuzz tuned set-point weighted PID controllers for unstable systems, Proc. NSIMC 2000, pp. 80–88 Feb., Madras.

Chen, C.-L. and P.-C., Chen, Application of fuzzy logic controllers in single loop tuning of multivariable system design, Computers in Industry, 17 (1991) 33–41.

Chen, L.M. and H.C. Chang, Global stabilization of biological reactors by linear feedback control, Chem. Engg. Commun., 27 (1984) 231–254.

Chidambaram, M., Set point weighting for PI/PID controllers, Chem. Engg. Commun., 179 (2000) 1–13.

Chidambaram, M., Set point weighting for unstable first order plus time delay systems, IEEE CCSP-2000, Bangalore, July, 2000.

Chidambaram, M., Set-point weighting parameter for integrating plus time delay systems, Proc. NSIMC-2000, 25-26th Feb., Madras, 2000.

Chidambaram, M., Control of unstable systems: a review, J. of Energy, Heat and Mass transfer, 19 (1997) 49–56.

Chidambaram, M., Applied process control, Allied Pub., New Delhi, 1998, pp. 68.

Chiu, S., Using fuzzy logic in control applications: beyond fuzzy PID control, IEEE Control Sys., Oct., (1998), 100–105.

Daley, S. and H.F. Gide, A justfication for the wider use of fuzzy logic algorithms, Proc. IME., 199 (1985) 43–49.

DeCarli, A.P. Liguori and A. Marroni, A fuzzy PI control strategy, Control Engg. Practice, 2 (1994) 147–153.

DePaor, A.M. and M. O'Malley, Controllers of Ziegler-Nichols type for unstable process with time delay, Int. J. Control, 49 (1989) 1273–1284.

Dimitratos, J., G. Ellicabe and C. Georgakis, Control of emulsion polymerization reactors, AICHE J., 40 (1994) 1993–2021.

Driankov, D., H. Hellendoorn and M. Reinfrank, An Introduction to Fuzzy Control, Springer-Verlag, Berlin, 1993.

Hang, C.C, K.J. Astrom and W.K. Ho, Refinement of the Ziegler-Nichols tuning formula, IEE Proc., D, 138 (1991) 111–118.

Hang, C. C. and L. Cao, Improvement of transient response by means of variable set-point weighting, Trans. IEEE, IE-43, (1996) 477–484.

He, S.Z., S. Tan, F.L. Xu and P.Z. Wang, Fuzzy self tuning of PID controllers, Fuzzy Sets and Systems, 56 (1993) 37–46.

Jacob, E.F. and M. Chidambaram, Robust control of unstable bioreactor, Process Contr. and Quali., 8 (1996) 147–157.

Kobayashi, K., H. Ogata and R. Murai, A method of inducing fuzzy rules and membership functions, Contr. Eng. Pract., 1 (1993) 283–290.

Lee, C.C. Lee, Fuzzy logic in control systems: Fuzzy logic controllers Part-I, IEEE Trans. Sys. Man. Cybern., SMC-20 (1990) 404–418.

Lee, C.C. Lee, Fuzzy logic in control systems: Fuzzy logic controllers Part-II, IEEE Trans. Sys. Man. Cybern., SMC-20 (1990) 419–435.

Mahesh, M.M. and M. Chidambaram, Fuzzy state estimation for a semi-batch copolymerization reactor, Chem. Eng. Tech. 15 (1992) 395–400.

Mahesh, M.M., M. Chidambaram and K.P. Madhavan, Fuzzy control of a semi-batch copolymerization reactor, Chem. Eng. and Process., 32 (1993) 327–331.

Malki, H.A. and H. Li, New design and stability analysis of fuzzy PD control system. IEEE Trans. Fuzzy Syst., 2 (1994) 245–254.

Odian, G., Principles of Polymerization, McGraw-Hill, New York, 1970.

Ramkumar, K.B. and M. Chidambaram, Fuzzy self tuning PI controller for bioreactors, Bioprocess Engg., 12 (1995) 263–267.

Satishkumar, S. and M. Chidambaram, Control of unstable bioreactor using fuzzy tuned PI controller, Bioprocess Engg., 20 (1999) 127–132.

Satishkumar, S. and M. Chidambaram, Fuzzy self-tuned PID control of unstable systems, Proc. 21st National system conference, Hyderabad, Allied Pub., New Delhi, (1997) 292–297.

Schork, F.J., Reactor operation and control, in Polymer Reactor Control (Ed.) C. McGreavy, Blackie Academic and Professional, New York (1994) 148–200.

Shimizu, K., An overview of control system design of biological reactor, Adv. Biochem. Biotechnol., 50 (1993) 66–81.

Sugano, M., An Introductory survey of fuzzy logic, Inf. Sci. 36 (1985) 59–83.

Van der Wal, A.J., Application of fuzzy logic in industry, Fuzzy Sets and Systems, 74 (1995) 33–41.

Leva, A. and A.M. Colombo, Method for optimizing set-point weights in ISA-PID autotuners, IEE Proc. CTA, V 146 (1999) 137-146.

Visioli, A., Fuzzy logic based set-point weight tuning of PID controllers, IEEE Trans. Sys. Man. Cyber., SMC-29, (1999) 587–592.

Wang, L-X, A course in fuzzy systems and control, Prentice-Hall Int. ed., New Jersey, (1997).

Wang, P.P. and C.-Y., Tyan, Fuzzy dynamic system and fuzzy linguistic controller classification, Automatica, 30 (1994) 1769–1774.

Xu, C.W. and Y.-Z., Lu, Fuzzy model identification and self learning for dynamical systems, IEEE Trans. Sys. Man. Cyber., SMC-17 (1987) 683-389.

Yi, S. and S. Park, A new fast algorithm, Fuzzy Sets and Systems, 60 (1993) 33–40.

Ying, H., W. Sider and J.J. Buckley, Fuzzy control theory: A nonlinear case, Automatica, 25 (1990) 513–520.

Ying, H., A nonlinear fuzzy controller with linear control rules is the sum of global two-dimensional relay and a local nonlinear PI controller, Automatica, 29 (1993) 499–505.

Zadeh, L.A., Fuzzy sets, Inf. Control., 8 (1965) 338–346.

Zhao, Z.Y., M. Tamizuka and S. Isaka, Fuzzy gain scheduling of PID controllers, IEEE Trans. on SMC, 23 (1993) 1392–1398.

Review Questions

1. What are the advantages of FLC over PID controllers?
2. What are the basic steps involved in designing FLC?
3. Draw the block diagram of a FLC.
4. Redo the example given in Section 14.4 for the instantaneous values of C and T as 0.61 and 85°C.

5. Draw the block diagram of multi-level control by using fuzzy logic.
6. Ketonen (Control Engg. Practice, 1993, V1, 273–281) has given the results on implementing FLC in a Phosphoric acid plant. Give a summary of that work.
7. Parekh et al. (IEEE Trans. Pack and Manuf. Tech. Part A, V17, 1994. 192–201) have presented an experimental implementation of FLC on a pH control problem. Give a report on this work.
8. Hanakuma et al. (Intern. Chem. Engg., 1994, V34, 91–96) have reported experimental application of self tuning fuzzy control of a demethanizer column in an ethylene plant. Give a summary of this work.
9. Kim and Kim (Jl. of Chemical Engg. Japan, 1995, V28, 617–620) have reported an experimental application of combined fuzzy and predictive multivariable control to a binary distillation column. Give the details of this study.

Index